CONTENTS

Volume I

KEYNOTE ADDRESSES	1
SMART MACHINES	47
BLADES	141
ANALYTICAL METHODS	201
EXPERIMENTAL METHODS	303
INSTABILITIES	345
ELECTRICAL MACHINES	393
AUTHOR INDEX	

CASE STUDIES

C663/022/08 Centrifugal pumps uncoupled shafts laser alignment and its influence on vibration level
D Litvinov, Riga Technical University, Latvia — 489

C663/026/08 Model and experiment for vibration reduction of a single cylinder reciprocating compressor
N Levecque, J Mahfoud, G Ferraris, R Dufour, INSA Lyon, France; D Violette, Danfoss Commercial Compressors, France — 497

C663/035/08 Experimental rotordynamic investigations of a flexible low-pressure-spool
D Peters, Gleason Germany (Holdings) GmbH, Germany; C Kaletsch, R Koehler, R Nordmann, Technische Universität Darmstadt, Germany; B Domes, Rolls-Royce Deutschland Ltd. & Co. KG, Germany — 511

C663/087/08   Dynamic effects in rotors of large horizontal-axis   523
wind-turbines with free-moving masses in blades
S D Garvey, University of Nottingham, UK;
J E T Penny, Aston University, UK

FAULTS

C663/036/08   Improved simulation of faults in rolling element bearings   539
in gearboxes
N Sawalhi, R B Randall, University of New South Wales,
Australia

C663/058/08   Diagnostics of a spiral bevel gearbox using signals of the   549
motor inverter
R Ricci, P Pennacchi, Politecnico di Milano, Italy

C663/076/08   Bi-spectrum of a composite coherent cross-spectrum for faults   565
detection in rotating machines
J K Sinha, University of Manchester, UK

C663/080/08   Detection of rolling element bearing faults by analysis of   573
the motor current
A Ibrahim, F Guillet, M El Badaoui, IUT de Roanne,
France;
R B Randall, University of New South Wales, Australia;
D Rémond, INSA Lyon, France

C663/081/08   Bearing condition monitoring using multiple sensors and   585
integrated data fusion techniques
S L Chen, M Craig, R J K Wood, L Wang,
University of Southampton, UK;
R Callan, H E G Powrie, GE Aviation, UK

CRACKS

C663/018/08   Stochastic approach for crack identification in rotating shafts   603
using Monte Carlo simulations and a hybrid mechanical model
T Szolc, P Tauzowski, R Stocki, J Knabel, Polish Academy
of Sciences, Poland

C663/023/08   A shaft crack identification technique based on vibration   619
measurements
M Karthikeyan, R Tiwari, S Talukdar, Indian Institute of
Technology Guwahati, India

C663/028/08   Turbo-generator groups affected by transverse cracks: 631
a sensitivity analysis of vibrations versus crack position
and depth
*N Bachschmid, E Tanzi, P Pennacchi, Politecnico di
Milano, Italy*

C663/054/08   Detecting cracked rotors using an active magnetic actuator   645
*M I Friswell, Bristol University, UK;
Y Y He, Tsinghua University, PR China;
J E T Penny, Aston University, UK;
J T Sawicki, Cleveland State University, USA*

C663/056/08   Flexibility coefficients of a shaft with an elliptical front crack   657
*L Rubio, B Muñoz-Abella, University Carlos III of Madrid,
Spain*

TURBOCHARGERS

C663/006/08   Transient modal analysis of the nonlinear dynamics of a   671
turbocharger
P Bonello, University of Manchester, UK

C663/008/08   Experimental study of high speed turbocharger dynamic   685
stability
*R G Kirk, Virginia Polytechnic Institute and State
University, USA;
J C Nicholas, Rotating Machinery Technology, Inc., USA*

C663/072/08   An investigation of gyroscopic effects on the dynamic   697
behaviour of a turbocharger
*P Kamesh, M J Brennan, R Holmes, University of
Southampton, UK*

DIAGNOSTICS

C663/102/08   Bearing condition monitoring using feature generated   709
by genetic programming based on Fisher criterion
H Guo, Q Zhang, A K Nandi, University of Liverpool, UK

C663/097/08   Fan diagnosis in the field   719
A El-Shafei, RITEC, Egypt

DAMPING

C663/049/08   Active bearing unit for damping of bending vibrations in   733
rotating machinery
B Petermeier, H Springer, Vienna University of
Technology, Austria

C663/024/08   Damping properties of syntactic foams with nanoparticulate   745
additives
J A Rongong, G R Tomlinson, University of Sheffield, UK;
I Stepanova, A Ivanenko, S Panin, Tomsk Polytechnic
University, Russia

C663/057/08   Design of electromagnetic damper for aero-engine applications   761
A Tonoli, M Silvagni, N Amati, Politecnico di Torino, Italy;
B Staples, E Karpenko, Rolls-Royce plc, UK

C663/045/08   Damping of a flexible rotor by time-periodic stiffness and   775
damping variation
F Dohnal, B R Mace, University of Southampton, UK

C663/055/08   Application of squeeze-film dampers with a centrifugal   787
compressor
E A Memmott, K Ramesh, Dresser-Rand Company, USA

C663/098/08   Time-domain simulation of rotors with hysteretic damping   799
G Genta, Politecnico di Torino, Italy

MISALIGNMENT

C663/013/08   Influence of machine alignment and load on steam turbine   813
vibrations
P Pennacchi, A Vania, Politecnico di Milano, Italy

C663/021/08   The development of harmonics in rotor misalignment   825
A W Lees, Swansea University, UK;
J E T Penny, Aston University, UK

FLUID-FILM

C663/019/08   The procedure for numerical investigation of vibration of   837
a rotor with fluid film bearings and a disc submerged in
inwettable liquid
J Zapoměl, Technical University of Ostrava,
Czech Republic

C663/029/08   Shape optimisation of a labyrinth seal: leakage minimisation   849
and sensitivity of rotordynamic coefficients
*A O Pugachev, M Deckner, Technische Universität
München, Germany*

C663/033/08   Effect of cavitation on identification of the configuration   861
state of statically indeterminate rotor bearing systems
*N Feng, E Hahn, University of New South Wales, Australia;
W Hu, N Zhang, The University of Technology, Sydney,
Australia*

C663/079/08   A study on the oil whip of an elastic rotor supported in   871
tilting-pad journal bearings
*H Taura, Nagaoka University of Technology, Japan;
M Tanaka, Toyama Prefectural University, Japan*

BALANCING

C663/030/08   Multi-objective genetic algorithm application in unbalance   885
identification for rotating machinery
*H Fiori de Castro, K Lucchesi Cavalca,
L Ward Franco de Camargo, University of Campinas,
Brazil*

C663/062/08   Device asymmetries and the effect of the rotor run-up in a   899
two-plane automatic ball balancing system
*D J Rodrigues, A R Champneys, M I Friswell, R E Wilson,
University of Bristol, UK*

C663/067/08   Balancing of a north seeking device using partial, noisy and   909
delayed measurements
*I Bucher, A Elka, Technion, Israel Institute of Technology,
Israel*

C663/085/08   Use of the co-variance matrix in rotor balancing   919
*A I J Rix, Rolls-Royce plc, UK;
S D Garvey, S Jiffri, University of Nottingham, UK*

AUTHOR INDEX

CASE STUDIES

Centrifugal pumps uncoupled shafts laser alignment and its influence on vibration level

D Litvinov
Riga Technical University, Latvia

ABSTRACT

In this article the analysis of laser shaft alignment method practical application for the uncoupled shafts of centrifugal pumping unit is conducted and given recommendations for shaft alignment maximal exactness providing of the uncoupled shafts. The analysis of vibration level (vibration velocity) dependence from alignment exactness is also resulted.

1.   INTRODUCTION

During rotating machinery shaft alignment works conducting very often there is a situation when it is necessary to conduct laser shaft alignment for uncoupled shafts (Fig 1). Once it is necessary to be done because missing places for the laser shaft alignment equipment fastening. Though, the uncoupled laser shafts alignment main task is maximal exactness providing, and also the repetition of results.

Most laser shaft alignment systems have the specially developed method for conducting such of works. However, very low repetitions of alignment results during these methods application do not allow make alignment with proper accuracy. Accordingly low quality shaft alignment of the uncoupled shafts conduces to the vibration level increase of whole rotating aggregate.

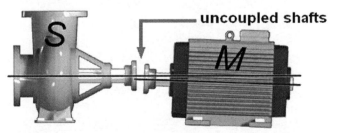

Fig 1 Rotating machinery with uncoupled shafts.

1.1. Laser shaft alignment peculiarities

The shaft alignment significant feature through the laser systems is that during such kind of works conducting the angular backlash of coupling is not allowing in relation to each other parts of gear coupling. In other words, it is necessary, that between itself, coupling must be hardly fastened. On small gear coupling with a diameter to D ≤ 200 mm, the hard connection during shaft alignment possible to provide by adhesive tape application. For gear coupling by a diameter D ≥ 200 mm (Fig 2), it is properly to make alignment with uncoupled shafts.

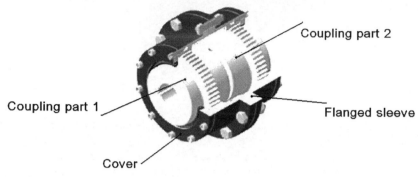

Fig 2 Gear coupling.

For alignment works conducting on the uncoupled shafts it is necessary that stationary unit (S) and movable unit (M) (Fig 3) was equipped with built-in inclinometer. It is the sensor for the rotating angle evaluation. If the inclinometer missing, an error even in 1° at the turn of the S and M units gives considerable distortions of results. Shaft alignment repetition of results goes down directly proportional to the S and M units setting angular error. Consequently further shaft alignment works are senseless.

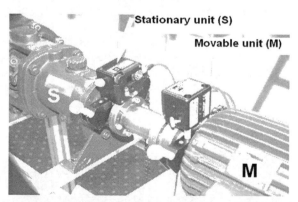

Fig 3 The units S and M with electronic 360° inclinometer inside.

As a result of the low quality conducted shaft alignment works there is a vibration level noticeable increasing, first, in the high-frequency range, and then passing to the low-frequency range. Thereupon as a result are the rolling bearings damage, and then the machinery failure.

2. EXPERIMENT

In order to do the analysis of pumping unit shaft alignment quality with coupled and uncoupled shafts, it is necessary to conduct an experiment. For the experiment the centrifugal pumping unit of K150-125-250a is used (Fig 4). Basic parameters of the unit:

Fig 4 Centrifugal pumping unit K150-125-250a.

- Rotation frequency of working shafts: 1450 r/min;
- Feed: 180 m^3 / h

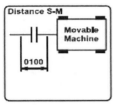

Fig 5 Distances inputting.

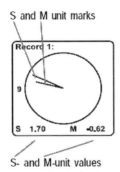

Fig 6 First reading.

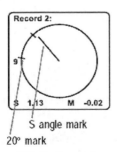

Fig 7 Second reading.

An experiment consists that at first is conducted pumping unit shaft alignment coupled, and then uncoupled coupling. Shaft alignment is carried out using the Swedish laser alignment system *Easy Laser*® D-525. Applying program Easy Turn ™. Totally are conducted 5 shaft alignment operations for coupled shafts and 5 shaft alignment operation for uncoupled shafts. The offset values and angular value during the alignments were set identically for different cases. After every shaft alignment procedure vibration velocity measuring is conducted. Information is summarized in tables. On the basis of tabular information the dependence graphs of vibration velocity value are built.

For the pumping unit shaft alignment works conducting for coupled clutch it is necessary to make the followings operations:

1. To set S (stationary machine) and M (movable machine) sensors. During shaft alignment of pumping unit, usually, a pump gets out as a stationary machine (S), because pipes are already supplied to it and to move it more difficult, than electric motor. A movable machine (M) is an electric motor.

2. Enter the distances, as prompted by the system (Fig 5).

3. Place the measuring units so that the marks are on top of each other (or almost on top). Adjust the laser beam to the closed targets. Open the targets. Record the first measurement value (Fig 6).

4. Second reading (Fig 7). Turn the shafts at least 20° in any direction (displayed as small marks on the circle).

5. Third reading. Turn shafts beyond the 20° mark (Fig 8).

Fig 8 Third reading.

6. The measurement results are displayed on the display of the measurement system. Horizontal and vertical positions for the movable machine are displayed both digitally and graphically (Fig 9).

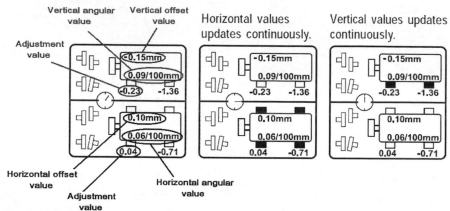

Fig 9 The example of the measurement results which could see on the display.

For the pumping unit shaft alignment of the uncoupled shafts conducting it is necessary to make the same operations as well as at the pumping unit with the coupled shafts. A difference consists only that at the turn of shafts at first it is needed to close targets on both sensors (S and M), than to turn a shaft with the S sensor, after to turn a shaft with the M sensor. After it through the program VALUES, included in the kit of the device, to check up exactness of angular setting of the sensors. Exactness must be ± 0.1°. Setting angles of sensors during an experiment were:

- First reading: 0.0°;
- Second reading: 20.1°;
- Third reading: 40.5°

During the first reading is not necessary to set sensors in 0.0° position. An angle can be any. The main thing is to take an installation angle error of sensors from each other to a minimum.

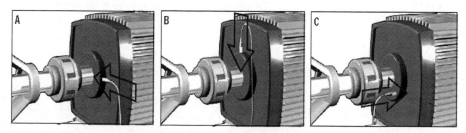

Fig 10 Mounting the vibration sensor.

After every shaft alignment procedure vibration velocity measuring is conducted, using program VIBROMETR, in three directions: horizontal (Fig 10, a), vertical (Fig 10, b) and axial (Fig 10, c). The normal vibration velocity value is indicated in the tables of the results (Table 1 and Table 2).

3. THE RESULTS OF EXPERIMENT

During an experiment of the pumping unit shaft alignment with coupled shafts (Table 1) and the pumping unit with uncoupled shafts (Table 2) information was got on. And the graphs of the vibration level dependence from offset value (Fig 11) and angular value (Fig 12) in the coupled shafts case are built. And also the graphs of the vibration level dependence from offset value (Fig 13) and angular value (Fig 14) in the uncoupled shafts case are built. For the results comparison the comparison graph of the vibration level at the offset value of the coupled and uncoupled shafts (Fig 15), and also the comparison graph of the vibration level at the angular value of the coupled and uncoupled shafts, are built (Fig 16).

Table 1 Vibration level after the pump unit shaft alignment with coupled shafts.

Number of the test	Revolutions of the electric motor (r/min)	Offset value (mm)	Angular value (mm/100mm)	Angular backlash max. (degree)	Vibration velocity (mm/s)
1.	1450	0.03	0.01	1	1.1
2.	1450	0.04	0.02	1	1.2
3.	1450	0.05	0.04	1	1.3
4.	1450	0.06	0.05	1	1.5
5.	1450	0.07	0.07	1	1.6

Table 2 Vibration level after the pump unit shaft alignment with uncoupled shafts.

Number of the test	Revolutions of the electric motor (r/min)	Offset value (mm)	Angular value (mm/100mm)	Angular backlash (degree)	Vibration velocity (mm/s)
1.	1450	0.03	0.01	0	0.5
2.	1450	0.04	0.02	0	0.6
3.	1450	0.05	0.04	0	0.7
4.	1450	0.06	0.05	0	0.9
5.	1450	0.07	0.07	0	1.0

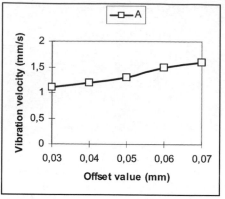

Fig 11 Vibration velocity dependence from offset value (coupled shafts).

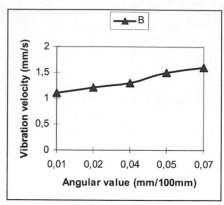

Fig 12 Vibration velocity dependence from angular value (coupled shafts).

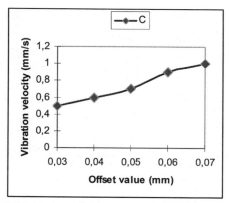

Fig 13 Vibration velocity dependence from offset value (uncoupled shafts).

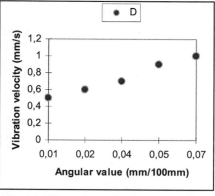

Fig 14 Vibration velocity dependence from angular value (uncoupled shafts).

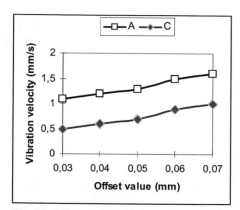

Fig 15 Vibration velocity comparison A and C.

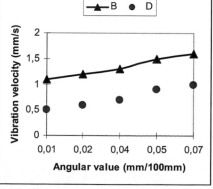

Fig 16 Vibration velocity comparison B and D.

4. CONCLUSION

Reasoning from the conducted experiment it is possible to make a conclusion, that at gear coupling presence on the pumping unit it is more expedient to conduct shaft alignment at the uncoupled shafts. It is justified, that even if a coupling is not worn down, it have a construction angular backlash (angular backlash of gear coupling it is free motion on a circumference about general axis of flanged sleeve in relation to a coupling part, Fig.17) approximately 1°, however, that negatively influence on the shaft alignment quality (in coupled shafts case) and accordingly on the vibration of whole aggregate. Especially it touches gear couplings by diameter above 200 mm and at revolutions of workings shafts over 500 r/min. As a result of the shaft alignment faulty works there is a vibration level increasing. It happens usually, at first, in a high-frequency range, and then passing to the low-frequency range. As a result is rolling bearings damage, and then the pumping unit failure.

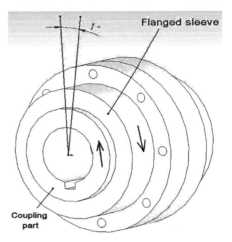

Fig 17 Angular backlash of gear coupling.

To avoid a similar situation it is necessary strictly abide by indications of the inclinometers during the shaft alignment works conducting for uncoupled shafts. Mounting the S and M sensors is necessary with the exactness ± 0.1°. Precision laser alignment of uncoupled shafts can be applied for any rotation machinery with gear couplings.

5. REFERENCES

1. Measurement and alignment. Training book. Basic and advanced course// Damalini AB, - 2004. p.33.
2. Measurement and alignment systems «Easy Laser». Manual 05-0215 Rev.7// Damalini AB, - 2005.p. A2-F8.

Model and experiment for vibration reduction of a single cylinder reciprocating compressor

N Levecque*, J Mahfoud*, D Violette**, G Ferraris*, R Dufour*
* LaMCoS, INSA-Lyon, CNRS UMR5259, France
** Danfoss Commercial Compressors, France

ABSTRACT

Basically, a reciprocating compressor consists of three main mechanical subsets: the hermetic housing on external suspension made of grommets, the crankcase mounted inside the housing on an internal suspension composed of springs, and the rotor-crankshaft assembly supported by the crankcase through two fluid film bearings. The counterweight mass located on the rotor-crankshaft is designed to balance the eccentric masses of the slider-crank mechanism. However excessive vibration levels can be observed due to fluctuation of the crankshaft angular velocity, alternative masses of the slider-crank mechanism, resulting pressure forces on the piston, assembly process, etc.

The objective of the study is to perform a reliable model for balancing the compressor by taking account the dynamic behaviour of the three subsets. A finite element model associated with the rotor-dynamic theory is developed in order to predict the steady-state unbalance response of target planes situated on the three compressor subsets. The rotor-crankshaft assembly is considered as a flexible body, while the crankcase and the housing are assumed to be rigid. The rotor-crankshaft model is adjusted by using experimental modal analysis at rest. The characteristics of the fluid film bearings that are speed of rotation dependant are considered. The forces of the pressure and of the slider-crank mechanism are expanded by using the Fourier transformation. By using the Influence Coefficient Method several balancing solutions were investigated for reducing the vibratory levels of the three main subsets. The carried out experiments showed that this multi-stage balancing procedure is rather more efficient than a classical dynamic balancing.

1   INTRODUCTION

Single cylinder reciprocating compressors are widely used in several types of refrigerant applications. They are driven by an asynchronous electrical motor and their operating speed depends on the power supply frequency (50 or 60 Hz).

Generally speaking, they consist of three main subsets: (1) the slider crank mechanism composed of the piston, the connecting rod, the crankshaft equipped with a counterweight mass and of an electrical rotor, (2) the crankcase equipped with the electrical stator and (3) the hermetic housing. These subsets are linked altogether by different types of suspension: the crankshaft to the crankcase by two fluid film bearings and air gap, the crankcase to the housing by springs and the housing to the frame by grommets and by suction and discharge pipes. To sum up a single cylinder compressor is a multi-stage system subject to vibration even if a balancing has been carried out.

Static balancing consists to adjust the counterweight mass designed in a plane to balance the eccentric masses (crankpin and crank arm masses, rotating mass of the connecting rod), located, due to technological requirements, in another plane. A dynamic balancing avoids the excitation of the moment due to the offset of these two planes, but it requires a second balancing plane classically located on the electrical rotor. Therefore a dynamic balancing has to be carried out on the crankshaft equipped with the electrical rotor and with a ring which mass is equivalent to the rotating masses. Unfortunately, such a balancing does not take into account the dynamic behaviour of the three main subsets. Consequently the subset responses can be over-pronounced. Noise and mechanical problems can occur such as rotor-to-stator or bearing rubs and failures at the weld spots of the pipes.

Complementing the balancing techniques, several technological solutions have been proposed for obtaining a best vibration reduction. It can be cited the optimization of the locations of the counterweight and the internal suspensions [1], and the introduction of a piston axis offset [2]. Moreover the multistage balancing method presented in [3, 4] was applied to rotary compressors made of two subsets.

In this paper a numerical technique is presented for balancing single cylinder reciprocating compressors considering the vibration levels on the three subsets. This can be done because the initial unbalances are fairly well-known. They are mainly due to eccentric masses of the crankshaft (crank-pin, crank-arms, counterweight ...) and of the rotating part of the connecting rod. The Finite Element (FE) models of the three subsets are combined with the rotordynamics theory, see Section 2. Moreover the constant and synchronous terms of the forcing excitations are taken into account. The static effort due to the slider-crank mechanism permits evaluating the bearing characteristics which are speed of rotation dependent and the synchronous component is taken into account in the masse unbalance response and in the balancing presented in Section 3. The influence coefficient technique yields a balancing, so-called multi-stage balancing, based on two speeds of rotation, two correction planes and several target planes located on the three subsets, especially at the anchorage points on the housing of the suction and discharge pipes. The proposed multi-stage balancing based on a model and the classical dynamic balancing based on experimental tests are implemented in two identical compressor prototypes. The proposed balancing efficiency is investigated by analyzing the measured steady state mass unbalance responses of the two prototypes for several operating conditions. Then the advantages and the limitations of the proposed multi-stage balancing are dressed and discussed.

2   FINITE-ELEMENT MODEL

The compressor is composed of three subsets: the rotor-crankshaft assembly which is a rotating part, the stator-crankcase assembly and the hermetic housing which are non rotating parts. The rotor-crankshaft assembly is connected to the stator - crankcase assembly through the fluid film bearings and the magnetic attraction between the electrical rotor and the stator. The crankcase is connected to the hermetic housing by an internal suspension made of springs and the discharge pipe, and the housing is mounted on an external suspension composed of grommets and the suction and discharge pipes. The stiffness of the pipes is assumed to be neglected regarding the stiffness of the springs and of the grommets.

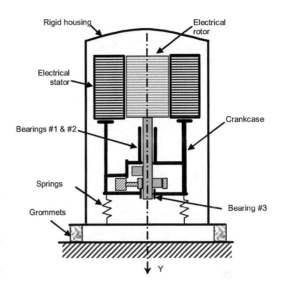

Fig. 1 The compressor's components

2.1   Whole compressor

The FE model, shown in Fig. 2, is governed by the rotordynamics theory presented in [5]. Each node contains the four degrees of freedom (DOF) of the bending motion: two lateral translations and the two associate rotations. The rotating part, considered as flexural, is modelled with two node beam elements while non-rotating parts are assumed to be rigid bodies modelled with rigid beam elements with no mass on internal and external suspension, respectively. The mass properties are modelled with additional mass elements located at the centre of inertia. The bearings are modelled with two-node bearing elements whose stiffness and damping parameters are speed of rotation dependent. Side-pull forces between the electric rotor and the stator are taken into account by using distributed additional two node elements with negative stiffness. Two-node elements, located on the rotor axis and containing transverse and angular parameters permit modelling the stiffness and the damping of the springs and of the grommets. After assembly, the unbalance response of the compressor is governed by the matrix equations:

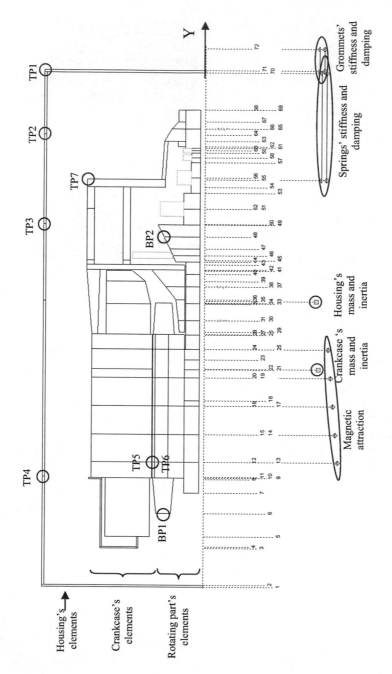

Fig. 2 Finite-element model for the whole compressor

$$[\mathbf{M}_R + \mathbf{M}_{NR}]\ddot{\mathbf{X}} + [\mathbf{C}(\Omega) + \mathbf{C}_B(\Omega) + \mathbf{C}_S]\dot{\mathbf{X}} + [\mathbf{K}_R + \mathbf{K}_B(\Omega) + \mathbf{K}_{SP} + \mathbf{K}_S]\mathbf{X} = \mathbf{F}(\Omega), \qquad (1)$$

with $\mathbf{X}$ being the displacement vector containing all the bending DOF of the assembly, $\mathbf{M}_R$ and $\mathbf{K}_R$ the classical mass and stiffness matrices of the rotating part; $\mathbf{C}(\Omega)$, the non-symmetric gyroscopic matrix; $\mathbf{C}_B(\Omega)$ and $\mathbf{K}_B(\Omega)$, the damping and stiffness matrices due to the bearings; $\mathbf{M}_{NR}$, the mass matrix of the non-rotating parts; $\mathbf{K}_S$ and $\mathbf{C}_S$, the stiffness and damping matrices associated with the suspensions; $\mathbf{K}_{SP}$, the anti-stiffness matrix associated with the side-pull forces, and $\mathbf{F}(\Omega, n\Omega)$, the external force vector which is the following summation:

$$\mathbf{F}(\Omega, n\Omega) = \mathbf{F}_C(\Omega) + \mathbf{F}_{CM}(\Omega) + \mathbf{F}_{P+CR}(\Omega, n\Omega), \qquad (2)$$

with $\mathbf{F}_C(\Omega)$, the force vector due to the eccentric masses of the crankshaft, $\mathbf{F}_{CM}(\Omega)$, the force vector of the correction masses, and $\mathbf{F}_{P+CR}(\Omega, n\Omega)$ the slider-crank force vector related to piston, connecting rod and cylinder pressure.

The industrial application concerns a refrigerant compressor which rotor-crankshaft assembly, stator-crankcase assembly and housing have roughly the following masses: 3.5 kg, 15 kg, and 7 kg, respectively. In what follows a particular attention is paid on rotating part, slider-crank forces and bearings.

2.2   Rotating part

The rotor-crankshaft assembly is mainly modelled with shaft elements. The crankshaft dissymmetry is taken into account by applying mass unbalances in response calculation. Rigid disk elements are used for modelling, with a mean radius, the presence of eccentric masses such as the counterweight, the crankpin... The piston axis is the angular reference. Unbalance masses, with a 180° phase, are applied on nodes (13, 15) to consider the counterweight dissymmetry while unbalance masses, with a 0° phase are located on nodes (16, 19, 21, 23) to model the crankpin and crank-arm dissymmetry. The sum of the unbalance masses situated at 0° is 3983 g.mm, and those situated at 180°, 4458 g.mm.

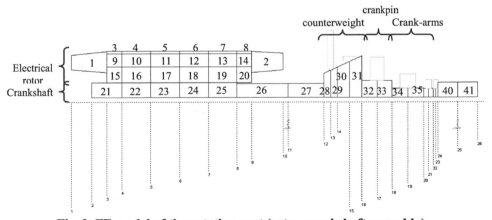

Fig. 3 FE model of the rotating part (rotor-crankshaft assembly).

The electrical rotor, made of steel laminations joined together with aluminium bars, is fitted onto the crankshaft, made of cast iron. The material properties are presented in Table 1.

Table 1 Material properties of the crankshaft and electrical rotor

Parts of the model		Young's modulus (GPa)	Mass density (kg/m³)	Poisson's ratio
Crankshaft	Elements 21 to 41	180	7200	0.3
Electrical rotor	Elements 1 and 2	70	2700	0.3
	Elements 3 to 8	5	7800	0.3
	Elements 9 to 14	37.5	5250	0.3
	Elements 15 to 20	5	7800	0.3

The FE model has been adjusted by carrying out an experimental modal analysis on a free-free rotor crankshaft assembly. The rolling hammer technique was used for obtaining the bending mode shapes with the data acquisition system LMS. To adjust the natural frequencies, diameters of the crank-pin and crank-arm elements were reduced to make the elements more flexible. It should be mentioned that this modification had no influence on the mass properties of that elements which are modelled with disk elements.

2.3 Slider-crank forces

The pressure force P_r is applied on the piston and on the crankcase and transmitted to the crankshaft by the connecting rod. Let P, B, M, and O be the centres of the piston, of the connecting rod, of the crank-pin and of the crankshaft, respectively (Fig. 4). Let m_{CR}, l, I_{CR}, ϕ be the mass, length, inertia, and auxiliary angle of the connecting rod. Consequently its alternative and rotating masses are $m_{re} = \dfrac{b}{l} m_{CR}$ and $m_{ro} = \dfrac{a}{l} m_{CR}$. Let r be the eccentricity of the crank-pin, x_p, and m_p, the position and the mass of the piston. The angular position of the crankpin is ϑ. The variable x_p is linked to ϑ the main variable of the mechanism [6, 7] and depends on the cylinder volume.

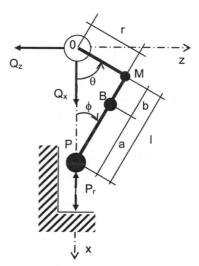

Fig. 4 Efforts transmitted by the slider-crank mechanism

Assuming a constant speed of rotation, the transmitted force components Q_x and Q_z have the following expressions:

$$Q_x = (m_{re} + m_p)(-\ddot{x}_p) + m_{ro} r \dot{\vartheta}^2 \cos \vartheta - P_r, \tag{3}$$

$$Q_z = \frac{1}{l\cos\phi}(I_{CR} - m_{CR}ab)\ddot{\phi} + \left[(m_p + m_{re})(-\ddot{x}_p) - P_r\right]\tan\phi - m_{ro}r\dot{\vartheta}^2\sin\vartheta. \qquad (4)$$

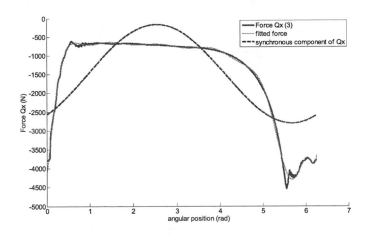

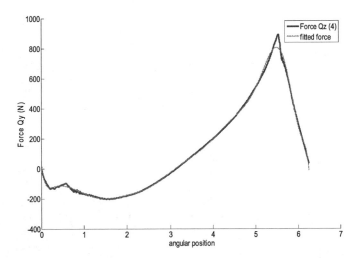

Fig. 5 Evolution of the forces with the rotation angle

The pressure diagram taken into account corresponds to the operating condition corresponding to evaporation and condensation temperatures: -10°/45° C. Each force component is fitted by fifteenth order polynomials and then expanded in a Fourier series:

$$Q(\vartheta) = a_0 + \sum_{k=1}^{N} a_k \cos(k\vartheta) + b_k \sin(k\vartheta), \qquad (5)$$

where the coefficients a_k and b_k are calculated by the coefficients of the polynomials. The constant terms a_{x0} and a_{z0}, corresponding to static forces, are calculated for both the x and z-components: $a_{x0} = 2969$ N, $a_{z0} = 218$ N. The crankpin is loaded by the static force F_p, given by:

$$F_p = \frac{1}{2}\sqrt{a_{x0}^2 + a_{z0}^2} = 1489N$$

(6)

For the higher order, the x-component coefficients are the most important and are the only ones used in the compressor dynamics, see Table 2. The compressor balancing is based only on the first order coefficients.

Table 2. Coefficients of the Fourier series

	i=1	i=2	i=3	i=4	i=5
a_{xi} (N)	-1087	-2172	-123	-23	-14
b_{xi} (N)	748	374	463	210	35

2.4 Bearing characteristics

Stiffness and damping coefficients are evaluated by using the tables proposed by Someya [8]. First of all, the Sommerfeld number S should be estimated. Let μ be the oil film viscosity, $\Omega = \dot{\theta}$ the speed of rotation, L, D, C_p, F the length, the diameter, the gap, and the static load of the bearing, respectively.

$$S = \frac{\mu\Omega LD}{F}\left(\frac{D}{2C_p}\right)^2$$

(7)

The upper bearing is made of two bearing models (L/D~1 and L/D~0.5) while the lower bearing has a ratio L/D~1. The reactions F on each bearing are deduced from relation (6). The stiffness and damping coefficients of the three bearings are calculated for 3000 and 3600 r/min.

3 PROPOSED MULTI-STAGE BALANCING

3.1 Influence Coefficient Method

The influence coefficients (IC) method is a well known experimental balancing method [9]. It consists of evaluating the influence of trial masses on the displacements at given planes, called measuring planes. The method assumes that the displacements are linearly proportional to the trial weights and that the initial unknown unbalance can be represented by a discrete finite number of unbalance moments that are placed on chosen balancing planes [10, 11]:

$$\mathbf{V}_{ini} = \mathbf{C} \, \mathbf{B}_{ini} \, . \tag{8}$$

$\mathbf{V}_{ini}$ is the vector of radial displacements due to the initial unknown unbalance $\mathbf{B}_{ini}$. The elements of $\mathbf{V}$ and $\mathbf{B}$ contain magnitude and phase information with respect to the reference phase, previously defined. $\mathbf{C}$ is the influence coefficient matrix. The method aims at determining balancing weights $\mathbf{B}_c$ to be placed at the chosen balancing planes so that the magnitudes of radial displacements, which are measured at the measuring planes, are minimized for different speeds of rotation:

$$\mathbf{C}\left(\mathbf{B}_{ini} + \mathbf{B}_c\right) = 0 \tag{9}$$

Here, the IC method follows a numerical approach. The IC matrix represents the system and is determined by using trial weights B_T and by predicting the resulting displacements V_R at given planes, so-called here target planes:

$$\mathbf{C}\left(\mathbf{B}_{ini} + \mathbf{B}_T\right) = \mathbf{V}_R \tag{10}$$

$$\mathbf{C} = \mathbf{B}_T^{-1}\left(\mathbf{V}_R - \mathbf{V}_{ini}\right). \tag{11}$$

The correction weights are calculated either by direct inversion if the numbers of balancing planes and of target planes are equal:

$$\mathbf{B}_C = -\,\mathbf{C}^{-1}\mathbf{V}_{ini}, \tag{12}$$

or by least squares technique, if not [12, 13]:

$$\mathbf{B}_C = -\left[\overline{\mathbf{C}}^t\,\mathbf{C}\right]^{-1}\overline{\mathbf{C}}^t\,\mathbf{V}_{ini}, \tag{13}$$

where $\overline{\mathbf{C}}$ represents the complex conjugate of matrix $\overline{\mathbf{C}}$.

3.2 Balancing procedure

The IC method with a numerical approach is applied on the compressor's model to determine the weights that reduce the vibrations on several target planes situated on different subsets of the compressor. The crankshaft dissymmetry and the slider-crank mechanism generate unbalance forces so-called initial unbalances. Only the synchronous forces can be balanced, that corresponds with the first order of the Fourier series (Fig. 5). The planes at the top of the electrical rotor (BP1) and at the counterweight (BP2) are available technologically for placing corrective weights. The aim is to reduce vibrations levels especially in the target planes (TP1 – TP7), where there are specific connections (pipe-housing, grommets-housing, springs-crankcase) and the air gap, (Fig. 2). The calculation of the corrective weights is done successively for a single-plane balancing and for a two-plane balancing, for one speed (3000 or 3600 r/min) and for two speeds (3000 and 3600 r/min). The model predicts successively the responses to the initial unbalances, and to the trial weights added to the initial unbalances. The initial unbalances and the calculated corrective weights permit providing predicted responses at the target planes which are compared to responses with the initial unbalances only, see Fig. 6.

The multi-plane balancing is more efficient than the single-plane balancing. The two-speed balancing, more convenient for 50 and 60 Hz operating conditions, is not really less efficient than the one-speed balancing. The responses calculated at 3000 r/min and at 3600 r/min are similar. Consequently only the results at 3000 r/min for the two-plane and two-speed balancing are presented (Table 3, Fig. 6).

Table 3 : Corrective weights for a two-plane and two-speed balancing.

Position	Modulus (g.mm)	Radius (mm)	Mass (g)	Phase (°)
Plane 1 (node 6)	824.6	26	31.7	-3
Plane 2 (node 48)	456.8	26	17.6	-177

Considering the responses predicted by the model, the proposed balancing sounds efficient, especially for the planes located on the housing. Responses for the planes TP 2, TP 3 and TP 7 are similar to those for TP 1 and TP 4 and are not presented here.

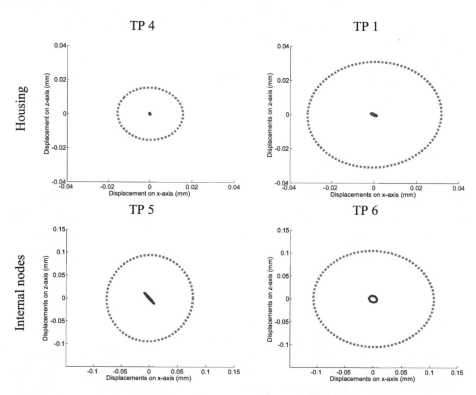

Fig. 6 Predicted responses at target planes #1, 4, 5 and 6 at 3000 r/min before (••••••) and after (▬▬▬) the proposed balancing

4   EXPERIMENTAL INVESTIGATIONS

The objective of the experimental investigation is to implement the proposed multi-stage balancing in a compressor prototype #1 for evaluating its efficiency regarding the classical dynamic balancing implemented in another similar compressor prototype #2.

The classical dynamic balancing is carried out on the rotor-crankshaft assembly by using, a balancing machine, the two balancing planes described in Section 3.2, and one speed (600 r/min). The obtained balancing quality corresponds to the G6.3 class at 3000 r/min of the ISO standardization.

The prototypes are mounted on a rigid frame by their grommets and by high flexible pipes. Their mass unbalance responses are measured for different operating conditions by using three tri-axial accelerometers stuck onto the hermetic housings at measurement points: MP1, MP3 (corresponding to target point TP4 and MP2 (corresponding to target point TP1), see Fig. 7. The piston axis is collinear to the Galilean x-axis, sketched on Fig.7.

Fig. 7 Experimental set-up showing the measurement points MP1, MP2 and MP3

The operating conditions took into account, the two nominal speeds of rotation corresponding to the 50 and 60 Hz frequencies, a constant condensation temperature (+45°C) and an evaporation temperature varying from -20°C to +10°C. To sum up the selected temperature operating conditions were: -20/45°, -15/45°, -10/45°, -5/45°, 0/45°, 5/45° and 10/45°. Fig. 8 collects the X-Y vibration magnitudes measured on Prototypes #1 and #2. Amplitude values were normalized with respect to the maximum obtained value.

The vibration levels of MP3 and MP2 along X and of MP1 along Y are low. The vibration levels of MP3 along Y and of MP2 and MP1 along X are consequent. Therefore it can be established that the steady state housing response is mainly governed by the torsion mode shape around the vertical axis Z. Prototype #2 provides high vibration levels at the bottom of the housing (MP2). To sum up it can be stated that Prototype #2 housing motion is composed of torsion and strong conical mode shapes while Prototype #1 housing motion is composed of torsion and low cylindrical mode shapes. Prototype #2 produces a high vibration level in the plane located at the bottom where grommets and discharge pipe are connected. Moreover the operating conditions have roughly no influence on the motion of the Prototype#1 housing. Broadly speaking Prototype #1 provides a more satisfactory dynamic behaviour than Prototype #2.

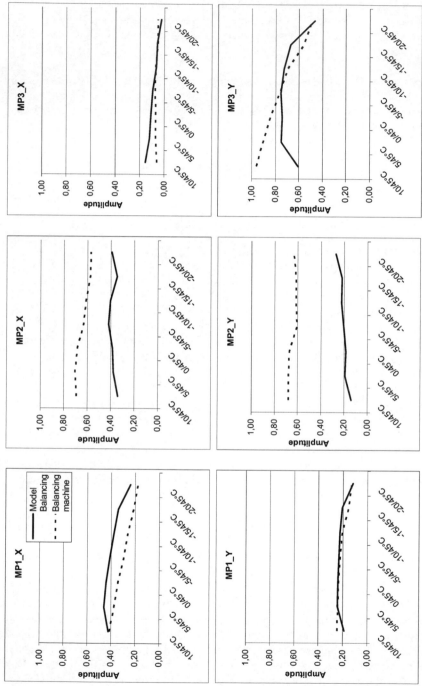

Fig. 8 Dimensionless vibration levels versus operating conditions – X and Y directions at measurement points MP1, MP2 and MP3. Prototype#1 with proposed balancing (⸺), Prototype#2 with classical balancing (······)

5   CONCLUSION

A numerical approach is presented for balancing a single cylinder reciprocating compressor. Vibration levels of the three subsets were considered. The constant and synchronous terms of the excitations were taken into account. The proposed numerical balancing is available for rotating machinery which unbalance masses are fairly well known. This is true for reciprocating compressors. Vibration levels stemming from the proposed balancing were compared with those obtained with dynamic balancing on a balancing machine. It was shown that this multi-stage balancing is rather more efficient than a classical dynamic balancing focusing only on rotating part. It can be an alternative solution for saving time and reducing the costs. In this case, it is important to study its sensitivity to the production tolerances.

6   REFERENCES

[1]   K. Kjeldsen, P. Madsen, 1978, *Proceeding of Purdue Compressor Technology Conference, edited by James F. Hamilton 55-59.* Reduction of compressor vibration by optimizing the locations of the counterweight and the internal springs

[2]   A. Kubota, T. Nagao, K. Tsuboi, T. Kakiuchi, 2006, *International publication number WO 2006/049108 A1.* Reciprocating compressor

[3]   F. Sève, M.A. Andrianoely, A. Berlioz, R. Dufour, M. Charreyron, 2003, *Journal of Sound and Vibration, 264(2), 287-302.* Balancing of machinery with flexible variable-speed rotor

[4]   G. Ferraris, M.-A. Andrianoely, A. Berlioz and R. Dufour, 2006, *Journal of Sound and Vibration, 292, 3-5, 899-910.* Influence of cylinder pressure on the balancing of a rotary compressor

[5]   M. Lalanne, G. Ferraris, 1997, *Rotordynamics prediction in engineering,* 2nd edition, John Wiley & Sons

[6]   N. Ishii, K. Imaichi, N. Kagoroku and K. Imasu, 1975, *ASME Paper 75-DET-44.* Vibration of a small reciprocating compressor

[7]   R. Dufour, J. Der Hagopian, M. Lalanne, 1995, *Journal of Sound and Vibration, 181(1), 23-41,* Transient and steady state dynamic behavior of single cylinder compressors: prediction and experiments

[8]   T. Someya, 1991, *Journal-Bearing Databook,* Springer-Verlag

[9]   W. C. Foiles, P. E. Allaire, E. J. Gunter, 1998, *Shock and Vibration, 5, 325-336.* Review: Rotor Balancing

[10]   R.E.D. Bishop, G.M.L. Gladwell, 1959, *Journal of Mechanical Engineering for Science 1, 66–77.* The vibration and balancing of an unbalance flexible rotor

[11]   J.W. Lund, J. Tonnesen, 1972, *ASME Journal of Engineering for Industry 94 233–242.* Analyses and experiments on multiplane balancing of a flexible rotor

[12]   T.P. Goodman, 1964, *ASME Journal of Engineering for Industry 8 273–279.* A least squares method for computing balance correction masses

[13]   J. Mahfoudh, J. Der Hagopian, J. Cadoux, 1988, *Mécanique, Matériaux, Electricité, Vol. 427, 38-42.* Equilibrage multiplans-multivitesses avec des contraintes imposées sur les déplacements

Experimental rotordynamic investigations of a flexible low-pressure-spool

D Peters[1], C Kaletsch, R Koehler, R Nordmann, B Domes[2]
[1]Gleason Germany (Holdings) GmbH, Germany
Department of Mechatronics in Mechanical Engineering, Technische Universität Darmstadt, Germany
[2]Rolls-Royce Deutschland Ltd. & Co. KG, Germany

ABSTRACT

According to a Low Pressure Rotor of a three-spool engine a test rotor with a long, flexible shaft, which operates in a super critical speed range was tested. The rotor is supported with a ball bearing in the front and two roller bearings equipped with squeeze film dampers in the rear. The vibration amplitudes at the resonances will be reduced by means of the squeeze film dampers in a sufficient manner for normal operation and low unbalance levels. In the case of a high unbalance, e.g. core blade off, the squeeze film damper will bottom out or will be highly non-linear with unknown dynamic behaviour. Therefore a test rig was build up at the chair of Mechatronics in Mechanical Engineering in cooperation with Rolls-Royce Deutschland. It consists of a rotordynamic similar dummy rotor in order to investigate mainly the non-linear vibration behaviour of the rotor-bearing-system under extreme operation conditions, such as out of balance of the shaft and the turbine dummy, oil off conditions of the squeeze film dampers and additional tests.

In this paper the design of test rig will be described. One of the important tasks concerning the engine is to develop an efficient balancing procedure in order to avoid high speed balancing of the whole rotor or trim balancing in the engine environment. For this investigation different levels of unbalances can be applied at locations along the test rig shaft, as well as at the turbine dummy and also by using special balancers, which give the opportunity to change the unbalance level electronically during operation. A comparison to the linear rotordynamic analysis and the non-linear vibration effect with different unbalance levels will be presented as well as the comparison of the test results with a non-linear rotordynamic analysis. The significant non-linear rotordynamic components are the squeeze film dampers, calculated by a one-dimensional SFD-law.

1   INTRODUCTION

The reduction of vibrations in aero engines is an important part in the field of rotordynamics. This problem requires a well known simulation model for the determination of the system behaviour. Measurements at test rigs are important for the validation of engine models which are developed by the engine suppliers. This paper is about a full scaled test rig for the investigation of the rotordynamic behaviour of the low-pressure spool (LP-spool) of a three spool engine. This rotor has a long, thin, flexible shaft, which operates in a supercritical speed range. The rotor is supported by a ball bearing in the front and two roller bearings equipped with squeeze film dampers (SFD) in the rear. The SFDs are the main reason for the non-linear behaviour of the spool. Furthermore the test rig is equipped with a similar dummy rotor compared to the engine in the face of the rotordynamic behaviour. Due to the elastic support structure in the engine, the bearings are supported by adjustable circular springs to simulate the engine casing stiffness as good as possible.

Test rig sensors are available for displacement, velocity, oil temperature, oil pressure and oil flow, together more than 60 sensors. This great number of sensors required a sophisticated data acquisition and evaluation system.

For model validation experimental and simulation results of rotor displacements along the spool will be compared. One of the important tasks of Rolls-Royce is to develop an efficient balancing procedure in order to avoid high speed balancing of the whole rotor for the engine. Therefore different levels of unbalances are applied at several axial locations along the low pressure shaft, at the turbine dummy and by using special automatically balancing devices. The letter gives the opportunity to change the unbalance level electronically during operation. Here a sensitivity investigation of unbalances for determination of the influence coefficients is applied. Another task is the behaviour at higher unbalances at the dummy turbine, in the face of a prediction for high unbalances at the engine. The simulation is realised by a FEM-program based on Timoshenko beam elements combined with a non-linear one-dimensional SFD-law, calculated in the time domain.

2   TEST RIG MOTIVATION AND OBJECTIVES

The maximum speed of the rig is about 9000 r/min. The LP-spool is very flexible, due to the long, thin, hollow shaft. Two bending modes have to be passed up to the maximum speed, leading to the supercritical behaviour. For this reason the rotordynamic behaviour of the rotor is very sensitive to any disturbances like unbalances. This is the reason for the unbalance sensitivity study. At low unbalances a linearisation of the SFDs around the operating point is meaningful. For higher unbalances a linearisation is not correct anymore, due to the strong progressive SFD-forces characterised by increasing of the SFD eccentricity. An important part is the behaviour during the operation with higher unbalances. The behaviour of the SFDs depends on several geometric features, the oil viscosity (oil temperature) and the oil supply pressure. Consequently the prediction of the damping forces of the

SFDs is very difficult, due to the complex non-linear relationship between rotational speed, eccentricity, damping behaviour, design of the oil inlet configuration and other time dependent parameters. Therefore for vibration analysis a simplified theoretical model was used, which could consider only the main non-linear effects of the SFD. The oil pressure and the oil temperature influencing the SFD behaviour are also adjustable at the test rig and were measured and recorded.

3   LOW-PRESSURE SPOOL

Subsequent the mechanical parts then the data acquisition and in a final step the used sensors will be described. A short overview is given in Figure 1. Beside the LP-spool on the top the position of important sensors and other units are presented here.

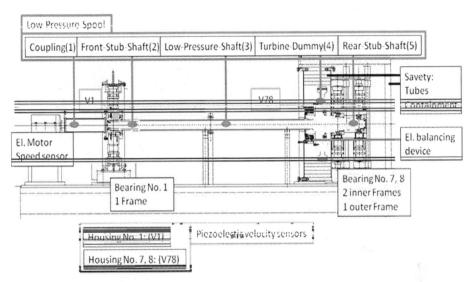

Figure 1: Test rig design and components

3.1   Test Rig Rotor
The test rig shown in Figure 1 is subdivided into a double membrane coupling, a front stubshaft (FSS), a low pressure shaft (LPS), a turbine dummy (TD) without blades and a rear stubshaft (RSS). Its overall length is approximately 3 m and its weight is higher than 140 kg, which nearly equals the real sizes and weights of a corresponding aero engine rotor. The rotor is driven by an electric motor via a double membrane coupling. Three bearings, bearing No.1 and the double bearings No. 7 and 8 behind the TD support the spool. The connections between every part of the assembly are realised by flanges. All parts together have similar rotordynamic behaviour like the engine rotor.

The locating ball bearing No. 1 is mounted on the FSS(2) with a relatively high axial tolerance of 0.5 mm to give enough freedom in the case of shaft bending. The connection to the LPS is realised by a shrink fit and an axial interlock. On each side of the FSS labyrinth sealings should avoid oil losses. Because of the expected large radial displacements of the LP-spool in the resonance the labyrinth sealings have a relatively high radial gap of 0.2 mm.

The TD(4) is flanged between the LPS and the RSS. To minimize the air resistance the blades are omitted. However, the real turbine is represented by its rotordynamic attributes: The centre of gravity, the mass and the moment of inertia. The whole part has a weight of about 100 kg and a diameter of 0.55 m. In addition two electric balancing devices are flanged in the front of the TD. Unbalances up to 22400 gmm can be put on the LP-spool during a test run without stopping of the machine.

The conical RSS(5) belongs to the end part of the LP-spool with two roller bearings, whereas the inner ring of the Squeeze Film Dampers (SFD) is shrunk onto the outer ring of the bearing. Different oil circuits supply the bearings and the Squeeze Film Dampers with oil.

3.2   Connections

The bearings are mounted in housings, one for bearing No. 1 and one for bearings No. 7 and 8. Accessories in these housings are: The outer sealing rings, the injectors, the oil outlet and the connection modules for the circular springs. The springs consist of closed outer and inner rings with high stiffness and small dimensions. Thus enables to create test rig support stiffness similar to the support stiffness of the real engine. The springs fix the housing in horizontal and vertical direction. Every spring unit has its own frame with a connection to the foundation. The rear construction is different to the front: For advanced applications two inner frames are placed in one outer frame.

All frames are fixed to the foundation plate. The first eigenfrequency is at 284 Hz, which has practically no influence to the test runs. The test runs go up to 150 Hz.

For safety the outer frames 7 and 8 are covered with steel plates of 51 mm thickness and the TD with additional three rings of 25 mm wall thickness.

3.3   Data Acquisition and Sensors

To control the test rig, display and store sensor values, a data acquisition is needed. The used one is from "National Instruments" while the software tool is LabVIEW. It allows a very flexible usage of the programmable user interface. Different graphic tools allow for controlling and displaying, for example the electrical power transmission speed und the displacements in micrometers of the eddy current distance sensors. About 60 analogue voltage in- and outputs can be controlled.

PCI-Measurement-cards with connector blocks digitise the sensor values. A sampling rate of 10000 Hz is possible for the main sensors, for example the eddy current distance sensors or the velocity sensors. Less important data like the temperatures, pressures and flow rates of the oil system are sampled with 100 Hz.

Eddy current distance sensors allow a contact-free measurement of the displacements of the spool in three axial planes shown in Figure 2. One plane is defined by 2 sensors arranged in 90 ° to each other. Hence it is possible to evaluate the vibration orbit of the shaft.

The first plane detects mainly the first bending mode (Figure 2) near the highest amplitudes. In front of the turbine the second plane is placed in expectation of characteristic 2nd bending mode (Figure 2) amplitudes.

Both planes show the absolute displacements relative to the foundation. In the third plane four sensors are necessary to measure the absolute displacement: Two for the displacement between the foundation and the housing of bearing No. 7 and 8 as well as two for the displacement between the housing and the RSS behind the turbine. The latter gives an indication of the vibration of the SFDs.

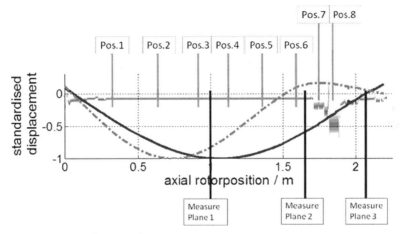

Figure 2: 1st and 2nd bending modes of the non-rotating shaft and the locations of artificial unbalance positions

Two velocity sensors, presented in Figure 1, are mounted on each housing to measure the acceleration, which is integrated to velocity, in three directions. Maximal magnitudes of 1000 mm/s can be measured. These sensors are comparable to those used in a real engine.

4   MEASUREMENTS

This paper presents displacements measured by eddy current displacement sensors at the three sensor planes along the LP-spool as well as velocities at the rear bearing housing. The measured data are compensated by the vibrations caused by the residual unbalance, for a better comparability with the simulations without a residual unbalance.

4.1   Sensitivity Study of Unbalances

The determination of the influence of the unbalances along the LPS is very important for the development of efficient balancing strategies by the manufacturer. The calculation of the unbalance sensitivity is not easy, due to the long, thin, hollow

LPS and its supercritical behaviour and the additional nonlinearity of the SFDs. The following measurements provide a basis for the determination of influence coefficients.

The unbalance tests are subdivided into 6 planes along the LPS shown in Figure 2 and three different relatively low unbalance levels. Level 1 is the lowest and level 3 the highest. For each position tests are carried out with all the three unbalance levels at the same angle position at the rotor circumference. Overall 18 test runs have been performed for this study. The positions of the unbalances are shown in Figure 2. The vibrations were measured in the middle of the rotor according to Figure 1. Of particular interest is the linear behaviour of the spool in the low unbalance range. As an example for this investigation, Figure 3 shows the standardised amplitudes of the 18 runs at the 2nd bending eigenfrequency (2nd EF). The amplitudes are standardised by a unit unbalance.

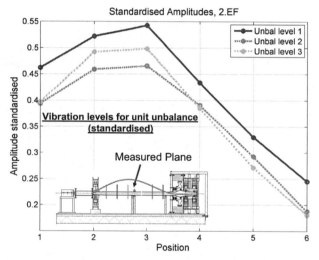

Figure 3: Displacements of unbalance levels standardised for unit unbalance, at 2nd EF

The highest sensitivity exists at the position 3 in the middle of the LPS, due to the influence of the 2nd bending mode, as shown in Figure 2. The influence of the excitation position 3 is 2.5 times higher as at the position 6, due to the influence of the node of the 2nd bending mode to position 6 as well as the mass of the dummy turbine. At position 1 decreased the influence in opposite to position 3, based on the bearing close to position 1. Due to the standardisation of the displacements to a unit excitation, a level of linear behaviour is interpretable. In the case of an ideal linear system, all curves must be the same. But the curves are different from each other, based on the nonlinear SFDs. Higher unbalance levels decreased the standardised displacements of the LPS, due to the increasing of the nonlinear SFD forces.

Figure 4 depicts the velocity response at the housing bearing No. 7, 8, due to the excitation at the 6 positions along the LPS as well as two positions (7, 8) in the front and rear of the turbine. The data are triggered by the 1st engine order (1. EO) and are standardised by the unit unbalance. The measured speed is at the 2nd bending eigenfrequency. The influence of the position 1 - 4 is nearly the same for the vibrations of the rear bearing housing. The influence of an unbalance at the turbine is for the velocities 7 times smaller as at the positions 1 - 4. The reasons are the high mass of the turbine and the closed position to the bearing No. 7. A point of interest is the behaviour of the phase of the velocities. The phase is skipping between the position 6 and 7, due to the vibration node of the 2nd bending mode between these axial locations.

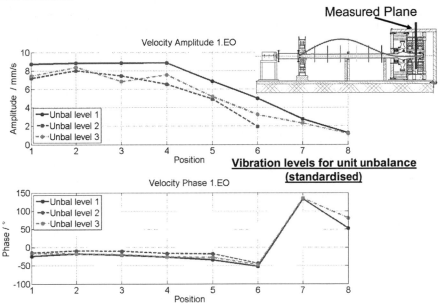

Figure 4: 1st EO velocities and phase angle for different unbalance levels, measured at the housing of the bearings 7 and 8, standardised by unit unbalance at 2nd EF

5   SIMULATION TOOL

The simulation has been carried out by the rotordynamic software tool Nirod (Non Linear Rotordynamics), a universal applicable, finite element method application. This program allows a linear and non-linear modelling under time domain. Nirod considers the non-linearities as external excitation forces. With this approach it is possible to compute rotating machines containing local non-linearities such as journal bearings and squeeze film dampers (SFD) efficiently. The SFD forces are calculated according to the models of [7-9]. The SFD model of [9] achieves the best simulation results regarding the experimental data of the test rig. This SFD- Model is based on the classical infinite short SFD Model. In contrast to the classical theory the model of M. Schwer includes the effects of end sealing and inertia forces. Furthermore the existence of cavitation is neglected. Due to this simplification, the model of M. Schwer is only applicable to well pressurised squeeze film dampers. The damping forces according to M. Schwer are computed after formula (1)

$$\begin{pmatrix} F_{r,Schwer} \\ F_{t,Schwer} \end{pmatrix} = -\begin{pmatrix} D_{rr} & 0 \\ 0 & D_{tt} \end{pmatrix}\begin{pmatrix} \dot{e} \\ e\dot{\gamma} \end{pmatrix} - \begin{pmatrix} M_{rr} & 0 \\ 0 & M_{tt} \end{pmatrix}\begin{pmatrix} \ddot{e} \\ e\dot{\omega} \end{pmatrix} \qquad (1)$$

F_r represents the radial and F_t the tangential damping force. The damping coefficients D_{rr} and D_{tt} are caused by viscous flow, M_{rr} and M_{tt} are inertial damping coefficients. Furthermore the viscous damping coefficients are in accordance to the infinite short and cavitation free (2π-theory) damper model except for the end sealing function $K_{endseal}$. The inertial coefficients are calculated by means of a perturbation method and a simplified two dimensional Navier-Stokes equation. Here the end sealing function depends on two parameters, which represents the flow resistance of the damper land without end sealing and the flow resistance of the end sealing itself. A detailed description of the force coefficients and the end sealing function can be found in the appendix.

6   COMPARISON WITH EXPERIMENTAL RESULTS

A point of interest is to compare a complete run-up of the test rig with a computed test run. Figure 5 contrasts the experimental displacements with computed results utilising the model of Schwer and a viscous damper. The data base on an unbalance of level 2 located at position number three near the centre of the LPS. The part on the top of Figure 5 shows the results of measurement plane 1 (ECS 12) and measurement plane 2 (ECS 34) according to Figure 2 and the corresponding calculations. The part on the bottom is the relative displacement at measurement plane 3 (ECS 78) between the RSS and the rear bearing housing. The dashed lines are based on the calculations by the model of Schwer [9]. The dotted are lines derived from calculations by a viscous damper and the drawn through lines are experimental results.

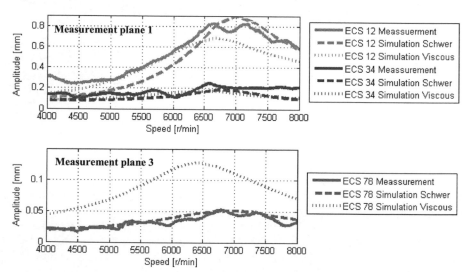

Figure 5: Comparison of experimental and computed run-up for an unbalance of level 2 at position number 3.

At measurement plane 1 (ECS 12) the simulation by the model of Schwer overestimates the experimental results in resonance speed about 8%. The computed data with the viscous damper are about 9% lower than the measured amplitudes. Both calculations are nearly equal and correspond approximately with the measured amplitudes at measurement plane 2 (ECS 34). Furthermore the computed results for the relative displacement of the RSS and the bearing housing (ECS 78) diverge from each other. Whereas the model of Schwer corresponds well with the experimental data, the viscous damper diverges for about 40%. The conclusion is that a linear modelling of a squeeze film damper may lead to completely erroneous results, even for moderate unbalances. Figure 6 shows two further run-ups. In these cases the unbalance is placed on balancer 1 (Pos. 7 of Figure 2). The unbalance of measurement A is slightly larger than the unbalance of measurement B. Experimental data are represented with drawn trough lines, computed results are represented with dashed lines. The model of Schwer is used again for the computation. The turbine including the balancers is modelled as a slim disk. In reality the point of unbalance force (balancer 1) and the base point of the turbine have different axial positions. This generates also different unbalance moments, which are transmitted into the shaft at the base point of the turbine. The rotating unbalance momentum of the TD is abstracted as two external momentums according to formula (2) and (3). ω is the angular velocity of the spool and M_{res} the resulting unbalance momentum.

$$M_y = M_{res} \cos (\varpi \, t) \tag{2}$$

$$M_x = M_{res} \sin (\varpi \, t) \tag{3}$$

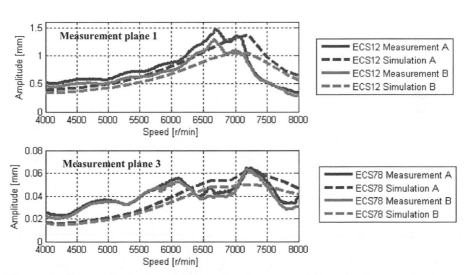

Figure 6: Comparison of experimental and computed run-ups for an unbalance at balancer 1 measured at meassurement plane 1 and meassurement plane 3. The unbalance of case A is bigger as in case B.

The upper part of Figure 6 shows the displacement amplitudes at measurement plane 1 in the middle of the rotor for both unbalances. The resonances of the computed curves are shifted to higher speed by less than 5 %. After the resonance speed the amplitudes decreases slower than in reality. However the maximal amplitudes of calculation match quite well with the experiments.

The lower part in Figure 6 depicts the relative displacement of the RSS between the SFDs and the rear bearing housing. The computed curve fits well together with the measured amplitudes. Furthermore the reduced measured amplitudes between 6000 and 7100 r/min are caused by a moving bending node of the mode shape of the rotor. In this speed range the node moves close to the measurement plane. The simulation shows the same effect less distinctively.

Figure 7 compares the experimental unbalance sensitivity study with a computed study. All amplitudes are standardised by a unit unbalance, so that in the case of a perfect linear system the standardised vibrations of the different unbalance levels must be the same.

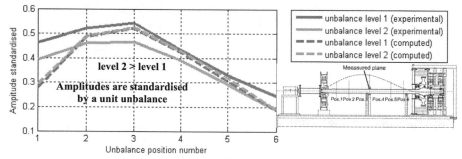

Figure 7: Comparison of experimental identified and the computed sensitivities at measurement plane 1.

Apart from position 1 the computed sensitivities correspond qualitatively well with the experimental sensitivities. In reality the SFD stiffens with increasing unbalance. This is explicit shown in Figure 7 because the standardised amplitudes for unbalance level 2 are smaller than the amplitudes for unbalance level 1. The computed results do not show this stiffening effect. The reason for this behaviour is that the real SFD is more strongly non-linear with increasing eccentricity than the SFD model by [9]. The variations at position 1 may be caused by an imprecise modelling of bearing 1 and the surrounding support structure.

7   CONCLUSIONS

The experimental results with a dummy rotor of real engine size shows the vibration sensitivity concerning unbalances at different axial locations. Although two resonances occurred in the speed range it could be shown that a low speed balancing is possi-ble and that the rotor could operate savely in the required speed range. The main reason for the mode-rate vibration behaviour is the installation of squeeze film dampers at the rear bearings.

With increasing unbalance the expected non-linearity appears. The comparison with the vibration analysis shows a good correlation for small unbalance levels as expected. For medium unbalances the non-linear approach of [9] correlate well with the test results. This is valid for the precondition of an oil filled squeeze film damper gap, which means sufficient oil supply pressure to avoid cavitation.

8   REFERENCES

[1]   Gasch R., Nordmann R., Pfuetzner H.: Rotordynamik , 2nd edition, Springer-Verlag 2002.
[2]   Knothe K., Wessels H.: Finite Elemente, Springer-Verlag, Berlin Heidelberg 1965, 2nd edition
[3]   Gasch R., Knothe K.: Strukturdynamik Band 1, Diskrete Systeme, Springer-Verlag Berlin Heidelberg 1987
[4]   Gasch R., Knothe K.: Strukturdynamik Band 2, Kontinua und ihre Diskretisierung, Springer-Verlag Berlin Heidelberg 1987
[5]   Federn K.: Auswuchttechnik, Band 1, Springer-Verlag Berlin Heidelberg New York 1977
[6]   Kraemer E.: Maschinendynamik, Band 1, Springer-Verlag Berlin Heidelberg 1984
[7]   El-Shafei, A., Cardall, S.H.: Fluid inertia forces in squeeze film dampers; Rotating Machinery and Vehicle Dynamics, DE-VOL.35, ASME 1991, Pages: 219-228.
[8]   Meng G., San Andres L.A., Vance J.M.: Experimental investigation on the dynamic pressure and force response of a partially sealed squeeze film damper, Rotation Machinery and Vehicle Dynamics, ASME 1991, Pages 251-256
[9]   Schwer, M.: Eigenschaften von Quetschöldämpfern – Ein Beitrag zur zuverlässigen Auslegung einer äußeren Lagerdämpfung; Dissertation; 1986
[10] Peters D., Kaletsch C., Nordmann R., Domes B.: Test Rig for a Supercritical Rotor of an Aero Engine, 12th IFToMM World Congress, Besancon (France), June 18 -21, 2007

APPENDIX

Squeeze film damper coefficients:

$$D_{rr} = -\frac{\eta B^3 R}{h_0^3}\left(\frac{\pi(1+2\varepsilon^2)}{(1-\varepsilon^2)^{5/2}}\right)K_{endseal} \qquad D_{tt} = -\frac{\eta B^3 R}{h_0^3}\left(\frac{\pi}{(1-\varepsilon^{3/2})}\right)K_{endseal}$$

$$M_{tt} = \frac{\rho\pi B^3 R}{h_0}\frac{1}{5(1+\sqrt{1-\varepsilon^2})}K_{endseal} \qquad M_{rr} = \frac{\rho\pi B^3 R}{h_0}\frac{M_{tt}}{\sqrt{1-\varepsilon^2}}K_{endseal}$$

End sealing function:

$$K_{endseal} = 1 + \frac{3}{1+R_0/R_S}$$

NOMENCLATURE

η	dynamic viscosity of the lubricating oil
ρ	density of the lubrication oil
B	damper land width
R	damper radius
h_0	damper radial clearance
e	damper eccentricity
$\dot{e}$	damper radial velocity
$\ddot{e}$	damper radial acceleration
$e\dot{\omega}$	damper tangential acceleration
ε	dimensionless damper eccentricity $= e / h_0$
R_0	flow resistance of the damper land without end sealing
R_s	flow resistance of the end sealing

Dynamic effects in rotors of large horizontal-axis wind-turbines with free-moving masses in blades

S D Garvey[1], J E T Penny[2]
[1] School of Mechanical, Manufacturing and Materials Engineering, University of Nottingham, UK
[2] School of Engineering and Applied Science, Aston University, UK

ABSTRACT

In several respects, larger is better for wind turbines. One reason is that wind-speeds increase appreciably at heights in the order of 200m above the surface so that power-density is greater. A second reason is that installation costs (for offshore turbines in particular) comprise a substantial proportion of total cost and this proportion decreases with size. The power from a horizontal-axis wind-turbine (HAWT) increases roughly in proportion D^2, rotational speed decreases in proportion to D and thus the turbine torque increases with D^3. The costs of both direct-drive generators and speed-increasing gearboxes are roughly proportional to the size of the input torque (not power). These components already account for substantial proportions of total system cost. The scaling laws clearly indicate that ultimately, as D increases further, the cost-per-kW of rated power of a conventional-design HAWT must increase again. This prompts consideration of different approaches to extracting the power from large HAWTs.

One possibility is to accommodate tubes within the blades and to allow large masses to move within these tubes under the influence of gravity. The power can then be extracted from the motion of the masses relative to the tubes. This paper considers only the possibility that air might be compressed by this relative motion but there are other possibilities. The movement of the free-moving masses excites several forms of vibration in the structure. As well as the moving gravitational load, there are parametric excitation effects due to moving mass, substantial Coriolis forces and substantial radial forces as the masses are decelerated near to end-of-stroke while air is compressed. This paper presents some simulations based on a notional 200m diameter machine rated at 18MW for a wind-speed of 15 m/s.

1   INTRODUCTION:
SOME LAWS OF SCALING APPLICABLE TO WIND-TURBINES

The power which can be extracted from a HAWT in a wind of velocity v is given by

$$P = \tfrac{1}{2}k\rho A v^3 = \tfrac{\pi}{8}k\rho D^2 v^3 \qquad (1)$$

where ρ represents the density of air, A ($= \pi D^2/4$) represents the area swept by the turbine and k is a constant which Lanchester [1] calculated to be not more than (16/27) (though many attribute the finding to Betz [2]). Most contemporary designs exhibit substantially lower k – typically 26% - 30% [3],[4]. Every wind-turbine has some rated wind-speed, v_{Rated} and produces a power P_{Rated} at this speed. When $v > v_{Rated}$, no more power is produced from the machine because the power-conversion equipment is at its limit but a power of P_{Rated} can be extracted at higher wind-speeds by allowing the aerodynamic surfaces to become deliberately less efficient – either by stalling or by furling [4]. If the projected area of the hub is significant, a correction for this must be made to (1) but where the hub diameter is small relative to blade tip diameter, we can consider that this correction is incorporated into k.

Most of the power is developed by sections of blade at largest radius. Thus aerodynamic efficiency of the blade tips is paramount (for $v \leq v_{Rated}$). For different designs of aerofoil (and different numbers of blades), there are different optimal tip-speed ratios, r. In general

$$r := \frac{\Omega D}{2v} \quad \text{and} \quad r_{optimal} \approx 5 \tag{2}$$

where Ω represents turbine rotational speed. For $3 < r < 9$, the efficiency curves are relatively flat. Turbines with higher numbers of blades tend to favour slightly lower values of r [4]. At higher tip-speeds, blade drag losses increase disproportionately. The rationale militating against lower tip-speeds is best understood from the aerodynamic torque, T

$$T := \frac{P}{\Omega} = \left(\frac{\pi \rho k}{4r}\right) D^3 v^2 \tag{3}$$

Evidently, at low values of r the torque must be higher and this torque causes significant swirl to be left in the tube of air which has passed through the turbine - carrying away kinetic energy. Equation (3) makes clear that although the power of a turbine rises with the square of its diameter, the aerodynamic torque developed must rise with the cube.

Now, for a cylindrical-airgap electrical machine, the torque developed across the airgap is

$$T_A = \tau_A \times \left(\pi D_A \times L_A\right) \times \left(\tfrac{1}{2} D_A\right) = 2\tau_A \times V_A \tag{4}$$

where $\{D_A, L_A, V_A\}$ represent the diameter, length and contained volume of the airgap middle surface. τ_A represents mean airgap magnetic shear stress and this value is fixed by the type of machine. Except in the case of superconducting machines, τ_A will never exceed 100kPa and more typical values are in the order 25kPa – 50kPa. What (4) essentially shows, therefore, is that the torque is proportional to contained volume. It is a relatively small stretch to see that cost of an electrical machine is proportional to its torque. It is more difficult to present this argument for the case of a gearbox but evidence bears this out. Thus, we may reasonably conclude that the cost of equipment in the nacelle of a large HAWT for transforming turbine shaft power rises with the D^3 – although output power rises only with D^2.

2 ENERGY-CONVERSION THROUGH MOVING MASSES WITHIN BLADES

The blades of large HAWTs are substantially hollow. Consider that a mass M is allowed to move within each blade such that its centre of gravity travels between diametral extremes D_I and D_O as Figure 1 indicates.

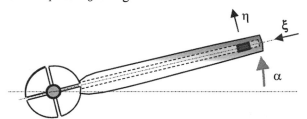

Figure 1. Energy conversion through moving masses in blades.

If the turbine rotates very slowly, then that mass might be allowed to fall through a net height of close to $(D_O - D_I)$ in each cycle and the maximum amount of energy done by gravity in that cycle is $Mg(D_O - D_I)$. This could be recovered from the movement of the mass if some energy conversion provision was present. In reality, the mass would not travel so far vertically and the work done by gravity in one cycle would be some factor, χ, times $Mg(D_O - D_I)$. Clearly, $0 \leq \chi \leq 1$. We will generally aspire to achieve $0.70 \leq \chi \leq 0.85$.

Denoting the number of blades on a HAWT as N, then the mean power which can be converted from allowing masses to move freely under the influence of gravity is

$$P = \left(\Omega / (2\pi) \right) . \chi N M g \left(D_O - D_I \right) \tag{5}$$

Combining equations (1), (2) and (4) yields the following expression for mass, M

$$M = \left(\frac{\pi^2}{8r} \right) \left(\frac{D^3}{D_O - D_I} \right) \left(\frac{k \rho v^2}{\chi N g} \right) \tag{6}$$

if all extracted power is converted through movement of mass relative to blades.

3 THE EFFECTS OF CENTRIPETAL ACCELERATION

Centripetal acceleration, a, at blade tips is significant compared to gravity in all current HAWTS. Centripetal acceleration at the maximum diameter position for a moving mass is $\Omega^2 D_O/2$ and using (2) to substitute for the rotational speed, Ω, produces

$$a = 2\left(\left(v^2 D_O r^2\right)/\left(D^2\right)\right) \qquad (6)$$

Evidently, $a \propto D^{-1}$. The proposal to use masses moving relative to wind turbine blades appears restricted to cases where $(a/g) < 1$. In fact this is not true – for two reasons: (a) we can employ *forced commutation* – allowing a small amount of energy back into the turbine blade approaching top-dead-centre in order to initiate downward motion of its mass and (b) we can arrange direct mechanical connection between radial motions of masses in different blades. Figure 2 illustrates the simplest case of such a connection – applicable only where N is even. Intriguingly, the tension in each tie will naturally remain constant and each piston can have *symmetrical* operation in the sense that it would convert the same quantity of energy in its inward and outward strokes. Tie length (between centres of gravity) is then $(D_O + D_I)/2$ and its tension is constant at $M\Omega^2(D_O + D_I)/4$.

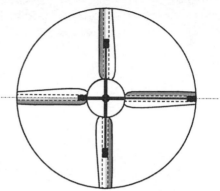

Figure 2. Mechanical connection of diametrically-opposed masses.

4   ENERGY CONVERSION VIA DIRECT COMPRESSION OF AIR

Consider that each blade has two controllable valves at each end: one linked to low-pressure air (possibly ambient) and the other linked to a high-pressure (and high temperature) manifold. We consider that adiabatic compression takes place taking some initial volume, V_0, of air at atmospheric pressure, P_0, up to some higher pressure, P_1 given by

$$P_1 =: bP_0. \qquad (7)$$

The total work done in adiabatically compressing a volume of air. V_0 at pressure P_0, and discharging it into a reservoir at pressure P_1 is given by

$$W = P_0 V_0 \left(b^{((\gamma-1)/\gamma)} - 1 \right) \times \left(\gamma/(\gamma-1) \right) \qquad (8)$$

where γ represents the usual ratio of specific heats (c_P/c_V) ($\gamma \approx 1.4$ for air), Given the pressure-ratio, b, and the work, W, enables us to determine what volume of air at pressure P_0 can be compressed and discharged. Selecting P_1 reasonably high has several advantages: (a) the minimum cross-sectional area of the tube within the turbine blade is reduced, (b) turbo-machinery used to re-expand the air can be relatively small and (c) if energy-storage using the compressed air is being considered, better energy-densities are obtained for high-pressure air. There are two disadvantages also: (a) leakage losses increase with P_1 and (b) high pressure air becomes hot and impacts on net efficiency. Sensible values for P_1 are in the order of 5MPa.

5   STATIC CALCULATIONS FOR A 200m DIAMETER HAWT

We consider a 4-blade turbine with tip diameter **200m** having rated wind-speed **15 m/s**. We take D_O = 170m, D_I = 40m, b = 70, k = 0.2715. From (1), the rated power is **P = 18.04 MW**. Taking tip-speed ratio s = 3.6, rated speed is Ω = **0.540 rad/s** (11.64 seconds per cycle). Estimating χ = 0.75 and applying (5) yields mass per blade is $M \approx$ **55,000kg** (55 tonnes).

At first this mass appears astronomical but compared with the scale of the object itself, it is modest. To give some perspective, three Boeing 747-400 aircraft would fit wing-tip to wing-tip inside a 200m diameter circle and the (take-off) mass of one of these aircraft is 380,000 kg. Steady bending moment in the wing root of a flying Boeing 747-400 is roughly 10 MNm. This compares with a steady bending moment of only **38 MNm** at the root of one of these huge blades due to the 55 tonne mass at radius 85m – if the blade is unsupported by wind.

The energy converted per blade per half-cycle is **26.22MJ**. Given a pressure-ratio of 70:1, equation (8) shows that the volume of air compressed and discharged in each half-cycle for one blade should be **31.62m^3**. This points to a minimum internal cross-sectional area for the compression tube of **0.4865m^2**. Here, rated power would involve compressing and expelling all of the air in the tube and allowing no re-admission. In the next section, we find that the actual compression area must be greater.

6   RIGID-BLADE CALCULATIONS FOR A 200m DIAMETER HAWT

In principle, the piston masses in the present turbine will begin to fall naturally since the peak centripetal acceleration at the mass centre of a diametric pair of pistons is only **9.477 m/s^2**. However, the piston-pair must travel a radial distance of 65m in a time of less than 5 seconds and the residual acceleration of 0.333 m/s^2 would cover a distance of only 4.2m in that time. Some re-admission of HP air is needed to induce a pair of pistons to move sufficiently quickly; how much is determined by a series of simulations tracking the radial motions of the piston pair and treating the **blades** as though they are **rigid** and rotate **at constant speed**. The main independent variable can be considered to be the length of piston travel over which HP is re-admitted. As this variable changes, the cross-sectional area of the compression-tube must also change to keep constant the net output-per-cycle of air to the HP reservoir. The ideal value is that which causes the radial velocity of the piston-pair to return to zero exactly as the pair reaches the end of stroke. Too little reflux will not allow the piston-pair to complete the stroke and too much will leave excess KE in the piston pair which will be dissipated as a very harmful impact. There is a second independent variable here – the angle (before top-dead-centre) at which readmission begins. The optimal value for this angle is that which minimises the reflux of HP air. In the present case, that happens to be close to 60° before top dead centre and the corresponding HP reflux length is **0.3518m** while the corresponding internal cross-sectional area is **0.5480m^2**. In each blade half-cycle, **11.231%** of the HP air discharged is re-admitted to initiate the piston motions - causing radial accelerations in the order of 7g. Figure 4 shows the radial displacement and velocity of one piston mass starting from -60°. The pistons come to rest again shortly before +80° and we can now evaluate the actual gravity utilisation factor as, $\chi = 75.88\%$.

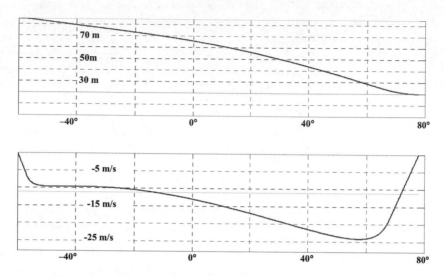

Figure 4. Piston mass position and velocity vs. blade angle.

7 FLEXIBLE-BLADE CALCULATIONS FOR A 200m DIAMETER HAWT

Present HAWTs use blade-root bearings to allow blade pitch adjustments. The requirement for an internal tube of significant diameter (~0.84m) appears to conflict with allowing pitch-control rotation of the blades. In fact, this requirement prompts an alternative construction for the blades where several degrees of pitch control freedom are achieved by articulating finite lengths of aerofoil section about a central *spine* serving as both compression tube and main blade structural member. Figure 5 here illustrates schematically (blade widths exaggerated deliberately).

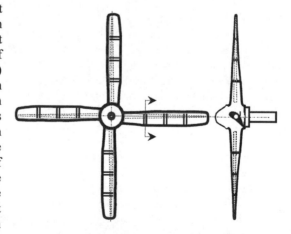

Figure 5. Blade aerofoil sections articulated off *spines*

The blade spines comprise an external "shell" to carry large bending moments and an internal constant-cross-section compression tube. Taking a nominal **internal diameter** of **0.85m** and its tube wall thickness to be **30mm**, the maximum hoop stresses in the internal tube would be in the order of a modest **99MPa** due to the (transient) **7MPa** pressure difference and maximum **longitudinal stress** sustained in this inner tube would be a modest **48MPa**. The **external shell** of each spine might have a thickness of **65mm** and might be **2.5m** in diameter at the root. Then with nominal root bending moment of **37.8MNm**, nominal bending stresses near to the hub would be around **115MPa**. The total mass for a single blade in this case might be 95 tonnes – concentrated mainly at low radii. The allowable stress for "E-glass" (the main-reinforcement material for composite wind-turbine blades) is around **181 MPa** [5].

8 ELEMENTS OF THE DYNAMIC CALCULATIONS FOR A 200m DIAMETER HAWT WITH INTEGRATED COMPRESSION

Equations of motion are prepared for the system of interest. It is assumed that the hub rotates at a fixed speed Ω = 0.540 rad/s about an axis fixed rigidly located in space. Each of the four blades is represented as a simple beam cantilevered off the hub from a radius r = 15m and stretching to r = 100m. The flexural stiffness of each blade is taken to be

$$EI(r) = 6\times10^9 + (120 - r)^3 \times 190,000 \ (\text{Nm}^2) \tag{9}$$

This has a peak value of 226×10^9 Nm^2 at r=15m and clearly falls to 6×10^9 Nm^2 near the tip.

The mass-per-unit-length of each blade is given by

$$\rho A(r) = 1400 + (120 - r) \times 20 \ (\text{kg/m}) \tag{10}$$

This has a peak value of 3500 kg/m at r=15m and falls to 1400 kg/m at the tip. It corresponds to a net blade mass of 257 tonnes. The moving mass in each blade is rounded to 90,000 kg and the centre of gravity of this mass is assumed to oscillate between r=20m and r=110m in each blade. The moving mass runs inside a circular tube of internal diameter 1.05m and it is guided within that tube by a system of bearings. The total friction/rolling-loss of the bearing system and piston-seal arrangements is estimated as 10kN. The masses in diametrically-opposite cylinders are tied together such that the distance between them remains constant.

Blade deflections are considered to comprise only flexural deformations in the rotation plane - measured relative to a rigid rotating reference frame. The deflection pattern of any one blade is expressed as a linear combination of the first ten natural mode-shapes of a straight cantilever beam having the properties described above (excluding the piston mass but including the effects of stress-stiffening. At any instant, the deflection state of each blade is then described by the ten deflection coordinates $\{q_1(t),\ q_2(t) \ ... \ q_{10}(t)\}$ and its velocity state is described by the ten velocity coordinates $\{v_1(t),\ v_2(t) \ ... \ v_{10}(t)\}$. An advantage of using these mode-shapes is that reasonable damping values can be assigned to the individual modes. Here, a simple 1% damping is applied to all of these modes and obviously this damping does not couple the modes. The radial position and velocity of the piston mass are included as further state-variables. The decision to exclude blade extension/compression is vindicated to a first order by recognising that even with the 55,000 kg piston mass locked at the largest radius, the first radial resonance frequency is around 75 rad/s (around 200 times fundamental frequency) and without the piston mass the first radial resonance is 100 rad/s. The tenth flexural resonance (excluding the piston mass) is just over 300 rad/s but including 10 flexural coordinates for the blade is justified because the presence of the piston mass at any one radial position has a dramatic influence on the mode-shapes. A formal criterion could be developed for how many blade flexural resonances to include but the intuitive test of including more mode shapes as displacement coordinates and checking whether this makes any noticeable difference is straightforward and shows that ten is more than sufficient here. Figure 6 shows the first four flexural mode-shapes of the blade given that properties follow (9) and (10).

The blade dynamics are considered to be influenced by a number of different effects here:

(i)   Gravity loads acting transversely on the blades change as the blade rotates. Gravity also acts on the piston mass in the circumferential direction.
(ii)   Centrifugal loads on the blades themselves cause a tension to exist in each one. This tension has the effect of stiffening the blade and it is present constantly.
(iii)   Gravity also contributes significantly to the tension in the blade and hence to the "stress-stiffening". This stiffening varies periodically. This tension (and the tension arising from centrifugal forces) introduces a strain-energy contribution proportional to the integral of $(T\alpha^2)$ [6] where T is the local tension and α represents the slope of the blade relative to the radial direction. In both cases the forces which originate the tension are "body-forces" which act in a specific instantaneous direction.
(iv)   Bearing/seal friction between the mass and the blade modifies the blade tension for all R between $R=15$m and the piston mass centre. This friction is taken to be independent of piston velocity relative to the blade except in its direction of action.
(v)   When the piston mass is compressing on the outward stroke, tension in the blade is increased. This presents a further time-varying stiffening effect
(vi)   When the mass is moving (relative to the rotating reference frame), Coriolis effects cause forces to exist on that mass in the direction perpendicular to the movement of the mass (within the rotating frame).

The effect of wind-shear is deliberately excluded here – not because it is thought unimportant but because the emphasis in this paper is to understand the large oscillatory components of loading caused by the motions of the piston masses.

530

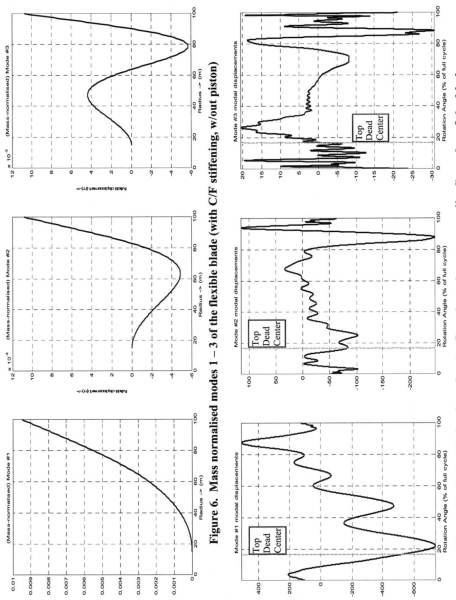

Figure 6. Mass normalised modes 1 – 3 of the flexible blade (with C/F stiffening, w/out piston)

Figure 7. Modal deflections for the first three (mass-normalised) modes of the blade.

9   CAUSING THE SIMULATION TO PRODUCE PERIODIC SOLUTIONS

In the real application, real-time controllers operate to control valves as the pistons approach end of stroke such that if the pistons have an excess of energy approaching the end, pressure is allowed to rise noticeably above the HP reservoir pressure in order to dissipate the residual energy smoothly before a major impact. Conversely, if the pistons had insufficient energy to complete the stroke, a small deliberate leakage of some HP air would reduce the demand on the piston so that it could, after all, complete the stroke. In the simulations conducted here, the ends of stroke were represented by equipping the ends with an artificial "attraction" for the pistons (when they are approaching end of stroke) and simultaneously endowing this interconnection with some damping. This artificial effect would not be needed if the flexure of the blades did not have a substantial effect on the fall of the piston masses but this is not the case. The attraction forces are restored to zero by the time that the piston pair is due to begin its next fall. Because the interaction between the piston masses and the stroke ends is dissipative, the quantity of re-admitted air must be increased to overcome the resultant (artificial) losses. The final result is a simulation which does achieve a periodic solution and which represents approximately what would occur in one of these large wind-turbines having integrated compression. An additional strategy had to be used also to achieve periodic solutions: the blade mass and stiffness matrices were multiplied by a large factor (~1000) during the simulation of the first cycle to suppress blade flexural response to very low levels. Then in subsequent simulations of individual cycles, the blade mass and stiffness matrices were gradually scaled back to their correct value and after achieving this value, a further 50 cycles were simulated to consolidate periodic behaviour. Non-periodic vibration behaviour in normal HAWTs has already been commented-upon [6].

10   RESULTS for the 200m DIAMETER WIND TURBINE

Figure 7 shows the modal responses of the first three (mass-normalised) modes. The dash-dot vertical line in each plot shows the angular instant at which the blade reaches top-dead-centre.

Note that since mode 1 achieves a peak instantaneous value of nearly 800 and since the circumferential displacement at the blade tip is 0.0095m for mass-normalised mode 1, we can infer that **blade tip movements** in the order of **7.5m** are occurring.

The natural frequencies of the first three modes were 0.3889Hz, 1.3394Hz and 3.2292Hz respectively. The modal deflections for seven other blade flexural modes were also calculated. Collectively, these modal deflections can be combined to determine what would be the instantaneous blade root moments and blade tip deflections. Figure 8 comprises plots of this blade root bending moment and blade tip deflection for a pair of diametrically-opposite blades. The periodicity of the behaviour is evident from the fact that the two blades undergo identical blade root moments and tip deflections – phase shifted by 180°. The maximum blade tip deflection of **7.48 m** occurs when the blade is roughly horizontal and rising. At this instant a piston mass is close to the tip of the blade and the aerodynamic power from four blades is being transformed into gravitational potential energy primarily in that one blade so it is not surprising that a visible downward deflection of the blade would occur.

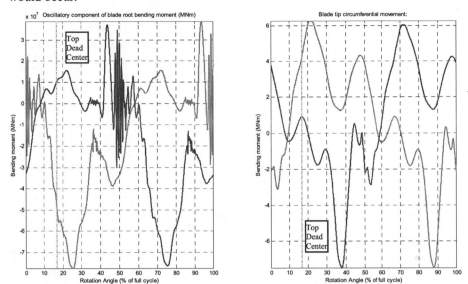

Figure 8. Blade root moment and blade tip deflection through one cycle for two opposite blades

The maximum blade root moment occurs at around 35° after top-dead-centre and has magnitude **77.5 MNm**. When a single blade of this machine is horizontal with the piston at its radial innermost position, the blade-root moment caused by gravity is **31.1 MNm**. For a horizontal blade with the piston mass at the radial outermost position, the gravitational blade-root moment is **68.9 MNm.** In this condition, the wind itself would usually be alleviating this moment with a steady blade-root moment of around **8 MNm**.

We now consider the importance of some of the effects which govern the actual dynamics of this wind-turbine rotor having integral compression. Table 1 below shows the resonance frequencies of one blade with and without the gravitational and centrifugal tension stiffening.

Table 1. Blade Resonances (Hz)

Mode #	C/F Off Gr. Off	C/F On Gr. Off	C/F Off Gr. On	C/F On Gr. On
1	0.3471	0.3889	0.3651	0.4050
2	1.2932	1.3394	1.3124	1.3579
3	3.1842	3.2292	3.2033	3.2480
4	6.0277	6.0726	6.0471	6.0919
5	9.8242	9.8692	9.8438	9.8887
6	14.571	14.616	14.591	14.636

Here, the effect of the gravity stiffening (which alternates sinusoidally with the rotation) is in the order of 50% of the effect of centrifugal stiffening. The latter more than 25% to the flexural stiffness of mode 1.

As a piston approaches the radial extreme of a blade, the blade tension rises by a further **3.79MN** due to the internal pressure. This compares with a blade-root tension due to gravity of a maximum of **0.90 MN** and a blade-root tension due to centrifugal force of **1.32MN**. Evidently, the internal pressure in the blade during an outward stroke has a very strong influence on the dynamics.

As one piston falls, the Coriolis forces exerted are very substantial. From Figure 4, peak radial velocities of **24.2 m/s** occur at one instant where the moment-arm on the blade root is **54 m**. If the blades were rigid, the instantaneous blade-root moment would be **77.6 MNm** due to the Coriolis forces alone. The flexibility of the blades alleviates this substantially. Some impression can also be obtained about the importance of the radial Coriolis forces arising because of changes in circumferential velocity. The first flexural mode strongly dominates blade deflections. This is as expected since the rotational speed is only around 20% of the first blade resonance frequency. Peak blade tip circumferential velocities of **13.9 m/s** occur indicating that a radial Coriolis component of force of **0.825 MN** is in place – equivalent to piston acceleration of **15 m/s²**. Although this is clearly very significant compared with gravity, it is less than 25% of the piston acceleration/deceleration in place while the blade is either discharging or inducting high-pressure air.

11   CONCLUSIONS

This paper has outlined a dynamic simulation for the rotor of an extremely large wind-turbine having a compression facility integrated within it to enable all of the power extracted from the wind to be converted without transmitting any net torque through the turbine shaft. The compression is based on heavy pistons falling within compression-tubes fitted within the blades themselves under the influence of gravity. It is abundantly clear that this concept can only work well at very large scales and this paper has provided the first analysis of such a possibility with a view to setting out approximately what bending moments and blade tip deflections can be expected. The paper shows that, though large, these are manageable.

The simulation itself is rich in features. It is shown that centrifugal stiffening, gravitational stiffening, stiffening due to blade tension when a piston is discharging outward and Coriolis forces all have a substantial influence on the blade dynamics. The analysis done here presumed that the rotor hub rotated at constant speed. Relaxing this assumption would make increase the level of complexity of the simulation further.

12 REFERENCES

[1] Lanchester F. W. "A contribution to the Theory of Propulsion and the Screw Propeller". Trans. Inst. of Naval Architects. vol. 57, p98-116, 1915

[2] Betz. A "Das Maximum der theoretisch möglichen Ausnutzung des Windes durch Windmotoren". Zeitschrift fur das gesamte Turbinewesen, vol. 26, p307-309, 1920

[3] Gorban A. N., Gorbov A. M. and Silantyev V. M. "Limits of the Turbine Efficiency for Free-Fluid Flow". ASME Journal of Energy Resources Technology, vol. 123, 2001, pp 311- 317.

[4] Hau E. "Wind Turbines: Fundamentals, Technologies, Application and Economics". Ed. 2. Springer. 2006

[5] Kong C., Kim, T, ., Han D. and Sugiyama Y "Investigation of Fatigue Life for a Medium Scale Composite Wind Turbine Blade". International J. of Fatigue, vol. 28, 1382-1388. 2006.

[6] Lin S.M. and Lee S.Y. "Modelling and Bending Vibration of the Blade of a Horizontal-Axis Wind Turbine". Computer Modelling in Engineering and Sciences, vol. 23(3), p175-186. 2008

FAULTS

Improved simulation of faults in rolling element bearings in gearboxes

N Sawalhi, R B Randall
School of Mechanical and Manufacturing Engineering,
University of New South Wales, Australia

ABSTRACT

Faults in rolling element bearings in a gearbox have previously been simulated using a detailed model of the gears, shafts and rolling element bearings, but only a simple model of the casing supporting the bearings. This included only one low frequency (rigid body) mode of the casing on its supports, and one High frequency to represent a typical high frequency response excited by the bearing faults. For signals bandpass filtered around this resonance, the results of envelope analysis were quite similar to measured signals, but the overall spectrum comparison over a wide frequency range was poor. In this paper, much improved spectral matching has been achieved by using a finite element model of the casing to determine its response over a much wider frequency range. Results are given for localized faults in the inner race, outer race and rolling elements, but it is intended to apply the model to extended faults in the inner and outer races, where an even greater improvement in the modelling is expected.

1. INTRODUCTION

There are three main advantages in simulating faults in rotating and reciprocating machines:

1) The ability to place and dimension the faults where desired, rather than having to wait until they occur naturally and quite randomly.
2) To aid in understanding the complex interactions between components, in particular where nonlinear interactions are involved. In gearboxes, information from bearing faults is often best carried by their effect on the meshing gears.
3) Perhaps most importantly, it is not viable to experience the number of failures required to collect sufficient data to train artificial neural networks to recognize the type, location and size of developing faults. Simulation gives this possibility.

Dynamic modelling of rolling element bearings has been extensively developed (e.g [1] and [2]). Different studies are emerging to combine the bearing model with other

---

rotating parts in the machine (rotor, gears, etc) in order to study the effect of the vibration induced by bearings on other components. For instance, reference [3] studied the vibration transmitted through rolling element bearings in rotor system and in geared rotors.

In gearboxes, faults in rolling element bearings manifest themselves in two ways, as additive (in the case of localized faults) series of impulse responses, typically at response frequencies much higher than the repetition rate of the rolling elements striking the fault in a race, and as modulation effects (in the case of extended faults), since the gearmeshing forces are modulated by the varying support given by the bearing in the case of a fault [4]. A simulation model of a gearbox with bearings should reproduce these two failure indication modes. In this paper only localized faults are considered, but in future work extended faults will also be modelled. An early simulation model was presented in [5], based on detailed modelling of the gears, shafts and bearings, but with only a primitive model of the gearbox casing. This added only two resonances to those of the internals (torsional and lateral vibrations), a low frequency rigid body resonance of the casing (or bearing pedestal) on its mounts and a high frequency (15 kHz) resonance to represent a typical resonance excited by localized bearing faults. The gearbox being modelled had a number of resonances in this range. Ref. [5] shows that by demodulating an optimum band around the high frequency resonance gave very similar results to measurements made on the modelled gearbox for localized faults in the inner race, outer race or rolling elements, but that for extended faults the matching was not so good, presumably because of the much wider frequency range over which the modulation action of the extended faults acted. This paper describes the development of a much better model of the gearbox casing, based on a finite element (FE) model, to give improved matching of the response signals over a wider frequency range. In this paper, only results for localized faults are given, but the same model will later be tested for extended faults as well.

2.   FE MODEL OF THE GEARBOX

The UNSW gearbox under investigation (Figure 1) was built by Sweeney [6] to investigate the effect of gear faults on transmission error. The gearbox is mounted to the rig bedplate through rubber pads to isolate the general vibration from the gear case [6]. In this test rig, the single stage gearbox (in this case a spur gear set with 1:1 ratio and 32 teeth on each gear) is driven by a 3-phase electric motor to make up losses and control the speed, but with circulating power via a hydraulic pump/motor set. The input and output shafts of the gearbox are arranged in parallel and each shaft is supported by two double row ball bearings (Koyo 1205). The flywheels are used to reduce the fluctuations of the input and output shaft speeds. The couplings are flexible in torsion and without stiffness in bending, making them very helpful for the attenuation of the shaft torsional vibration.

Figure 2 shows the meshing of the FE model (using shell elements) of the gear-case, which for simplicity was mounted on springs in the corners to represent the actual rubber pads a little in from the corners. The modal analysis was performed using

Patran®/ Nastran®. Mesh seeds were used to refine the mesh at the edges and ensure the connection between the elements on the different surfaces. This was also used to refine the mesh around the holes, and to connect the rolling element bearings to the gear-case. The stiffness values of the spring elements used to represent the rubber pads were tuned to update the model (see model update section below). The stiffness in the vertical direction was seven times greater than that in the axial and lateral directions for an optimum model update.

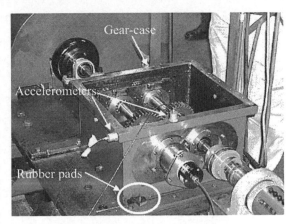

Figure 1 UNSW gearbox showing the internals, the gear-case and the rubber pads

2.1 Modelling the Internals

The gearbox internals (gears, shafts and rolling element bearings) were modelled using different elements. Gears were modelled using solid elements. The two gears were connected using spring elements. As the Line of action (LoA) for the two gears is at 20 degrees (pressure angle), the total stiffness was resolved in the z and x directions. The gears were connected to the shafts using multiple point connectors

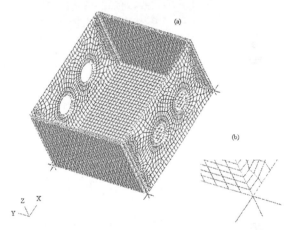

Figure 2 FE model of gearbox casing, showing rubber pad attachments in the corners

(MPS) elements to form a rigid link. Shafts were modelled using beam elements. Rolling element bearings were modelled using spring elements, which were connected the shafts on one side and to the gear-case on the other side.

2.2 FE Model Update

In order to improve the accuracy of the FE model, so that it closely matched the results of the experimental modal analysis (carried out previously by Endo, 2006 [7]), the stiffness values of the rubber pads in the horizontal, vertical and axial directions were updated so to match the low order mode shapes. The stiffness of the gearmesh and the bearings were also allowed to vary in the update. A comparison between the FEA and experimental modal frequencies (first ten, including some non rigid modes) is presented in Table 1. Figure 3 shows a typical comparison between experimental and FE mode shapes and frequencies (Mode 2).

Table 1 Comparison of Experimental and FE Modal Results

Mode #	Experimental modal frequency (Hz) [7]	FEA result (Hz)	Error (%)
1	63.9	71.4	11.7
2	100	98.7	1.3
3	174	174.6	0.3
4	251	234.4	6.6
5	319	314.6	1.3
6	344	320.8	6.7
7	404	417.4	3.3
8	414	428.6	3.5
9	445	476.2	7
10	543	508.1	6.4

Even though the updating was only carried out for frequencies up to about 500 Hz, the modes of the FE model up to 20 kHz were used in generating Frequency Response Functions (FRFs) for the later response analysis. Even though these could not be expected to be accurate, this was considered to be of lesser importance at high frequencies where the modal overlap would be quite high, and where typical responses would represent averages over several modes. The average modal spacing is approx. 40 Hz over the whole range, so for 1% damping the overlap ratio would be >1 above 2 kHz.

3.   GENERATION OF FRFs

Frequency Response Functions (FRFs) were extracted from the FE model by synthesizing them from the natural frequencies and mode shapes in the LMS Virtual Lab® environment. LMS Virtual Lab® allows the user to compute the FRF between selected input and output degrees-of-freedom (DOFs). The synthesis is based upon the modal superposition of the selected mode set. The results of the modal analysis (mode sets) were imported by LMS® where the FRF computations were performed. The FRFs were calculated between the bearing force locations (input points) and the accelerometer position (output point).

Since the connection between the internals and the casing is only at the bearings and their supports, which are relatively massive and stiff compared with the rest of the casing, it was assumed that the actual response of the casing would have little effect on the bearing forces, and it was decided to extract the latter from the original lumped parameter model (LPM) as described in [5]. These could then be applied to the FE model of the casing (plus internals) to determine the response at the vertical accelerometer mounting position shown in Figure 1. However in future work on this combination, a modal model including the casing within the closed loop will be considered.

3.1 Lumped Parameter Model (LPM)

The LPM is a 34 DOF model as shown in Figure 4, and described in detail in [5]. It models the gears, shafts and bearings in detail, and for example the gearmesh is modelled as a time (actually rotation angle) varying nonlinear stiffness, with the possibility of adding geometric (i.e. unloaded static) transmission error as one of the inputs to the system [7]. Similarly, the bearings are modelled as time varying nonlinear stiffnesses, and either localised or extended faults can be added to the raceways and rolling elements (localised only). An important part of the bearing model is that the small random variations in spacing of the rolling elements caused by slip are included. The way in which the faults are modelled is described in detail in [5]. As mentioned above, the 34 DOF model only includes a primitive model of the bearing supports. The system equations are solved using Simulink®.

3.2 Combining the LPM and the FE models

The bearing forces from the LPM model were convolved with the impulse responses corresponding to the FRFs of the casing. A flow chart showing the process involved in combining the two models is presented in Figure 5. The convolution was performed in Matlab® using the efficient overlap/add method based on FFTs.

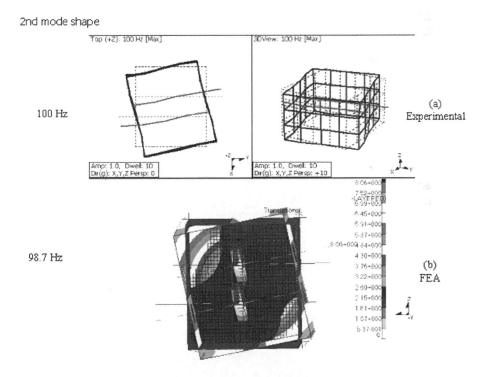

Figure 3 Comparison of experimental (a) and FE (b) mode shapes and frequencies for Mode 2.

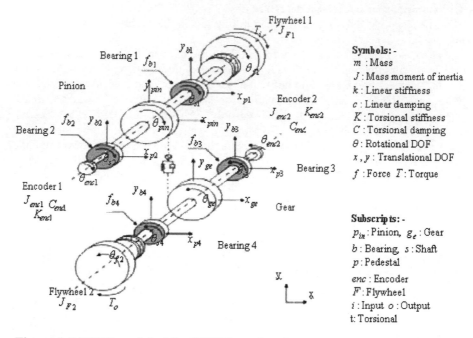

Figure 4 34 DOF model of the UNSW gearbox internals and bearing supports

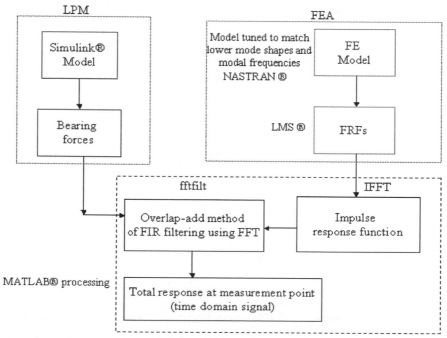

Figure 5 Schematic showing the generation of response signals at the accelerometer position

4. RESULTS FOR LOCAL FAULTS

4.1 Inner Race Fault

Figure 6 shows the time signals and spectra for a localised inner race fault as simulated in [5], Part 1. The figure compares measured signals with the more complicated simulations described in this paper and the simpler simulations from [5]. The time signals are bandpass filtered in the optimum band found using Spectral Kurtosis (SK) which maximises the kurtosis (impulsiveness) of the filtered signal, a complex Morlet wavelet filter bank being used for the filtration [7]. Broadband and envelope spectra are shown. The envelope spectra (g, h, i) correspond to the three filtered time signals of (a, b, c), this being the normal way of making the diagnosis.

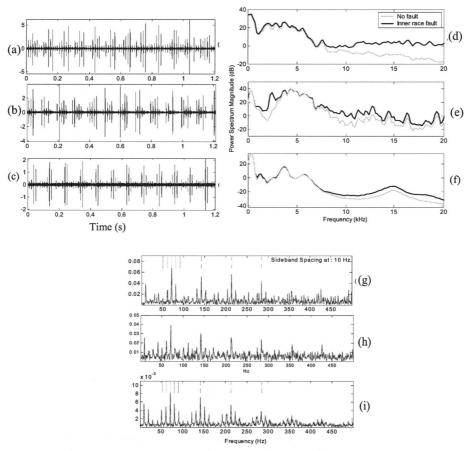

Figure 6 A comparison of filtered time signals (m/s²) (a, b, c) broadband spectra (d, e, f) and envelope spectra (g, h, i) for measured signals (a, d, g) and signals simulated using the FE model (b, e, h) and the lumped parameter model (c, f, i). The time signals and envelope spectra are with the inner race fault, and broadband spectra with and without the fault.

545

For the localised inner race fault, the results with the improved model are all closer to the measured results than those from the simple model, but in particular the broadband spectra are more similar, as expected. The only anomaly, where the improved model shows a small discrepancy, is that the response in the frequency range around 2.5 kHz has increased as a result of the fault, this not being the case for the measurements. An explanation of this is being sought, but it is still clear that the results of envelope analysis by filtering a high frequency band determined as having maximum spectral kurtosis give a clear diagnosis of the inner race fault, with harmonics of the BPFI (ballpass frequency, inner race) surrounded by sidebands spaced at the shaft speed 10 Hz, the rate at which the inner race fault passes through the load zone.

4.2 Outer Race Fault

Figure 7 shows the filtered time signals for a localised outer race fault, and also the envelope spectra, for the new model compared with the measurements and the simple LPM. In this case also there is a better match between the results of the new model and the measurements than for the simpler model. For an outer race fault with unidirectional load, the fault is always in the load zone, so there are no modulation sidebands around the harmonics of BPFO (ballpass frequency, outer race). The broadband spectra in this case were quite similar to those of Figure 6.

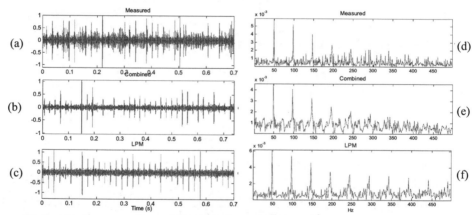

Figure 7 A comparison of filtered time signals (m/s²) (a, b, c) and envelope spectra (d, e, f) for a localised outer race fault. Measured signals (a, d) are compared with signals simulated using the FE model (b, e) and the lumped parameter model (c, f).

4.3 Ball Fault

Figure 8 shows the filtered time signals for a localised ball fault, and also the envelope spectra, for the new model compared with the measurements and the simple LPM. Once again there is a better match between the results of the new model and the measurements than for the simpler model. A rolling element fault passes through the load zone at cage speed, so in the envelope spectra, sidebands can be seen with this spacing.

5. CONCLUSION

Adding a finite element model of the gearbox casing has improved the agreement of simulated bearing faults and measurements, at least for the localised faults in inner race, outer race and rolling element presented in this paper. The same model will be used to check the response to extended faults in the inner and outer races, and it is expected that the results for this should be even better because of the wider frequency range over which they manifest themselves.

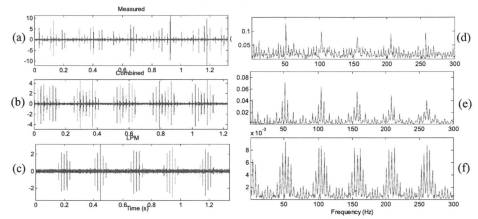

Figure 8 A comparison of filtered time signals (a, b, c) and envelope spectra (d, e, f) for a localised ball fault. Measured signals (a, d) are compared with signals simulated using the FE model (b, e) and the lumped parameter model (c, f).

ACKNOWLEDGMENT

Support for this study was given by the Australian Defence Science and Technology Organisation (DSTO) under their Centre of Expertise scheme.

REFERENCES

[1] Fukata, S., Gad, E.H., Kondou, T., Ayabe, T. and Tamura, H. (1985): On the vibration of ball bearings. *Bulletin of JSME*, vol.28 (239), pp. 899-904.

[2] Liew, A., Feng, N. S. and Hahn, E. J. (2002): Transient rolling element bearing systems, *Trans. ASME Turbines and Power*, vol. 124(4), pp. 984-991.

[3] Lim, T. C. and Singh, R. (1991): Vibration transmission through rolling element bearings, Part III: Geared rotor system studies. *Journal of Sound and Vibration),* vol. 151(1), pp. 31-54.

[4] Antoni, J, & Randall, RB, (2002) "Differential Diagnosis of Gear and Bearing Faults", *ASME Journal of Vibration and Acoustics*, Vol, 124, Apr 2002, pp. 165-171.

[5] Sawalhi, N, and Randall, RB, (2007) "Simulating gear and bearing interactions in the presence of faults, Part I. The combined gear bearing dynamic model and the simulation of localised bearing faults" and "Part II: Simulation of the vibrations produced by extended bearing faults", *Mechanical Systems and Signal Processing*. In press.

[6] Sweeney, PJ,(1994), "Transmission Error Measurement and Analysis". PhD Dissertation, UNSW, Sydney 2052, Australia.

[7] Endo, H (2006), "Simulation of Gear Faults and its Application to the Development of a Differential Diagnostic Technique". PhD Dissertation, UNSW, Sydney 2052, Australia.

[8] Sawalhi, N, and Randall, RB, (2007) "Semi-automated Bearing Diagnostics – Three Case Studies", Comadem Conference, Faro, Portugal, June.

Diagnostics of a spiral bevel gearbox using signals of the motor inverter

R Ricci
Dept. of Mechanical Engineering, Politecnico di Milano, Piacenza, Italy
P Pennacchi
Dept. of Mechanical Engineering, Politecnico di Milano, Milan, Italy

ABSTRACT

Condition monitoring and diagnostics of rotating machines have reached great importance as a consequence of both the increase of the system complexity and the necessity of reducing out of service time. The characterization of system health depends strictly on the effectiveness of signal processing techniques used for the analysis. Moreover, the cost of monitoring system should be consistent with that of the mechanical system taken into consideration: using an expensive diagnostic apparatus to monitor an economical mechanical component is a coarse error.

This paper analyses a spiral bevel gearbox; the aim of authors is the implementation of an economical monitoring system that allows the detection of the fault in early stage: the techniques commonly used for spur gearbox diagnostics are applied to signals of the motor inverters. No additional transducers are used.

1 INTRODUCTION

The great complexity reached by mechanical system and the necessity to know their condition during operation have increased the importance of diagnostics in recent years. Detection of both fault entity and position reduce the standstill of the plant. The reduction of the time needed for repairing and the consequent money savings have encouraged investments of resources in diagnostics.

Several types of diagnostics techniques based on experimental signal acquisition and analysis are proposed by different authors. Since each technique is based on different physical theory, also the results obtained are different; in other words, given a mechanical system to analyse, it is possible to know its condition using different methods. Therefore, for a reliable mechanical analysis it is important to chose techniques that are the most effective for the case and the situation under testing.

Gearboxes are widely used in several industrial fields, ranging from electric utilities to automotive, from ships to helicopters. While many techniques for spur gear monitoring have been proposed by different papers during the two last decades, no study has been developed for spiral bevel gears.

Boulahbal et al. [2] have suggested the use of wavelets for spur gear conditions monitoring: firstly the phase and secondly the amplitude analyses of wavelet transform of vibration signals acquired on the gearbox allow detecting the transients generated both by wear and by cracked teeth. A similar approach is proposed by Ohue et al. [7] for the evaluation of gear coupling dynamical behaviour: in this case wavelets are used first for the definition of the vibration level of sintered and steel gear and then for the pitting detection.

Second order cyclostationariety is the signal analysis technique taken in to consideration by Capdessus et al. [3] and by Zhu [8]: vibration signals, measured respectively on the bearing housing of the pinion and on the automobile gearbox, are used for the gear health monitoring at different operating condition.

Wang [6] has proposed the resonance demodulation technique: this methodology is based on the hypothesis that the defect, like a cracked tooth, excites the structural resonances. The residual signal, obtained by removing the gear meshing harmonics from the synchronous average signal [1], is band-pass filtered and then demodulated to extract the features. The resonance demodulation is suitable for spur gears, since the meshing harmonics are easily detected in the frequency domain. Spiral bevel gear could hide these components reducing therefore the effectiveness of the methodology.

Several suitable gear monitoring techniques are compared [4] by Dalpiaz et al., which take into consideration both conventional time-domain methodologies, like synchronous average signal, and the more complex wavelet transform and cepstral analysis.

The objective of this paper is therefore the application of several diagnostics techniques, proposed for the spur, to spiral bevel gearboxes. The results obtained in this case may be different from those achieved until now in literature; the comparison among the different techniques allows the definition of the most effective for the application considered.

Moreover the signal taken into consideration in the test-rig are not vibrations but rotating speed and torque signals, contrarily to the previous studies. The aim of this work is also the realisation of an economical condition monitoring system: signals analysed are acquired directly from the motor inverters without adding others transducers. This allows reducing the complexity of monitoring system and saving money.

The paper is organised as follows: the experimental test-rig is presented in the next section, while all the signal analysis techniques are introduced in section 3. The analysis using the synchronous average signal is reported in section 4, whereas results of the application of other and more complexes techniques are discussed in the next sections.

2 DESCRIPTION OF THE EXPERIMENTAL SET-UP

As stated in the introduction, signal analysis techniques have been applied mostly to spur gears in literature. In this case, the aim of the study is a spiral bevel gearbox with orthogonal axes, shown in Figure 1 (left); the gear ratio is set to 1:5 and the contact ratio is between 2 and 3.

Figure 1: spiral bevel gearbox (left), test-rig layout (right).

The test-rig of Piacenza Campus of Politecnico di Milano, shown in Figure 1 (right), is constituted by the driving asynchronous motor (DM in the following), inverter controlled, the gearbox and the user unit (UM in the following). This last unit is another asynchronous motor similar to the previous, which provides the resistant torque. Its inverter is connected to a personal computer. In this way, it is easy to impose the resistant torque applied by the motor, using the control software.

The considered gearing is characterized by a smooth behaviour due to its geometry (spiral bevel gears) and its contact ratio. Therefore, the application of smart analysis techniques, to early detect faults of the gearing, is even more challenging.

With the aim to explore different working conditions, several data acquisitions have been performed with different operating speeds (550 rpm, 1100 rpm, 1650 rpm, 2200 rpm), set in the driving motor inverter, and with different resistant torques of the user (0 Nm, 2 Nm, 4 Nm, 6 Nm, 8 Nm, 10 Nm).

In these conditions, the driving inverter control has a reference constant speed that leads to the equilibrium between driving and resistant torque.

A damage of the teeth will suddenly affect this equilibrium producing variations in the torque transmitted and in the rotating speed. When these variations fall outside of the inverter control bandwidth, then the system is not able to immediately compensate them and they can be measured. This hypothesis has been verified by applying different analysis techniques either to a new gearing and to a damaged one (one tooth artificially removed at about 210° with respect to the phase reference). The results of this procedure are presented in the next sections.

3 THEORETICAL BACKGROUND OF METHODOLOGIES

Direct comparison of experimental signals acquired on the same mechanical system but in different condition is a nonsense. Different length of acquisition records and the presence of noise above all do not allow the proper analysis of measured phenomenon. Therefore it is necessary to obtain a signal that actually reflects the dynamics taken into consideration. This is possible by realizing the *synchronous average* of the signal, under the hypothesis that the mean of the acquired signal is periodical. The hypothesis is generally confirmed in mechanical systems with rotating shafts.

In the present study, the synchronous average of the signal is based on the subdivision of acquired time histories into cycles of n user gear revolution and on their grouping in different vectors. Each vector $\mathbf{y}_k = \{y_k(0),...,y_k(m_k)\}$ can have a different time resolution: the cycle having the highest number of acquired points is the vector with the best time (and angular position) resolution.

In order to obtain cycles of the same length, each vector must be re-sampled by means of a linear interpolation. Assuming the continuity of the motion transmission in the spiral bevel gearbox, time can be substituted by angular position: in this way, also the angular localisation of faults is possible.

$$l_k = \text{length}(y_k) \tag{1}$$

$$da = \frac{360° \cdot n}{\max(l_k)} \tag{2}$$

$$\mathbf{\alpha} = \{0°, da, ..., 360° * n - da\} \tag{3}$$

$$\mathbf{y}_k \rightarrow \mathbf{y}'_k = \{y'_k(\alpha_1),...,y'_k(\alpha_p)\}, \quad p = \text{length}(\mathbf{\alpha}) \tag{4}$$

Synchronous average signal is obtained by means of the average of all cycles in each angular position, that is the average of the rows of the matrix for each column.

$$cycles \begin{bmatrix} y'_1(\alpha_1) & y'_1(\alpha_2) & y'_1(\alpha_3) & \cdots & y'_1(\alpha_p) \\ y'_2(\alpha_1) & y'_2(\alpha_2) & y'_2(\alpha_3) & \cdots & y'_2(\alpha_p) \\ \cdots & \cdots & \cdots & \cdots & \cdots \\ y'_r(\alpha_1) & y'_r(\alpha_2) & y'_r(\alpha_3) & \cdots & y'_r(\alpha_p) \end{bmatrix} \overset{position}{} \tag{5}$$

The average reduces the influence of random components: the resulting signal is therefore representative of the studied phenomenon with less noise and fluctuations than the original acquired signal. The synchronous average is both the simplest technique for fault detection in gearboxes and the reference signal for other techniques.

The *residual analysis* requires first the calculation of the differences between the synchronous average and the raw signal in the frequency domain: this way the fundamental frequencies are cancelled from the FFT of the synchronous average. The inverse transformation of the modified spectrum in the position (or in the time) domain provides the residual distribution. Undamaged gearboxes have typical uniform residual distribution, while sudden amplitude variations are symptoms of fault presence. Whereas for spur gear the frequencies to be cancelled are the meshing frequency and its multiples in the synchronous average spectrum, for spiral bevel gears these components are different and unknown. The motor rotation frequency and its multiples are considered in this case, because these components are those that introduce energy in the mechanical system.

The *demodulation analysis* is the third technique used: also this analysis is based on the knowledge of the reference signal spectrum. The signal of the synchronous average is transformed using FFT: the result is pass-band filtered in the frequency domain. Normally, the filtering bands are those centred on the meshing frequency for the spur gearbox. The part of the spectrum taken into consideration is then normalized with respect to the values (real and imaginary parts) assumed by the first filtered frequency: the result obtained from this operation is finally inverse transformed in the position domain. Differently to the approach proposed in [1], which foresees the amplitude (AD) and the phase (PD) demodulation separately, a unique normalization is used in this study. This is a consequence of a brief analysis that showed AD and PD as not effective for spiral bevel gears. Therefore, a partially different approach is applied in this paper: results obtained with this procedure are commented in the following. Like residual analysis, sudden variations of signal show an anomaly in the transmission of motion and allow the localisation of the fault position.

The *cepstrum analysis* is more complex than other methodologies taken into consideration before: cepstrum is obtained by inverse Fourier transform of the logarithm of the auto-spectrum:

$$C_{p_{xx}}(\tau) = F^{-1}\{\log_{10} G(\omega)\} \tag{6}$$

The logarithm applied to the auto-spectrum allows highlighting of low amplitude harmonics. The components of the resulting signal are now in the "quefrency" domain and are named "rahmonics"; the inverse of the spacing between rahmonics of the same family represents the distinctive frequency of the phenomenon. The position of the rahmonics along the horizontal axis shows the presence of frequencies taken into consideration in the signal. Therefore, cepstrum analysis allows fault detection in both frequency and position.

The *wavelet transform technique* provides similar information to the cepstrum analysis: this technique is based on the evaluation of the correlation between a wave and the synchronous average signal. The original wave (mother wave) can be rescaled and shifted in time. This approach is different from the techniques discussed before: the wavelet transform requires the knowledge of the frequencies of interest before the analysis. The resulting 3D graph highlights the possible presence of some frequencies in the signal; for example, a fault symptom can be shown by a sudden variation of the correlation between the wave and the signal at the user rotation frequency.

The last technique employed is the analysis of the *second order cyclostationariety*, which is possible giving that the autocorrelation of the synchronous average signal is a periodical function. This hypothesis is confirmed for the mechanical system considered. The spectral correlation density function (SCD) is defined as:

$$S_x^\alpha(f) = \int_{-\infty}^{+\infty} R_x^\alpha(\tau) e^{-2\pi j f \tau} d\tau \qquad (7)$$

and it can be calculated by means of the matrix used to define the synchronous average signal in eq. (5). The procedure is the following: 1) considering matrix in eq. (5), the autocorrelation function is calculated per each angular position; 2) then, the auto-spectrum of each autocorrelation is obtained by means of Fourier transform. The new matrix, which is the correlation spectrum among different cycles per each angular position, is known as the Wigner-Wille spectrum; 3) a further Fourier transform, in this case with respect to each cycle, is applied to the matrix and the SCD is finally obtained.

In order to obtain an easy to understand displaying of the results, the sum of different cycles with respect to the frequencies is evaluated. The most relevant frequencies of the original signal are those that have the highest peaks.

4 SYNCHRONOUS AVERAGE

Figure 2 shows the synchronous average, based on 5 user revolution, of the signal of UM speed, for undamaged (left) and damaged gear (right) respectively. The averages are relative to DM speed of 2200 rpm and resistant torque of 10 Nm.

By comparing the two signals, it is possible to note big differences during the transmission of the rotation; in fact, the variations in Figure 2-left are due to random irregularities, while those shown in Figure 2-right, in correspondence of about 210° of each user revolution, are closely related to the presence of defect. The presence of the broken tooth is therefore highlighted by a sudden amplitude variation of the UM rotating speed. These remarkable results are obtained for the maximum value of the resistant torque and of the rotating speed, but they can be very different in other operating conditions.

In case of not heavy duty, the synchronous average variations are less evident and are not shown for brevity.

The synchronous average analysis provides effective fault detection when the kinetic energy introduced in the mechanical system is high. Therefore, this methodology is suitable for tooth fault detection only when working conditions are heavy and the variations of the signal due to the damage are greater than the motion irregularities.

554

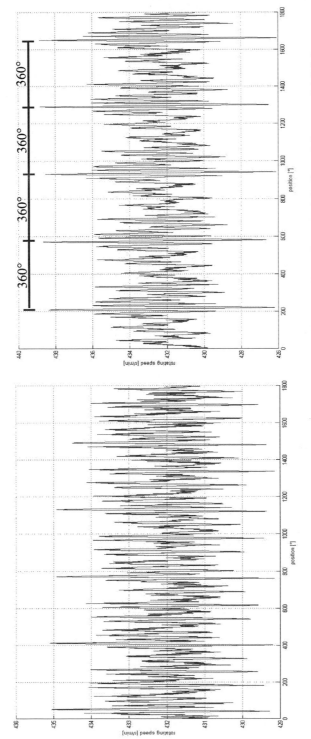

Figure 2: synchronous average of the signal of the user rotating speed: undamaged (left) and damaged (right).

5 RESIDUAL ANALYSIS

Figure 3-left shows the spectrum of the synchronous average signal of the UM rotating speed, when the DM operating speed is 550 rpm and the resistant torque is 8 Nm for the undamaged gearbox. The components to be cancelled from the spectrum are those closely related to the DM rotating speed and are marked in figure with *a*, *c*, *e* and the meshing frequencies (*b*, *d* in Figure 3-left).

After the inverse Fourier transform of the modified spectrum, the anomalies of the motion transmission are not emphasized (Figure 3-right). Residuals are randomly distributed around the mean. The greater amplitude in the centre of the diagram is due to the Hanning window employed for the first FFT of the synchronous average signal.

Now, the signal acquired using the damaged gearbox is analysed, in the same operating conditions. Figure 4-left shows the UM rotating speed spectrum: frequencies marked are those to be deleted. After the inverse Fourier transform, the signal in the position domain is obtained from the modified spectrum. Figure 4-right shows the residual sudden variations. Like in the synchronous average analysis, these variations are due to the presence of the fault. Also in this case, the damage is correctly localized at about 200°-210° with respect to the phase reference.

Therefore, the residual analysis allows the position of the defect to be precisely detected also in less heavy working conditions of the gearbox than those used in synchronous average analysis.

6 DEMODULATION ANALYSIS

Usually, in spur gearboxes, the filtering band taken into consideration for demodulation analysis must be centred on the meshing frequency. Because of the smooth behaviour of the spiral bevel gearbox, it is very difficult to detect the meshing frequency in the synchronous average spectrum. For this reason, the filtering band is different from that proposed in literature for spur gears. The operating conditions selected are 550 rpm for the DM rotating speed, with a resistant torque of 10 Nm. The signal analysed is that of the UM torque. The demodulation analysis, performed for the damaged gearbox, gives the results shown in Figure 5.

Conversely to the techniques applied previously, demodulation analysis allows the presence of a fault to be detected even if the DM rotating speed and the resistant torque are not of heavy duty. Sudden variations in the amplitude can be noted in Figure 5-right, emphasizing the gear damage. Naturally, the heavier are the working conditions, the clearer are the results.

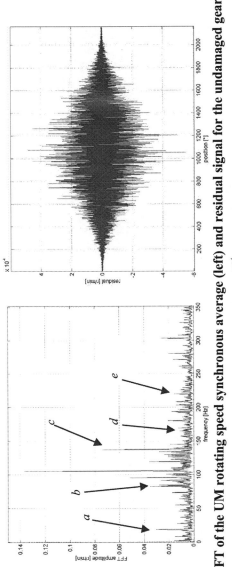

Figure 3: FFT of the UM rotating speed synchronous average (left) and residual signal for the undamaged gearbox (right).

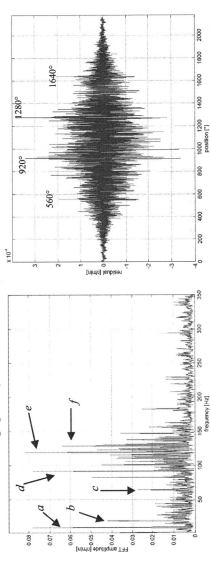

Figure 4: FFT of the UM rotating speed synchronous average (left) and residual signal for the damaged gearbox (right).

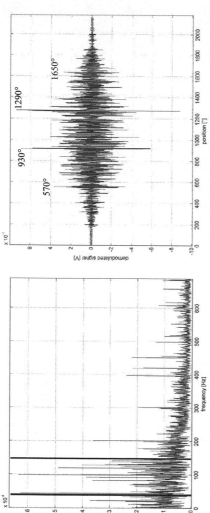

Figure 5: spectrum of the synchronous average of UM torque (left) and demodulated signal (right) for the damaged gearbox.

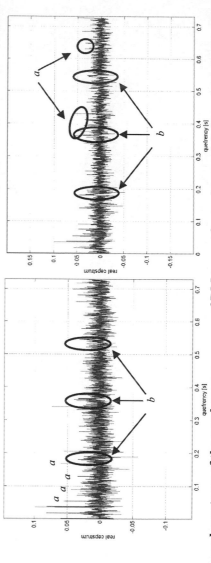

Figure 6: real cepstrum of the synchronous average of DM torque for undamaged (left) and for damaged (right) gearbox.

7 CEPSTRUM

Figure 6-left shows the real cepstrum of the signal of DM torque in the operating conditions of 1650 rpm and the resistant torque of 10 Nm, obtained from the synchronous average of 8 user revolutions for the undamaged gearbox. The family of rahmonics marked with "*a*" is closely related to the DM rotating speed: in fact the rahmonics are spaced by the quefrency of 0.0363 s corresponding to the frequency of 27.5 Hz. The three groups of rahmonics marked with "*b*" are spaced by the quefrency of 0.181 s: these peaks reveal the presence of the UM rotating speed.

The real cepstrum of the signal acquired in the same operating condition, but for the damaged gearbox (Figure 6-right) emphasizes the same structures shown by the gearing without faults.

The difference between the two cepstra is related to the distribution of the first family of rahmonics: in the damaged case, the DM rotating speed frequency appears starting from 0.4 s. For the second family of rahmonics it can be noticed a small growth of energy. The results of real cepstrum analysis do not allow detecting effectively the gear fault: signal obtained for undamaged gearbox is too similar to a signal related to a gear without one tooth.

8 WAVELET TRANSFORM

In this section the results of the wavelet transform, using the Morlet mother wavelet, are discussed. The signal taken into consideration is the synchronous average of the DM torque, in the operating conditions of 1650 rpm with the resistant torque of 8 Nm.

The wavelet transform of the DM torque of the damaged gearbox, Figure 7-left, does not highlight any variation of the correlation coefficients. The related 2D wavelet diagram is reported in Figure 7-right. The sinusoidal behaviour of 2D diagram is due to the low resolution in the time domain of the Morlet wavelet at low frequencies; also the evident higher peaks at both the begin and the end of the 2D diagram are a consequence of the wavelet used for the analysis: these portions are not taken into consideration in the discussion of the results. Neither the 2D section allows the presence of the damage to be found.

The wavelet transform is not an effective fault detection technique for the spiral bevel gearbox: it does not allow the fault to be recognized and located, even though in rather heavy operating conditions.

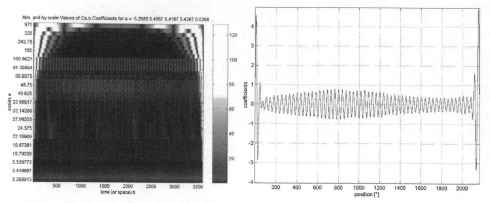

Figure 7: wavelet transform of the synchronous average of DM torque for the damaged gearbox: 3D (left) and 2D at 2× driving motor rotating speed (right).

9 SECOND ORDER CYCLOSTATIONARIETY

This technique has been applied to the synchronous average of the DM torque in the operating conditions of 1650 rpm for the DM rotating speed and the resistant torque of 10 Nm. Figure 8-left shows the results SCD amplitude obtained for the undamaged gearbox: the sum of the different cycles, along the frequency resolution, emphasizes the frequencies closely related to the motor rotating speed. These components are marked in Figure 8 from the second to the fifth harmonics with "*a*", "*b*", "*c*", "*d*" respectively. Some sidebands can be observed about 137.5 Hz, closely related to the UM rotating speed, since their spacing is 5.5 Hz.

This analysis concerns the frequencies only. The possible position of the damage cannot be found and another graphical output is considered for the second order cyclostationariety analysis, that is the Wigner-Wille spectrum. The signal obtained from the sum of each one of the components of the Wigner-Wille spectrum (see Figure 8-right), with respect to the position, allows the possible fault to be located. In this case, the randomness indicates no faults in the gearbox and the small variations of the amplitude are due to the rotating speed irregularities.

The results obtained for the damaged gearbox, in the same operating conditions (Figure 9) are extremely different. The value of the SCD amplitude is bigger than that of the undamaged gearbox per each frequency. The most relevant frequencies are included between 100 Hz and 160 Hz with the highest peak at about 132 Hz, the value of which is doubled with respect to the case previously discussed. The sidebands about this frequency have the spacing of 5.5 Hz. The presence of the fault is also emphasized by the absence of frequencies closely related to DM rotating speed in this case. These components are replaced by UM rotating speed frequencies. Moreover, the sidebands composed by similar peaks are due to collisions that repeat with the frequency of 5.5 Hz (Figure 9-left).

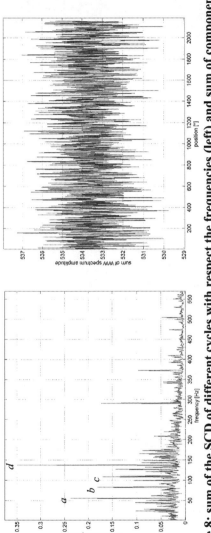
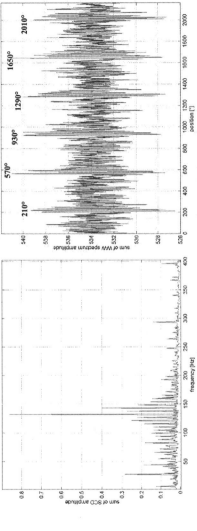

Figure 8: sum of the SCD of different cycles with respect the frequencies (left) and sum of components of the Wigner-Wille spectrum with respect to the position (right) for the undamaged gearbox.

Figure 9: sum of the SCD of different cycles with respect the frequencies (left) and sum of components of the Wigner-Wille spectrum with respect to the position (right) for the damaged gearbox.

The sum of the Wigner-Wille spectrum components are more effective than the SCD for the localization of the fault: the sudden variation in the signal shown in Figure 9-right allows detecting the position of the damage with respect to the phase reference. Therefore, the results obtained by means of this technique are similar to those of the residual and the demodulation analysis.

The second order cyclostationary analysis provides good results in this case. The presence of the damage in the tooth can be observed in the SCD by means of the increasing of the energy value and, above all, by the presence of sidebands closely related to the UM rotating speed. In the time domain, this is highlighted by the sudden variations of the sum of the Wigner-Wille spectrum components.

10 CONCLUSION

The aim of this work is to demonstrate the suitability of signals of control inverters for diagnostic purposes of mechanical systems. The system taken into consideration is a spiral bevel gearbox in two states: undamaged and faulty with a damaged tooth. The techniques used provide different results: the synchronous average signal is suitable for fault detection in heavy operating conditions, whereas the residual analysis and the demodulation analysis give good results (in the frequency domain but above all in the time-position domain) even in not heavy duty operating conditions. It is important to note that the demodulation analysis is realized by a different approach than that proposed by previous papers: the normalization used in this study allows reaching results more clear and effective than the amplitude and phase demodulation until now taken into consideration.

The SCD function, related to the second order cyclostationariety analysis, allows determining the most important frequencies in the signal: when the gearbox is damaged, the most evident components in the diagram are strictly related to the fault presence. Being the SCD relative to the frequency domain only, for the localization of the faults on the gears it is necessary to take in to consideration the Wigner-Wille spectrum: the combined use of the two representations allows characterizing completely the damage.

Cepstrum and wavelet analysis do not allow the fault detection for the considered type of gearbox: these techniques provide acceptable results only when the gearbox is so heavily damaged than the monitoring and diagnostics of the gearbox health is a nonsense, because the fault can be detect even by means of the noise generated during the rotation (or with the simpler synchronous average).

In conclusion, the experimental results demonstrate that it is possible to monitor the conditions and to detect the fault in spiral bevel gearbox, even only exploiting the inverter signals by means of suitable techniques. The monitoring system does not require any other transducer besides those used for the mechanical system motion control and the objective of this paper, i.e. an economical apparatus for diagnostic purpose, is reached.

REFERENCES

[1] P.D. McFadden, *Interpolation techniques for time domain averaging of gear vibration,* Mechanical System and Signal Processing, 3 (January 1989), 87-97.

[2] D. Boulahbal, M.F. Golnaraghi, F. Ismail, *Amplitude and Phase Wavelet Maps for the Detection of Cracks in Geared Systems,* Mechanical Systems and Signal Processing, 13 (May 1999), 423-436.

[3] C. Capdessus, M. Sidahmed, J.L. Lacoume, *Cyclostationary Processes: Application in Gear Faults Early Diagnosis,* Mechanical Systems and Signal Processing, 14 (May 2000), 371-385.

[4] G. Dalpiaz, A. Rivola, R. Rubini, *Effectiveness and Sensitivity of Vibration Processing Technique for Local Fault Detection in Gears,* Mechanical Systems and Signal Processing, 14 (May 2000), 387-412.

[5] J. Lin, L. Qu, *Feature Extraction Based on Morlet Wavelet and its Application for Mechanical Fault Diagnosis,* Journal of Sound and Vibration, 234 (June 2000), 135-148.

[6] W. Wang, *Early Detection of Gear Tooth Cracking Using the Resonance Demodulation Technique,* Mechanical Systems and Signal Processing, 15 (September 2001), 887-903.

[7] Y. Ohue, A. Yoshida, M. Seki, 2004, *Application of the Wavelet Transform to Health Monitoring and Evaluation of Dynamic Characteristics in Gear Sets,* Proceedings of the IMechE Journal of Engineering Tribology, 218 (February 2004), 1-11.

[8] Z.K. Zhu, Z.H. Feng, F.R. Kong, *Cyclostationarity Analysis for Gearbox Condition Monitoring: Approaches and Effectiveness,* Mechanical Systems and Signal Processing, 19 (May 2005), 467-482.

Bi-spectrum of a composite coherent cross-spectrum for faults detection in rotating machines

J K Sinha
School of Mechanical, Aerospace and Civil Engineering (MACE)
The University of Manchester, UK

ABSTRACT

Fault keeps generating in rotating machines. The identification of different faults in rotating machines by vibration techniques is mainly related to the machines RPM and its harmonics and sub-harmonics in the measured vibration responses from number of bearing pedestals. The higher order spectra (HOS) are tools to relate different harmonics and sub-harmonics of a frequency in a signal. Recently this feature of the HOS has been used to identify the crack and misalignment in a rotor. This concept has further been extended here in the present study. Here a composite coherent cross-power spectrum (CCCS) has been derived for the measured vibration responses from all the bearing pedestals of a rotating machine and then the HOS, namely the Bi-spectrum of this coherent spectrum has been computed to enhance the discriminating feature for the fault identification. This has been demonstrated through a numerically simulated example of a rotor with a small crack.

1. INTRODUCTION

Fault keeps generating in rotating machines. Identification of faults and their rectification is important from safety and productivity consideration. Vibration based condition monitoring is generally used for such requirement. Sinha [1] gave a detailed spectrum of the vibration based condition monitoring. In general, each fault can be identified by some kind of characteristic responses of the rotor either during steady state machine operation or during transient (runup or rundown) operation. For example, crack in a shaft or misaligned shaft generate 2X (twice the machine RPM) and higher harmonics in addition to 1X component in the shaft response during its rotation.

In recent studies, the higher order spectra (HOS)- Bi-spectrum and Tri-spectrum are proposed to identify the shaft crack and misalignment based on the observations made on a small rig [2-3]. It is because the HOS are capable to inter-relate number of harmonics of a frequency and their phase relationship in a signal. However the machine like Turbo-Generators (TG) sets, where there are number of bearings all along the rotor and so the number of vibration measurement locations. This simply

---

means that the identification of any fault generally becomes subjective when using the recently proposed HOS method if there are number of measurements along the rotor length like other vibration based condition monitoring techniques used in the TG sets. Hence a single composite coherent cross-spectrum (CCCS) is computed for the responses measured from all the bearing pedestals in the present study to overcome this limitation. The coherence between measurements at two bearings to remove uncorrelated signals has been used to compute the coherent cross-spectrum between them and these coherent cross-spectra between all bearings are then used to compute a single CCCS using the geometric mean. The Bi-spectrum has been then applied to this CCCS. It has been observed that the Bi-spectrum of the CCCS gives much better identification of a crack in a rotor compared to the Bi-spectrum at each bearing in a simulated example with added noise in their vibration responses. The paper gives the computational details of the CCCS and its Bi-spectrum and results of the simulated example.

2. THE BI-SPECTRUM

The conventional power spectrum density (PSD) provides information on the second-order properties (i.e., energy) of a signal whereas the bi-spectrum can provide information on the signal's third-order properties. In a physical sense, the bi-spectrum provides insight into non-linear coupling between frequencies (as it involves both amplitudes and phases) of a signal compared to the traditional PSD that gives only the content of different frequencies and their amplitudes in a signal. The PSD of a time series $x(t)$ is computed by the discrete Fourier transform (DFT) of the signal as

$$\text{PSD, } \mathbf{S}_{xx}(f_k) = E[\mathbf{X}(f_k)\mathbf{X}^*(f_k)], \ k = 1, 2, 3, ..., N \tag{1}$$

where $\mathbf{S}_{xx}(f_k)$ is the power density, $\mathbf{X}(f_k)$ and $\mathbf{X}^*(f_k)$ are the DFT and its complex conjugate at frequency f_k for the time series $x(t)$. N is the number of the frequency points. $E[.]$ denotes the mean operator here, it means that the PSD is the averaged spectrum over a time length of the signal, say, t. The bi-spectrum is the double Fourier Transformation of the second order moment of a time signal [4] that involves two frequencies components (both amplitudes and phases) of the signal, and computed by the signal DFT as [4]

$$\text{Bi-spectrum, } \mathbf{B}_{lm}(f_l, f_m) = E[\mathbf{X}(f_l)\mathbf{X}(f_m)\mathbf{X}^*(f_l + f_m)], \ l + m \le N \tag{2}$$

The bi-spectrum is complex and interpreted as measuring the amount of coupling between the frequencies at f_l, f_m, and $f_l + f_m$.

3. THE COMPOSITE COHERENT CROSS-POWER SPECTRUM (CCCS)

Similar to the PSD in the equation (1), the cross-power spectral density (CSD) between the two the signals, $\mathbf{x}_1(t)$ and $x_2(t)$ is computed using the discrete Fourier transform (DFT) of these signal as

$$\text{CSD, } \mathbf{S}_{x_1 x_2}(f_k) = E[\mathbf{X}_1(f_k)\mathbf{X}_2^*(f_k)], \ k = 1, 2, 3, ..., N \tag{3}$$

where $\mathbf{S}_{x_1 x_2}(f_k)$ is the CSD, $\mathbf{X}(f_k)$ and $\mathbf{X}^*(f_k)$ are the DFT and its complex conjugate at frequency f_k for a signal, $\mathbf{x}(t)$. Generally the CSD correlates the common features between the two signals in frequency domain, however in vibration analysis these signals are also contaminated by noises. Hence to remove the uncorrelated signals due to noises, the CSD can be modified as

$$\text{CCSD, } \mathbf{S}_{x_1 \gamma_{12} x_2}(f_k) = E[\mathbf{X}_1(f_k) \gamma_{12}^2 \mathbf{X}_2^*(f_k)] \tag{4}$$

where $S_{x_1 \gamma_{12} x_2}$ is the *Coherent Cross-power Spectral Density* (CCSD) and γ_{12}^2 is the ordinary coherence [5] between two signals averaged over the time length, t. So if the number of signals is n, then a *Composite Coherent Cross Spectrum* (CCCS) for all the signals can be computed as the geometric means of all the CCSD as below;

$$\text{CCCS, } X_{cccs}(f_k) = \sqrt{\left(S_{x_1 \gamma_{12} x_2} \quad S_{x_2 \gamma_{23} x_3} \quad \cdots \quad S_{x_{n-1} \gamma_{(n-1)n} x_n} \right)^{1/n}} \tag{5}$$

Obviously, $X_{cccs}(f_k)$ is a complex parameter which contains the only correlated information of all the signals at the frequency (f_k), and finally the Bi-spectrum in the equation (2) has been computed for this CCCS. The computation of the CCCS is nothing but like a Data Fusion of data from number of sensors in frequency domain. The advantage of the Bi-spectrum for this CCCS in the fault identification in rotating machines has been discussed in the Section 3.

4. THE NUMERICALLY SIMULATED EXAMPLE

The schematic of the numerically simulated experimental example is shown in Figure 1. It consists of a shaft made of the steel supported on 3 anti-friction bearings on flexible bearing pedestals. The length of the shaft is 1.2m and 20mm diameter, and the Young's modulus and the density of the rotor material are assumed as 210 GPa and 7800 kg/m³ respectively. The shaft carries 2 balance disks of thickness 25mm and the outer diameter 250mm, one between the Bearings 1 and 2, and other between 2 and 3 as shown in Figure 1. The locations of the bearings, disks and the stiffness of the bearing pedestals are shown in Figure 1.

An FE model using 2-node Timoshenko beam element was constructed for the shaft and stiffnesses of the bearing pedestals and the masses and moments of the disks have been added to the appropriate locations to represent the rig in Figure 1. The stiffness proportional damping matrix was used assuming the rotor damping of 0.5%. The first three critical speeds of the numerical rig have been estimated at 42.28Hz, 61.55Hz and 201.14Hz when there was no crack in the shaft, however these were estimated to be 41.66Hz, 60.55Hz and 198.20Hz when a small crack of 10% of the diameter (for fully open condition) was assumed at 250mm from the Bearing 1. Hence the identification of crack due to small change in the frequency is perhaps difficult. This is the reason why the orbit plots consisting of the 1X (1 times of the machine RPM) and 2X components are used to identify the crack in a rotor. However it requires vibration measurements at two orthogonal locations at each bearing and even then the judgment of the crack in a rotor is often subjective if there are number of bearings in a machine. Recently, the author has demonstrated the identification of a crack in a rotor using the HOS utilizing the vertical response only at the bearing pedestal on a small scale experimental rig [2-3]. The shaft in the experimental rig [2-3] was supported on two bearings only and crack was between the two bearings so the HOS for the vertical response at both the bearings gave the same results, however when the number of bearings are more than two which is generally the case with real machines like TG sets. The HOS at all the bearings may identify the crack, but the use of the CCCS in the HOS (here it is Bi-coherence) has also been tested here to confirm whether it enhance the discriminating feature and give a unique parameter for the crack identification.

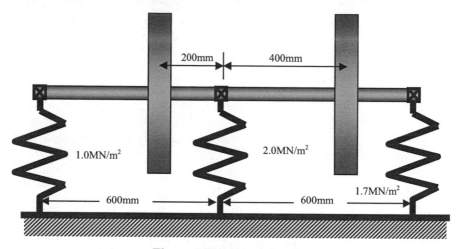

Figure 1 The Simulated Rig

The measured responses at all the bearing pedestals polluted with 30dB SN ratio were estimated for the machine run-up with the linear chirp rate of 0.5Hz/s. The measured responses were estimated for the shaft with a crack and without crack. The crack was modelled using the method proposed by Sinha *et al.* [6]. The breathing of the crack during the machine run-up was also considered [3].

5. RESULTS AND DISCUSSION

The order tracked responses at the bearings in the vertical direction for the cracked shaft are shown in Figure 2. The responses at the first critical speed and its harmonics for the crack shaft were seen between 51s to 55s, and the amplification at the sub-critical speed at a few bearings was also observed. However, no such observations were observed other than the responses at the critical speed for the no crack case as expected. The harmonics of the first critical speed after 4th machine order were not observed to be significant or rather not seen clearly.

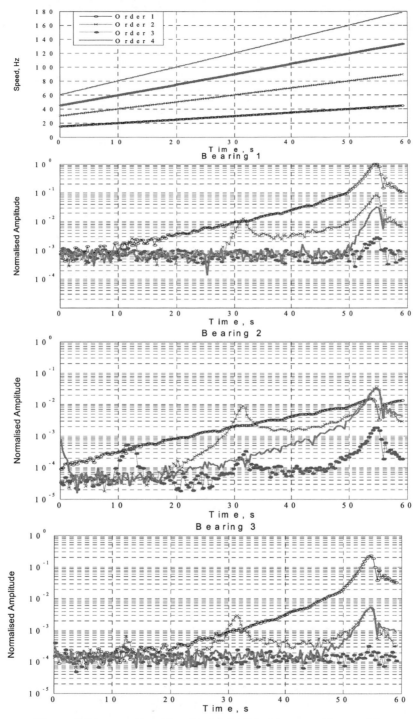

Figure 2 Responses at Bearings 1-3 with respect to the machine Orders during startup for the cracked shaft

569

The Bi-spectrum was computed first for both the cracked and no crack responses for all the three bearings. The components of the Bi-spectrum (B11 to B33) are shown in Figure 3. The Bi-spectrum component, B_{mn}, means the response at mth and nth orders related to the $(m+n)$th order response. Similarly, the component, B33, means that the 3rd order response (two times) related to a 6th order response. Definitely most of the individual components of the Bi-spectrum showing difference between the crack and no crack cases, however the discrimination was not that significant for the simple case considered here.

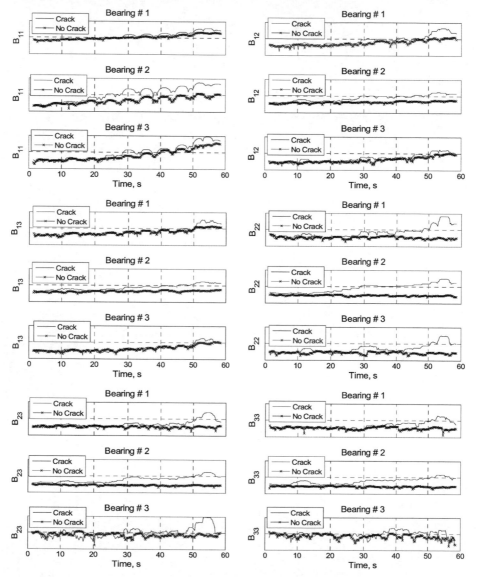

Figure 3 Different components of Bi-spectrum at Bearing Pedestals during machine startup

Now the Bi-spectrum components – B11 to B33 were computed again but using the CCCS for both the crack and no crack cases. These are shown in Figure 4 which shows significant improvement in the discrimination between the crack and no crack and gives a single parameter for all the bearings.

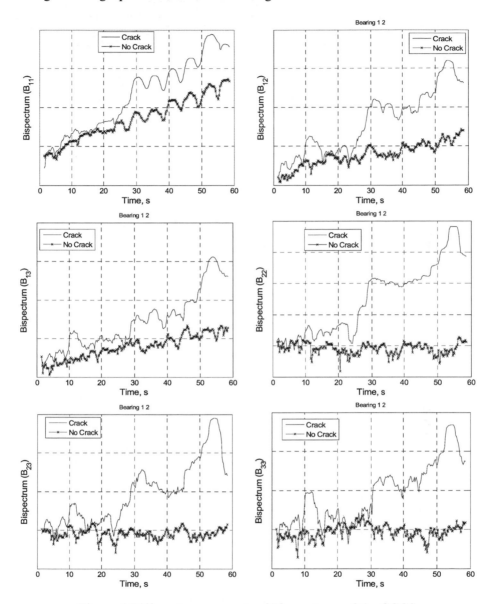

Figure 4 Different components of Bi-spectrum of the CCCS during machine startup

6. CONCLUSION

The computation of the *Composite Coherent Cross-Spectrum* (CCCS) has been suggested. The advantage of the proposed CCCS method is that it removes the noises between the signals and only relates the correlates signals in the frequency domain. The CCCS is nothing but the data fusion of number of signals in much improved way. The proposed CCCS was then used in the Bi-spectrum computation. The usefulness of the proposed Bi-spectrum of the CCCS in the crack detection in the rotating machine has been demonstrated. The greatest advantage of the proposed method is that it reduces the number of measurements at all bearings to one and may ease the faults detection process in the rotating machine. Here in the paper, only one fault – crack was used for the demonstration, however it is now plan to carry out number of experiments on a laboratory rig with different faults to further enhance the confidence level in the proposed HOS of the CCCS.

REFERENCES

[1] Jyoti K. Sinha, Health Monitoring Techniques for Rotating Machinery, Ph.D. Thesis. University of Wales Swansea, Swansea, UK, 2002.
[2] Jyoti K. Sinha, Bi-spectrum for identifying crack and misalignment in shaft of a rotating machine, *Smart Structures and Systems, An International Journal* 2(1) 2006.
[3] Jyoti K. Sinha, Higher Order Spectra for Crack and Misalignment Identification in Shaft of a Rotating Machine, *Structural Health Monitoring: An International Journal* 2007.
[4] Ewins, D. J., 2000, *"Modal Testing – Theory, Practice and Application,"* Research Studies Press, U.K., 2nd Edition.
[5] Collis, W.B., White, P.R., and Hammond, J.K., (1998). Higher-Order Spectra: The Bispectrum and Trispectrum, *Mechanical Systems and Signal Processing* 12(3): 375-394.
[6] Jyoti K. Sinha, M.I. Friswell, S. Edwards, Simplified Models for the Location of Cracks in Beam Structures using Measured Vibration Data. *Journal of Sound and Vibration* 251(1), pp.13-38, 2002.

Detection of rolling element bearing faults by analysis of the motor current

A Ibrahim[1], F Guillet[1], M El Badaoui[1], R B Randall[2], D Rémond[3]
[1]Laboratory LASPI, IUT de Roanne, France
[2]School of Mechanical and Manufacturing Engineering, UNSW, Australia
[3]Laboratory LaMCoS, INSA Lyon, France

1. INTRODUCTION

There are a number of papers giving evidence of the detection and diagnosis of faults in rolling element bearings, based on the analysis of the current of the induction motor driving the machine. The reason given for the effect of the faults on the current has so far been that the fault in the bearing causes the rotor to become displaced radially in the stator field. However, it can be shown that in the case of small local faults in the races, even if their actual depth is say 100 μm, the amount the shaft can actually move in the short time that the fault is being traversed is only a small fraction of this. Moreover, since the displacement takes place over a small fraction of the time between impacts (corresponding to the rolling element spacing) the total maximum displacement is divided over a large number of harmonics and thus the displacement attributable to the first harmonic (detected as a sideband in the motor current signature) would be approximately equal to 0.29 μm. This paper argues that this would have a negligible effect on the motor current, and instead argues that the more likely reason why the faults manifest themselves in the current is because they give a fluctuating resistive torque which acts immediately (in contrast to the radial displacement which takes time to integrate to a perceptible displacement even in response to sudden changes in acceleration). Evidence that even very small spalls in a bearing give torque fluctuations is given by measurements of torsional vibrations made using optical encoders. Experiments made on a test rig have given further confirmation of this interpretation of the effect of bearing faults on motor current.

2. BEARING FAULT SIGNATURE

Fig 1 shows the typical construction of a ball bearing and defines the dimensional parameters used in this paper. The balls are bound by a cage which ensures a

---

uniform distance between them and prevents any contact. Bearing defects can occur as a result of fatigue of their material under normal operational conditions. First, cracks will appear on the tracks and on the balls. Then, pitting and scuffing of material can quickly accelerate the wear of a bearing and intensive vibrations are generated as a result of the repetitive impacts of the moving components on the defect. For instance, when a rolling element contacts a defect on the inner or outer raceway, it produces an impact which in turn excites the structural modes of the bearing and its support [8]. In an operating bearing, a series of impacts occurs with a repetition frequency which depends on whether the defect is on the inner or the outer race, or on the rolling element. Therefore, the overall vibration signal measured on the bearing shows a pattern consisting of a succession of oscillating bursts dominated by the major resonance frequencies of the structure. Rolling element bearings experience some slip of the rolling elements, thus yielding as a consequence that the occurrences of the impacts never reproduce exactly at the same position from one cycle to another. Furthermore, when the position of the defect is moving with respect to the load distribution on the bearing, the series of impacts is modulated in amplitude. From these considerations, some randomness should be expected in the measured vibration signal as well as some amplitude modulation.

In addition, the bearings can be damaged by external causes such as [4] :
- contamination of the bearing by external particles: dust, sand
- corrosion induced by the penetration of water, acids
- inadequate lubrication which can cause heating and wear of the bearing
- bad alignment of the rotor
- current which passes through the bearing and which causes electric arcs

The defect will produce one of the four characteristic fault frequencies in the machine vibration depending on which bearing surface contains the fault; each bearing defect has its own signature and is characterized by a fundamental frequency (1)-(4) which can be calculated on the basis of the structure and dimensions of the bearing and the shaft speed (Fig 1) [1].

F_R rotor frequency;
F_C cage fault frequency;
F_I inner raceway fault frequency;
F_O outer raceway fault frequency;
F_B ball fault frequency;
D_b ball diameter;
D_c pitch diameter;
N_B number of rolling elements;
β ball contact angle.

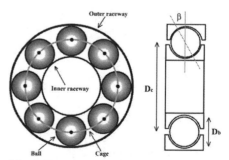

Fig 1 Structure and Dimensions of a bearing

$$F_C = \frac{1}{2} F_R \left[1 - \frac{D_b}{D_c} \cos \beta \right] \qquad (1)$$

$$F_O = \frac{N_B}{2} F_R \left[1 - \frac{D_b}{D_c} \cos \beta \right] \qquad (2)$$

$$F_I = \frac{N_B}{2} F_R \left[1 + \frac{D_b}{D_c} \cos \beta \right] \qquad (3)$$

$$F_B = \frac{D_c}{D_b} F_R \left[1 - \left(\frac{D_b}{D_c} \cos \beta \right)^2 \right] \qquad (4)$$

In this study, the defect results from a natural degradation of the outer raceway of the bearing which supports the rotor adjacent to a gear reducer.

3. LOAD TORQUE OSCILLATIONS DUE TO BEARING FAULTS

Single-point defects begin as localized defects on the raceways (or rolling elements) and as the rolling elements pass over these defect areas, small collisions occur producing mechanical shockwaves which then excite the frequencies of natural mechanical resonance in the machine. This process occurs every time that defect impacts with another part of the bearing, and its rate of occurrence is equal to one of the previously defined characteristic fault frequencies. Restated, the mechanical resonance frequencies (carriers) are modulated by the characteristic fault frequency (baseband signal) [7].

Bearing faults have a direct impact on the torque of the machine and cause load torque oscillations which can be detected by mechanical or electrical measurements.

3.1 Mechanical approach

There is a crack in the outer raceway (Fig 2) with a width of $\Delta l = 0.5\ mm$. The rotor shaft speed is $24.7\ Hz$. Knowing the dimensions of the bearing, we can calculate the time of passage past the damaged surface:

$\theta = \omega * t$

where θ is the angular position $= \frac{\Delta l}{r}$

r = outer raceway radius.

ω = shaft speed and

t = time.

Fig 2 Bearing fault

$$t = \frac{\theta}{\omega} = \frac{\Delta l / r}{\omega} = \frac{0.5mm}{24.7Hz * 20.74mm} = 9.76 * 10^{-4}s$$

If the defect causes a radial displacement, under gravity ($a = 10ms^{-2}$) the rotor shaft falls a maximum distance of:

$$s = \frac{1}{2} at^2 = 4.76 \mu m$$

This displacement can be modeled by a square pulse which lasts $0.976\ ms$ and is repeated for the passage of each ball ($16\ ms$). Thus, there will be approximately 16 harmonics in the Fourier series until the first zero passage in the spectrum. Since they are aligned in phase, the total displacement will be divided between 16 harmonics, therefore the size of each harmonic is approximately equal to 0.29 μm. It is not conceivable that such a displacement of the rotor would give a perceptible effect in the stator current and the reason proposed here for this effect, is that the defect in the bearing acts on the torque of the machine.

The following mechanical equation establishes the relation between the torque and the rotor angular velocity (ω_r)

$$T_{em} - T_{load} = J \frac{d\omega_r}{dt} \tag{5}$$

where J is the moment of inertia of the drive system. For motoring operation, load torque (T_{load}) is positive and for generating operation, load torque is negative.
The test bench is not equipped with a torque meter. Based on equation (5) and the close relationship between the torque and the speed, we will exploit the instantaneous frequency in order to find information related to the defect.

3.1.1 Instantaneous frequency
Instantaneous frequency (IF) is an important signal characteristic arising in many fields. It is a concept intimately linked to time-frequency analysis, where it can be obtained from a time-frequency distribution (TFD) as the first conditional moment in frequency, suggesting that the IF is the average frequency at each time. Conceptually the IF can be interpreted as the frequency of a sinusoid which locally adjusts the analyzed signal. Nevertheless, in the case of non-stationary signals with frequential properties varying in time, the introduction of the concept of instantaneous frequency can prove to be useful for the analysis.
Physically, it is significant to note that the analyzed signal must certainly be a signal with only one component, where there is only one frequency or a narrow band frequency varying according to the time; in the case of signals with multi-components, the concept of IF is no longer valid [2]. An example is given by the case of the stator current signal of an asynchronous machine because this signal contains the harmonics of the fundamental frequencies, including the harmonics of the slot passage frequency and other components. Two basic definitions are proposed in the literature to calculate instantaneous frequency; one is established starting from the derivative of the phase of the analytic signal associated with the studied signal, and the other is given on the basis of the time-frequency representation of the studied signal.
The instantaneous frequency $f_i(t)$ of a mono component analytic signal $Z(t) = a(t) * e^{j\emptyset(t)}$ is defined as being the derivative of its phase:

$$f_i(t) = \frac{1}{2\pi} \frac{d\emptyset(t)}{dt} \tag{6}$$

The analytic signal $Z(t)$ is a complex signal containing the same information as the real signal $x(t)$. In contrast to the real signal, the complex signal contains only

576

positive frequencies. It is built starting from the real signal by using the Hilbert transform symbolized by the operator H:

$$z(t) = x(t) + jH[x(t)]$$ (7)

where:

$$H(t) = \int_{-\infty}^{+\infty} \frac{x(t-\tau)}{\pi\tau} d\tau$$ (8)

3.1.2 Detection of fault through Instantaneous Frequency

The test rig is composed of an asynchronous motor, 1.1 kW; a double reduction speed reducer coupled to the rotor shaft which is carried by two bearings, where one of them is defective. This machine is supplied directly by the power supply network of the laboratory; it actuates a DC machine which outputs in a rheostat to apply a load. The mechanical vibrations are measured by an accelerometer attached to the machine above the bearing; the angular position of the rotor shaft is obtained by an optical encoder fixed at the free end of the shaft. The optical encoder delivers a square signal whose frequency is a multiple of the rotational frequency; this enables study of the speed fluctuations. We also make electrical measurements (voltages and currents) for the development of new methods of diagnosis based on the analysis of the electric signals [9].

Table 1 shows the bearing parameters taken from the data sheet, and $\cos\beta$ deduced from various measurements. The motor, with two pole pairs, operates at an average shaft speed of 1483 r/min ($F_R = 24.71\ Hz$). A real defect is located in the outer raceway. Using Equation (2), the data of Table 1, and shaft speed 24.71 Hz the outer race fault frequency is calculated to be 62.89 Hz.

The sampling frequency of the encoder signal used is 102400 Hz and the bandwidth of the pass band filter is 16000 – 34000 Hz. Fig 3 represents the power spectrum density of the instantaneous speed estimated using the optical encoder. The strongest components in this spectrum are related to the rotational frequency 24.71 Hz and its harmonics. The bearing fault frequency at 62.89 Hz is also detected with its harmonics but with lower levels. It should be noted that the speed signal also carries much other information.

This result is corroborated by the envelope analysis [6] of an acceleration signal demodulated around the high frequency 8.5 kHz which also confirms the presence of the defect at 62.9 Hz. Fig 4 is the spectrum of the squared envelope detected in the 600 Hz span around 8.5 kHz. It shows the first four harmonics of an outer raceway defect $F_O = 62.9\ Hz$ and the shaft speed at 24.71 Hz.

Table 1
Bearing parameters

Type	D_c	D_b	N_B	$\cos\beta$
SKF 6303-2Z	32mm	8.735 mm	7	1

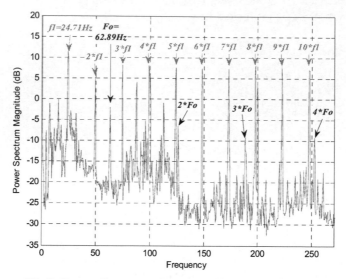

Fig 3 Power Spectrum Density of instantaneous speed

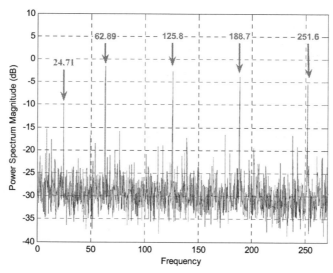

Fig 4 Power Spectrum Density of the squared envelope of vibratory signal

Faults in rolling element bearings are often low-energy signals, which are difficult to detect until they become severe. Envelope analysis enables detection of these signals, and gives information about the precise fault frequencies. The envelope analysis constitutes a reference but requires an accelerometer sensor as well as a high frequency analysis (because of amplification of the weak bearing signals by high frequency resonances). At low frequency, the effects of a bearing fault are masked by the large number of much stronger mechanical events.

3.2 Electrical approach

3.2.1 Relation between torque and current

The induction machine model is developed from its fundamental electrical and mechanical equations [3]. The voltage equations in d-q frame (Park's Transformation [5]) for an induction motor are given by

$$V_{qs} = R_s I_{qs} + \omega \psi_{ds} + \frac{\psi_{qs}}{dt} \tag{9}$$

$$V_{ds} = R_s I_{ds} - \omega \psi_{qs} + \frac{\psi_{ds}}{dt} \tag{10}$$

The stator flux linkages can be defined using stator leakage inductance (L_{ls}) and magnetizing inductance (L_m) as

$$\psi_{qs} = (L_{ls} + L_m).I_{qs} + L_m.I_{qr} \tag{11}$$

$$\psi_{ds} = (L_{ls} + L_m).I_{ds} + L_m.I_{dr} \tag{12}$$

The flux linkages also can be defined as the integral of the applied voltage less the resistance drops.

$$\psi_{qs} = \int (V_{qs} - R_s I_{qs})dt \tag{13}$$

$$\psi_{ds} = \int (V_{ds} - R_s I_{ds})dt \tag{14}$$

The rotor flux linkages can be defined by the following equations:

$$\psi_{qr} = \frac{L_m.R_r}{L_r.s + R_r}.I_{qs} + \frac{L_r.\omega_r}{L_r.s + R_r}.\psi_{dr} \tag{15}$$

$$\psi_{dr} = \frac{L_m.R_r}{L_r.s + R_r}.I_{ds} - \frac{L_r.\omega_r}{L_r.s + R_r}.\psi_{qr} \tag{16}$$

where L_r is the rotor inductance, s is the slip, ω_r is the rotor angular speed and R_r is the rotor resistance. The stator currents and the electromagnetic torque can be solved using the following equations:

$$I_{ds} = \frac{V_{ds}}{L_{sm}.s + R_s} - \frac{\psi_{dr}}{L_r}\frac{L_m.s}{L_{sm}.s + R_s} \tag{17}$$

$$I_{qs} = \frac{V_{qs}}{L_{sm}.s + R_s} - \frac{\psi_{qr}}{L_r}\frac{L_m.s}{L_{sm}.s + R_s} \tag{18}$$

$$T_{em} = \frac{3}{2}\frac{P}{2}\frac{L_m}{L_r}(\psi_{dr}.I_{qs} - \psi_{qr}I_{ds}) \tag{19}$$

where P is the number of pole pairs;

Equation (19) highlights the relationship between torque and current and justifies the effect of the defect on the current.

The relationship between electric power and the electromagnetic torque is given by the following relation

$$P_{abc} = T_{em} * \frac{\omega_s}{P} \tag{20}$$

Where ω_s is the current pulsation.

In the following, we will treat the instantaneous power which will be an image of the torque.

3.2.2 Instantaneous Power

In induction motors the simple voltage V_a has a lead of angle α over the current, and the power factor angle is seen as the angle between I_a and V_a. In addition, because the induction motor is powered by a balanced three phase source, then the simple voltage V_b lags the simple voltage V_a by an angle $2\pi/3$. The simple voltage V_c has a lead of $2\pi/3$ over the simple voltage V_a. The same reasoning can be employed for the three currents.

$$V_a = U_m \cos(wt + \alpha) \tag{21}$$

$$V_b = U_m \cos(wt + \alpha - 2\pi/3) \tag{22}$$

$$V_c = U_m \cos(wt + \alpha + 2\pi/3) \tag{23}$$

$$I_a = I_m \cos(wt) \tag{24}$$

$$I_b = I_m \cos(wt - 2\pi/3) \tag{25}$$

$$I_c = I_m \cos(wt + 2\pi/3) \tag{26}$$

where U_m, I_m and w are respectively the maximum voltage, maximum current and power supply pulsation.

The phase-to-phase voltages are computed by (27)-(28):

$$U_{ab} = V_a - V_b \tag{27}$$

$$U_{cb} = V_c - V_b \tag{28}$$

Partial instantaneous input powers P_{ab} and P_{cb} are computed by multiplying U_{ab} by I_a and U_{cb} by I_c:

$$P_{ab} = U_{ab} * I_a \tag{29}$$

$$P_{cb} = U_{cb} * I_c \tag{30}$$

Finally the total instantaneous input power P_{abc} is then computed by adding the two partial powers:

$$P_{abc} = P_{ab} + P_{cb} \tag{31}$$

In presence of a fault, torque oscillations involve variations in the power-factor and in this case α must be replaced by $\alpha(t) = \omega_{osc}t + \alpha$
where ω_{osc} is the fault frequency.

Finally, the instantaneous power will be:

$$P_{abc}(t) = \frac{3}{2}U_m I_m(\cos(\omega_{osc}t + \alpha)) \tag{32}$$

3.2.3 Processing of electrical signals
There are a number of papers giving evidence of the detection and diagnosis of faults in rolling element bearings, based on the analysis of the current of the induction motor driving the machine.
This paper proves that bearing faults give rise to torque oscillations, which appear directly as a phase modulation of the stator currents.
These oscillations thus generate stator current harmonics and then lead to the creation of current sideband components at predictable frequencies, F_{BE}, related to the vibration and electrical supply frequencies by

$$F_{BE} = |F_E \pm F_{osc}| \tag{33}$$

where F_E is the power supply frequency $= 50Hz$, $k = 1; 2; 3; \cdots$ and F_{osc} is one of the characteristic bearing fault frequencies $(F_C, F_O, F_I$ and $F_B)$, being equal to F_O in this study.

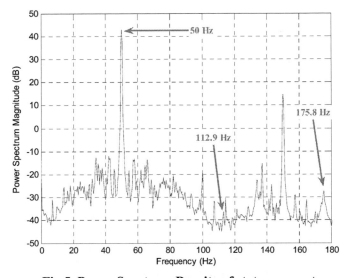

Fig 5 Power Spectrum Density of stator current

Fig 5 is the power spectrum density of a measured stator current. This spectrum detects a small peak at 112.9 *Hz*, equal to 50 + 62.89 *Hz*, and at 175.8 *Hz* corresponding to the second harmonic (50 + 2*62.9). It is important to note that the frequency components produced by the bearing defect are relatively small when compared to the rest of the current spectrum. The largest components present in the current spectrum occur at multiples of the supply frequency and are caused by saturation, winding distribution, and the supply voltage. This large difference in magnitude (Fig 5) can make the detection of the current spectrum bearing harmonics a significant problem. Therefore, we need another processing technique to precisely detect and estimate the bearing fault signature.

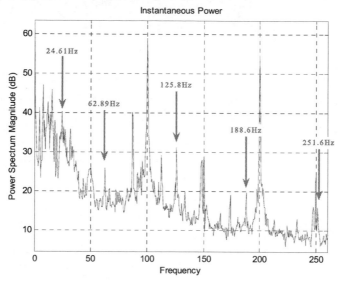

Fig 6 Power Spectrum Density of instantaneous power

With reference to equation (32), this information can be extracted from the instantaneous power spectra (Fig 6) at the fault characteristic frequency $F_O = \frac{\omega_{osc}}{2\pi} =$ 62.89 *Hz* and its harmonics or at frequencies $2F_E \pm F_O$. This spectrum detects clearly the outer raceway bearing fault and proves again that the bearing fault is transmitted to the stator current via the torque.

4. CONCLUSION

This study contributes to the diagnosis of bearing faults in rotating machines in general and more particularly in the diagnosis of asynchronous machines on the basis of electrical measurements. The effect of this type of defect on the stator current was for a long time modeled by a radial motion between the rotor and the stator of the machine, interpreted as an eccentricity. This paper argues, starting from two approaches: mechanical and electrical, that a bearing fault does not cause a radial displacement of the rotor, but generates torque oscillations which will be reflected in the stator currents. Experimental results based on mechanical and electrical measurement showed good agreement with the theoretical results of modelling the fault.

The use of signal processing is invaluable to detect the faults, and a good method must be chosen to allow a reliable and precise diagnosis.

5. REFERENCES

[1] R. A. Collacott, *Vibration Monitoring and Diagnosis.* New York: Wiley, 1979. pp. 109-111.

[2] P. J. Loughlin and B. Tacer, "Instantaneous frequency and the conditional mean frequency of a signal." *Signal Processing,* Vol. 60, no. 2 (July 1997): pp. 153-162.

[3] P. Vas, *Sensorless vector and direct torque control.* 1st edition. Oxford: Oxford university press, 1998.

[4] S. Nandi and H. Toliyat, "Condition monitoring and fault diagnosis of electrical machines - a review." *IAS Industry Applications Conference, Thirty-Fourth IAS Annual Meeting.* In proceedings of IEEE, 3-7 Oct. 1999. pp. 197-204.

[5] S. Leva and A. P. Morando, "Park equations for distributed constants line." *COMPEL J.,* Vol. 20, no. 4 (Sep./Dec. 2001): pp. 291-297.

[6] R. B. Randall, J. Antoni, and S. Chobsaard, "The relationship between spectral correlation and envelope analysis in the diagnostics of bearing faults and other cyclostationary machine signals." *Mechanical Systems and Signal Processing,* Vol. 15, no. 5 (September 2001): pp. 945-962.

[7] J. R. Stack, R. G. Harley, and T. G. Habetler, "An amplitude modulation detector for fault diagnosis in rolling element bearings." *IEEE Trans. Industrial Electronics,* vol. 51, no. 5 (October 2004).

[8] J. Antoni, and R. B. Randall, "On the use of the cyclic power spectrum in rolling element bearings diagnostics." *Journal of Sound and Vibration,* Vol. 281, no. 1-2 (March 2005): pp. 463-468.

[9] A. Ibrahim, M. El Badaoui, F. Guillet, and W. Youssef, "Electrical signals analysis of an asynchronous motor for bearing fault detection." *In proceedings of IEEE, IECON'06,* November 2006: pp. 4975-4980.

Bearing condition monitoring using multiple sensors and integrated data fusion techniques

S L Chen[1], M Craig[1], R J K Wood[1], L Wang[1], R Callan[2], H E G Powrie[2]
[1]Surface Engineering and Tribology Group, University of Southampton, UK
[2]GE Aviation, Digital Systems, UK

ABSTRACT

Recently, at the University of Southampton, a series of taper roller bearing tests have been conducted to evaluate the effectiveness of using on-line sensing technologies to detect incipient faults. The test rig instrumentation included vibration and electrostatic sensors as well as off-line processes such as debris analysis and tribological assessment of bearing condition, which are used for corroborative purposes. The test results indicate that the combination of these techniques are capable of detecting bearing deterioration shortly before complete failure, but the expected precursor events related to fatigue failure initiation are indistinguishable from the conventional univariate signal plots. Therefore, more advanced data fusion techniques have been developed to extract further information and enhance the detection of abnormal trends. This paper describes work on intelligent processes of both training and testing data and demonstrates how the prognostic window is significantly improved relative to original processed features. The approach also identifies the main variables which drive the anomaly detection so as to provide diagnostic information.

Keywords: Rolling Element Bearing, Condition Monitoring, Vibration Monitoring, Electrostatic Monitoring, Data Fusion.

1. INTRODUCTION

Vibration analysis has been used in bearing condition monitoring for over 50 years, and has achieved various degree of success. Electrostatic charge detection was originally developed by GE Aviation (formerly Smiths Aerospace) for the identification of debris in the gas path of jet engines [1]. Further research at the University of Southampton has shown its potential to detect precursor processes that indicate contact distress and wear in lubricated environments [2]. Recently, a series of taper roller bearing tests have been conducted to evaluate the effectiveness of using on-line sensing technologies to detect incipient faults [3]. The test rig instrumentation included vibration accelerometer and electrostatic sensors as well as off-line processes such as debris analysis and tribological assessment of bearing condition, which are used for corroborative purposes. Although the test results indicate that the combination of these techniques is capable of detecting bearing deterioration shortly before complete failure, expected precursor events related to initiation of bearing fatigue failure are indistinguishable from the conventional univariate signal plots. Moreover, these techniques require a lot of manpower to analyze the results in order to achieve detection and diagnostic information. Thus, developing a systematic approach that can automatically extract bearing abnormal events and diagnostic information is crucial.

So far, various approaches have been tried, including artificial neural networks (ANNs) [4], fuzzy logic [5], evolutionary algorithms [6] etc. to develop automatic fault detection and diagnosis systems. Although the performances of these systems are consistently improved with the adaptation of the algorithms or implementation of the data pre-processing [7] and feature selection [8] techniques, there is still a major concern over this development strategy: it is not easy to obtain multi-class training data covering all aspects of bearing symptoms that can appear very different and confusing. Nevertheless, data representing the normal condition of the bearings is the most easy-to-obtain dataset, and the systems can be built to detect whether the bearing condition deviates from the 'normal' region rather than mapping collected data into several established categories of the bearing conditions. This idea has been used as the central development strategy by novelty detection [9, 10] and statistical process control (SPC) [11, 12] techniques in image processing and chemical plant applications.

In this paper, a Gaussian mixture model (GMM) is proposed to summarize the probabilistic density of the training data (normal condition) and to build the normal model. However, the true condition of the training data is unknown and it might contain a number of anomalous data due to the sensor malfunction or environmental effects etc., so the GMM could not be expected to respond to these anomalies [13]. Therefore, the model is adapted with a novel approach to suppress regions with anomalies. Further, Hotelling's T-squared statistic is applied to extract abnormal information by measuring the distance between the test sample and the origin of the adapted normal model. Finally, the contribution values to the calculated T-squared statistic are calculated for each extracted feature, which are related to bearing components as well as to the sensor locations.

This paper is organized as follows: a brief introduction of the proposed approach is presented in section 2. Section 3 describes the experiments, extracted signals in both time and frequency domains and applied datasets. Section 4 provides the training process of the normal model, and the anomaly detection and diagnosis results of the test data is described in the section 5.

2. METHODOLOGY

Fig. 1 shows the flow chart of the proposed scheme. At the beginning, multiple features from time domain, i.e. RMS value and frequency domain, i.e. energies at bearing defect frequencies are processed for both the training and testing data. Next, the GMM approach is applied to describe the training data. Thirdly, the innovative methods are adopted to locate and trim Gaussian components in the existing GMM associated with anomalies. The scheme then utilizes principle component analysis (PCA) in each of the remaining Gaussian components, so that the multiple PCA model can be achieved. Finally, the multivariate testing data is fed into the multiple PCA models, so the Hotelling's T-squared statistic and variable contribution values can be calculated to obtain the anomaly detection and diagnosis information respectively.

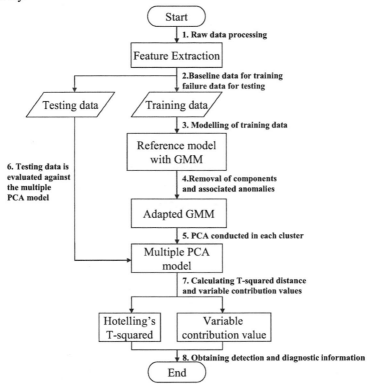

Fig. 1 Methodology flowchart of the proposed intelligent scheme

2.1. Gaussian Mixture Model

Often it is assumed that normal data may be described by a single Gaussian distribution. However, real world data can be distributed in a more complicated way. Data from different time intervals might occupy their own space of normality (e.g. vibration levels varies at different times due to a different load regime) or it may contain unknown anomalies. Clearly data with these elements will not follow a single Gaussian distribution. In this case, the data may be characterized by a combination of Gaussian distributions, or a Gaussian Mixture Model. Details of the GMM and its associated Expectation-Maximization (EM) algorithm used for clustering the training data can be found in [14]. It should be noted that the Bayesian Information Criterion (BIC) [15] was utilised in the current study to indicate the optimized number of clusters.

2.2. Adaptation of GMM to remove anomalies in the training data

As indicated in the introduction section, the true condition of the training data is unknown and it might contain a number of outlying data which could affect the fault detection results. In this paper, a novel approach is introduced to identify areas of the cluster space that might be associated with anomalies.

To define the proposed approach, two assumptions need to be made according to the general distribution of the anomalies: 1) The anomalies are occurring at occasional time intervals in the time series; 2) The clusters associated with anomalies are distant from the clusters with normal data.

- Entropy based method

To find out anomalies under the first assumption, the Entropy statistic from the information theory is applied. For n partitioned categories (equal time intervals) in the time series, and for a training vector x with all samples having a cluster ID v, the Entropy E is defined as [16]:

$$E(x = v) = \sum_{i=1}^{n} -p_i \log_2 p_i \qquad (1)$$

Where p_i is the probability of training vector x coming from cluster v of category i occurring.

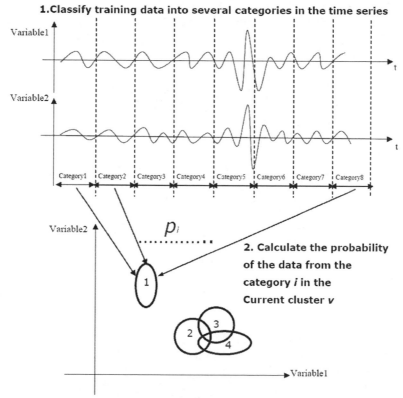

1.Classify training data into several categories in the time series

2. Calculate the probability of the data from the category i in the Current cluster v

3. Calculate the Entropy statistic for the current cluster v $\quad E(x = v) = \sum_{i=1}^{n} -p_i \log_2 p_i$

Fig.2 Steps of the entropy based approach to remove anomalies in the training data

Fig. 2 shows the concept and steps of using the Entropy statistic for finding the anomalies embedded in the GMM subspace. Hence, the main idea of the entropy based anomaly detection method is to explore whether the support cases in a particular cluster are from occasional time intervals or multiple time intervals. If data in a cluster is just from one time interval, the probability of the data points from this time interval in the current cluster will be as high as 100%, and the probabilities of the data points from other time intervals are 0%, so that the entropy score for the current cluster will be as low as zero. Regarding the first assumption, the current cluster could be recognized as a region associated with anomalies.

- **Distance based method**

Although anomalies are found occurring occasionally in the time series of the training data in most situations, however anomalies are also appearing frequently throughout the test, and the entropy based method will be insensitive to these anomalies. Hence, a distance based method is developed to measure the data dispersion-clusters with distant points from other regions being eliminated from the calculation. In this study, the Hotelling's T-squared distance is calculated between two clusters [17], using equations (2) and (3):

$$T^2 = \frac{n_A n_B}{n_A + n_B}(\mu_A - \mu_B)^T \hat{\Sigma}_{AB}^{-1}(\mu_A - \mu_B) \tag{2}$$

$$\hat{\Sigma}_{AB} = \frac{1}{n_A + n_B - 2}((n_A - 1)\hat{\Sigma}_A + (n_B - 1)\hat{\Sigma}_B) \tag{3}$$

Where n_A and n_B are the number of support cases in clusters A and B; μ_A and μ_B are sample means of clusters A and B, respectively; $\hat{\Sigma}_A$ and $\hat{\Sigma}_B$ are estimates of covariance matrices of the clusters A and B, respectively.

2.3. Anomaly detection and diagnosis

After characterization of the training data and adaptation of the GMM, the PCA is applied in each of the remaining Gaussian component, so the multiple PCA models can be generated. For detailed introduction of PCA, please refer to [18]. And then, for the given testing vector y_i, the sum of normalized squared scores, known as Hotelling's T-squared statistic is calculated to measure the variation of testing vector y_i within the PCA model. The T-squared statistic is defined as:

$$T_i^2 = t_i \lambda^{-1} t_i^T = y_i P_c \lambda^{-1} P_c^T y_i^T \tag{4}$$

where t_i refers to the ith row of Tc , the m by c matrix of c scores vectors from the PCA model, and λ is a diagonal matrix containing the eigenvalues ($\lambda 1$ through λc) corresponding to c eigenvectors (principal components) retained in the model. On the other hand, the anomaly detection threshold is defined by maximum Hotelling's T-squared distance of the training data.

It should be noted that Hotelling's T-squared statistic is calculated based on a single component within the multiple PCA models for each testing vector. The component is chosen with the minimum Mahalanobis distance, d_k, from its centre to the testing vector [9], given by equation (5):

$$d_k = \sqrt{(y_i - \mu_k)^T \sum_K^{-1} (y_i - \mu_k)} \tag{5}$$

Where μ_k and Σ_k are mean vector and covariance matrix of kth Gaussian component respectively.

Furthermore, the T-squared contribution values describe how individual variables contribute to the Hotelling's T-squared value for a given testing sample.

$$t_{con,i} = t_i \lambda^{-1/2} P_c^T = y_i P_c \lambda^{-1/2} P_c^T \tag{6}$$

Thus, diagnostic information is achieved by calculating the variables contribution value, to identify which variables are the main driven factors for the detected anomalies.

3. EXPERIMENT AND DATASETS

3.1. Experiment

The bearing rig tests discussed in this paper form part of a series of tests undertaken at the University of Southampton. Details of the rig may be found in other references (see [3] for example), so only a brief overview is provided here.

The bearing rig comprises four taper roller bearings housed in a chamber, as shown in Fig. 3. The bearings are mounted on a shaft driven by an electric motor. The outer two bearings (#1 and #4) are support bearings, whilst the inner bearings (#2 and #3) are test bearings. The two test bearings have a radial load applied to accelerate their failure. In addition, for the failure test analyzed in this paper, bearing #2 was pre-indented on the inner race. The baseline test used non-indented bearings. The bearing types were LM67010 (cup) and LM67048 (cone), which are all steel bearings. The rig was run at a constant speed of 2500 rpm under ambient conditions and fully flooded with the Shell Vitrea 32 lubricant at the oil flow rate of 4 litres min⁻¹.

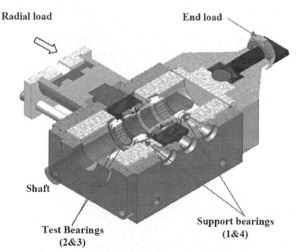

Fig.3 Bearing rig chamber showing main test components

The rig was instrumented with a number of condition monitoring sensors. In the test chamber there was one vibration sensor (mounted externally) and three electrostatic wear-site sensors (WSS) - one each on bearings #1 and #4 and one looking at bearings #2 & #3. The oil re-circulation pipework included an electrostatic oil-line debris sensor. Additional oil-line debris sensors included an inductive debris sensor, which provided wear particle counts for > 100 microns.

The testing also included a number of complimentary off-line analyses. Oil samples were taken during various stages of the testing and these were analyzed for sub-100 micron debris content. Tribological analyses of the bearing condition pre and post-test were conducted and included photographic evidence and mass loss calculations. Where applicable, the oil-line sensors and post-test / off-line analyses were used to help interpret the responses of the vibration and WSS sensors during the tests.

3.2. Feature extraction

The post-processed signal features from the three types of sensors (vibration, WSSs and OLS) were extracted to construct both training and testing data set for the developed intelligent scheme.

To form the vibration and WSSs training set, six dimensional features were selected for each of the two sensing technologies. In theses features, the first and second dimensions were chosen as the RMS value of the time domain and the energy at the rotational frequency respectively. The third to sixth features are based on the energies at the bearing defect frequencies, i.e. races, rollers and cage. These features are used to reveal the information on the bearing components. Each energy was externally clocked using once per revolution signal from the shaft tachometer. Data sets of RMS value and energies at bearing defect frequencies were produced at two points per minute. The bearing defect frequencies are listed as the equations (7) ~ (10) given by [19],

−Cage frequency (Hz)   −Roller frequency (Hz)

$$\omega_c = \frac{\omega_s}{2}\left(1 - \frac{d}{D}\cos\alpha\right) \quad (7) \qquad\qquad \omega_{re} = 2\omega_b = \frac{D\omega_s}{d}\left(1 - \frac{d^2}{D^2}\cos\alpha^2\right) \quad (8)$$

−Outer race frequency (Hz)   −Inner race frequency (Hz)

$$\omega_{od} = \frac{Z\omega_s}{2d}\left(1 - \frac{d}{D}\cos\alpha\right) \quad (9) \qquad\qquad \omega_{id} = z(\omega_s - \omega_c) = \frac{Z\omega_s}{2}\left(1 + \frac{d}{D}\cos\alpha\right) \quad (10)$$

Where ω_s is the shaft rotational speed in rad /s, d is the diameter of the roller, D is the pitch diameter, Z is the number of rolling elements and α is the contact angle.

For the OLS analysis, three time domain parameters including RMS values of the two oil line sensors and the cross-correlated value that couples two OLSs to identify charge moving at the oil flow rate, were utilised to generate data set. This training and testing sets were produced one point every two seconds. Table 1 tabulates the selected features for the bearing wear detection.

Table 1 Extracted features for the multivariate data analysis

Vibration	WSS1	WSS2	WSS3	OLS
RMS	RMS	RMS	RMS	RMS1
Tacho Energy	Tacho Energy	Tacho Energy	Tacho Energy	RMS2
Cage Energy	Cage Energy	Cage Energy	Cage Energy	Indicator
Roller Energy	Roller Energy	Roller Energy	Roller Energy	
OuterRace Energy	OuterRace Energy	OuterRace Energy	OuterRace Energy	
InnerRace Energy	InnerRace Energy	InnerRace Energy	InnerRace Energy	
Front housing	**Vicinity of Bearing1**	**Vicinity of Bearings2&3**	**Vicinity of Bearing4**	**Oil-line**

3.3. Bearing test datasets

In this paper, 2 sets of data are utilized to validate the proposed anomaly detection and diagnostic scheme. Table 2 summarizes the datasets and their test conditions and functions. From the test condition perspective, there are two modes of test, the first one is the baseline test, in which bearings are free from defects; the second one is the defect test, in which the inner race of one of the test bearings was pre-indented to accelerate the fatigue failure process. According to the investigation of the proposed condition monitoring scheme, the central idea of utilization of these data is: the baseline datasets are collected to build the reference model representing the normal operating condition, while the defect test datasets are evaluated against the established reference model to detect anomalies as well as diagnosing detected anomalies.

Table 2 Bearing test data used in this paper

Test No.	Test load/kN	Test duration/hours	Defect bearing	Speed/rpm	Application
1	20	80	none	2500	Training data
2	20	62	#2	2500	Testing data

4. TRAINING OF THE NORMAL MODEL

4.1. Construction of the classic mixture model

The first step is to build the suitable GMM to correctly describe the training data. Fig. 4(a) shows the Bayesian Information Criterion (BIC) for model-based method applied to test 1 data. It was found that the local minimum for the mixture model occurs with eleven clusters, hence, this was the number of clusters chosen and the trained normal model is illustrated in Fig. 4 (b).

4.2. Adaptation of GMM

It was indicated in section 2.2 that the training data (test 1 in this case) might contain a number of anomalies, and need to be eliminated by the developed entropy and distance based approaches.

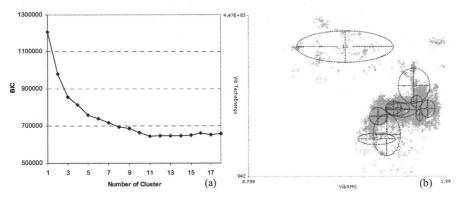

**Fig.4 BIC score for GMM applied to test 1 data (a)
Trained GMM for the test 1 data (b)**

At the beginning of the approach, the training data in the time series needs to be partitioned into several categories. The difficult task here is to determine the number of partitions. Mostly the partition needs to be assigned properly, as it directly affects the anomaly detection result. Thus the parameter of number of partitions has to be set explicitly. In this paper, the number of partitions is varied, and cross validation helps to find an appropriate parameter.

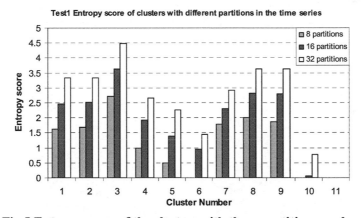

Fig.5 Entropy score of the clusters with three partition numbers

For the test 1 data, three partition numbers 8, 16 and 32 were assigned accordingly. Fig. 5 shows the calculated entropy score for each of the 11 trained clusters with 3 different partitions in the time series. It can be seen that the entropy score for each cluster increases lineally with increasing number of partitions. From the Fig.5, the entropy scores of clusters 10 and 11 are seen as significantly lower than the other scores of clusters, no matter how many partitions are assigned. Thus, these two clusters are highly suspected to be the regions associated with anomalies, and are suggested to be removed from the classical GMM.

Fig. 6 shows vibration RMS value before and after the removal of clusters 10 and 11 and their support cases. It was found a distinctive running-in region existed in the original plot (Figure 6(a)), if these running-in data points are used to build the GMM, they would lead to a coarser view of 'normality' which may not prove so effective when trying to detect real faults. After the removal of clusters 10 and 11 and their support cases as the Figure 6(b) shows, the running-in region data points have been successfully trimmed, and a new GMM can be generated for the anomaly detection.

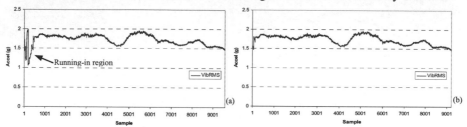

Fig.6 Vibration RMS plot before (a) and after (b) the removal of clusters 10, 11 and their support cases

However, the anomalies might be occurring frequently or distributed throughout the dataset, and the entropy based method would be ineffective in detecting these anomalies. This situation also occurred with the test 1 data. Drilling down to the extracted features in the frequency domains, as Fig. 7(a) illustrates, a significant number of anomalies still exist throughout the test, although clusters 10 and 11 and their associated support cases were eliminated.

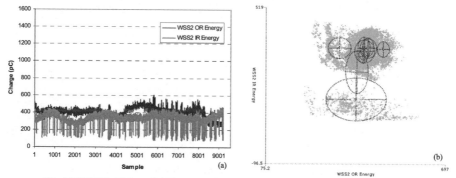

Fig.7 WSS2 OR and IR plot after the removal of cluster 10, 11 (a) GMM in the space of WSS2 OR and IR energy (b)

If the clustering space is plotted with the dimensions of WSS2 OR and WSS2 IR in Fig. 7(b). The cluster 9 might be the region with the anomalies discovered in Fig. 7(a). Going back to the entropy score plot (Fig. 5), cluster 9 is the region with high entropy score with 3 types of partitions. Hence, this cluster should not be removed.

Under this circumstance, the second strategy to remove anomalies in the training data is triggered. With the second anomaly detection strategy, Hotelling's T-squared distance is calculated between each of the Gaussian components. Fig.8 (a) illustrates the calculated result. It can be seen that clusters 3 and 9 are recognized as distant from the other clusters, so the support cases associated with these two clusters are identified as anomalies. Fig.8 (b) shows the WSS2 OR and WSS2 IR data after the removal of clusters 3 and 9 and their associated data points. Most of the defined anomalies have been filtered out from the original data plot, and shown a reasonable flat trend for the usage of the normal training data.

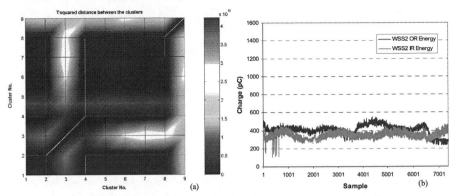

Fig.8 T-squared distance between the trained clusters (a)
WSS2 OR and IR energy plot after the removal of clusters 3 and 9 (b)

4.3. Principle Component Analysis in the remained Gaussian components

After the construction and adaptation of the normal reference model, the Principle Component Analysis (PCA) is applied in each of the remaining Gaussian component, so that the multiple PCA subspace can be built. Table 3 shows the number of the Principle Components (PCs) and their captured variance for each of the remained Gaussian components of the test 1 data.

Table 3 Summary of number of extracted PCs in the remained clusters

Remained clusters	Number of PCs	Cumulative Variance (%)
1	14	82.61
2	11	82.04
4	14	81.18
5	13	80.42
6	13	82.49
7	12	80.73
8	13	80.40

5.   ANOMALY DETECTION AND DIAGNOSIS RESULTS

5.1.   Anomaly detection

The effectiveness of using equipped multiple sensing technologies to detect impending failure of the bearings, and the multivariate dataset, marked as test 2 in this paper are reported in [3]. There are two aims behind the further analysis of the data. First to determine whether an earlier (prognostic) indication of the failure can be achieved, i.e. can the initiation of fatigue failure be detected in advance of the finial few hours? Second, the capability for determining what and where the fault is i.e. is there clear diagnostic information available? In the field, both of these are important if goals such as optimized maintenance planning, minimum logistic footprint and maximum equipment are to be achieved.

Fig. 9(a) shows the Hotelling's T-squared statistic with the test 2 data, which has been run through the normal reference model trained with test 1 data. From this figure there is evidence of abnormal behaviour at the start of the test which, is due to the bearing running in. The run out to failure from about 54 hours, dominates. However, it is encouraging to find evidence of change in advance of 54 hours: an overall increase at around 36 or so hours, with significant peaks between 43 and 45 hours. It is also noted that these detected prognostic activities could not be seen clearly in the original extracted features, which proves the advantage of the proposed anomaly detection approach.

So far, it is not clear the cause of the period of activity between 43 and 45 hours in Fig. 9(a), they may result from the generated wear debris by delamination from the pre-indent on the inner race of bearing #2 or surface debounding by subsurface cracks. It is also found in Fig.10, that the off-line debris counter could detect increases in debris production at 43 hours. This correlates with the detected activities by the T-squared statistic over the similar period of time. Hence, these abnormal events are invaluable, as they could be used to detect debris related abnormal conditions that result from the initiation of fatigue failure.

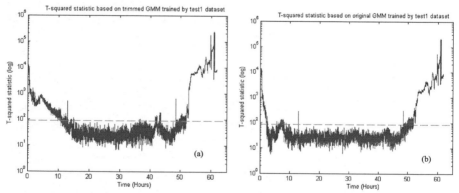

Fig.9 T-squared statistic with the test 2 data based on trimmed GMM (a)
T-squared statistic with the test 2 data based on original GMM (b)

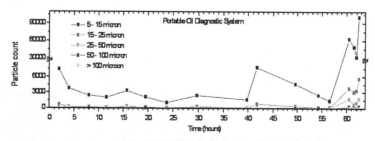

Fig.10 Off-line debris counter

In order to verify the importance of the model adaptation technique, the Hotelling's T-squared statistic is also calculated against the un-trimmed original GMM as shown in Fig. 9(b). The prognostic activities between 43 and 45 hours can not be visualized, this is due to a number of anomalies contained in the training data (test1) which is used to build the model of normality, and the abnormal events in the testing data which are similar to the anomalies in the training data become undetectable.

5.2. Anomalies diagnosis

The anomaly detection results indicate two periods of interest as shown in Fig. 9(a). In the field, it might be insufficient to conduct appropriate maintenance plan with only the knowledge of detected anomalies, as it is meaningful to understand where these anomalies occur. As stated before, the applied features could reveal the condition of different bearings and components within the bearing. Hence, there is a need to investigate which variables are the main driving factors for the detected anomalies to achieve a diagnostic function. In this study, contribution values for each of the variables are calculated to identify the main influencing factors that drive the detected anomalies of interest.

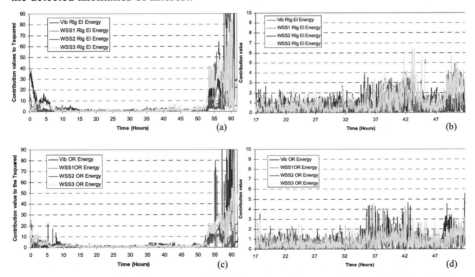

Fig.11 Contribution charts of the applied variables for the bearing diagnosis

Fig. 11(a) shows the absolute contribution values of the rolling element energy for the four applied sensors. Over the first period of interest (between 37 and 43 hours, as Fig. 11(b) shows), the WSS3 rolling element energy dominates the trend with additional peaks, while the other rolling element energies show insignificant contributions to the detected abnormal trend. On the other hand, it is clear to see that the vibration and WSS3 rolling element energies are the main driving factors to the increasing trend during the second period of interest (54-63 hours). For the contribution value of outer race energies (see Fig.11 (c) and (d)), vibration and WSS3 elements dominate both periods of interest, but do not seem as significant as the rolling element energies. For the contribution value of inner race energies (not shown here), the difference from the previous two bearing elements is that contribution values of inner race energies can not be distinguished to see which variables are the major contribution resource for both two areas of abnormal, and have the relatively weak impact. Apart from the contribution values of the bearing element energies, the oil-line variables are also examined, and found that the OLS variables were seen to be significant in the second period interest.

Therefore, it can be seen that the WSS 3 rolling element and vibration outer race energies obtain the highest impact on the detected prognostic events (between 37 and 43 hours), from which two diagnostic information is generated: 1) abnormal conditions occurred within the bearing #4, since it is monitored by WSS3. 2) As the WSSs are designed to detect charge generated between the contacting surfaces and wear debris, it can be inferred that the delamination or initial spallation occurring between the contacting surfaces of rolling elements and outer race of bearing #4, and amount of debris generated which is detected by off-line debris counter.

On the other hand, nearly all the variables have the impact on the abnormal condition at the end stage (54-63 hours), but the rolling element and outer race energies of vibration and WSS3 are the strongest. This could be explained as spallation occurred within the bearing #4 due to rolling contact fatigue, hence large amounts of debris and hugely modified contact surfaces generate high charge and strong energies which are reflected by the WSS and vibration sensors, respectively. Furthermore, OLS variables also show a significant trend during this period. This is because the OLSs are installed in the oil-line, and the generated debris was brought by the oil flow passing through the OLSs, causing significant charges.

5.3. Bearing inspection
In order to relate the physical condition of the bearing elements to the diagnostic results, both these faults are corroborated by post-test inspections of the bearings. It is clear to see in the Fig. 12 that there is significant damage to the outer race of bearing #4 and several rolling elements of the same bearing suffered material loss. This confirms that, for the two periods of interest, the fatigue initiation and spallation occurred between the contacting surfaces of the rolling elements and the outer race of bearing #4.

Fig.12 Post-analysis of the failed bearing #4 of outer race and rolling elements

6. CONCLUSIONS AND FUTURE WORK

A new intelligent scheme has been developed to detect early signals of bearing distress. From the work conducted to date the following conclusions can be made:

1) A novel approach based on entropy statistic and T-squared distance has been developed to effectively remove the anomalies in the training data, and the demonstration results show the importance of adaptation of the normal model.
2) The T-squared distance has been utilized for extracting abnormal events from the multivariate data. Prognostic information has been extracted approximately 10 hours before the severe wear occurred.
3) The contribution values of the extracted features can assist operators locate faulty bearings as well as indicate that the vibration features have the strongest impact on the T-squared statistic at the second period of interest while the electrostatic features are more sensitive to the precursors. Furthermore, off-line debris analysis and final bearing surface inspection also provide strong physical evidence for the precursors and predicted faulty bearing locations.
4) The vibration features were examined by the developed scheme for the first time to obtain the physical understanding of the vibration sensor. From the testing results, the vibration sensor could be used to detect and locate prognostic events that are related to fatigue initiation, this capability was complementary with the electrostatic wear-site sensors which also provide contribution to detect prognostic events. On the other hand, vibration sensor is found to be more sensitive for monitoring severe wear of the bearing (e.g. large amounts of wear debris were entrained between the raceways and rolling elements due to the spallation giving rise to significant shocks).

Although the test results have shown the effectiveness of the scheme, there are still several concerns that need to be addressed by future work:

1) In the approach of reference model adaptation, number of the partitions assigned in the time series is set explicitly. A mechanism is required to set this parameter by investigating number of the data points and dimensionality in the training data.
2) The threshold value for the anomaly detection needs to be optimized.
3) To achieve the automated reasoning, the system needs to be assigned with the knowledge that represents different conditions of the bearings. The current diagnostic method using contribution values is not an automatic approach. Hence, it is important to develop diagnostic training data representing different bearing symptoms (bearing fault mechanism, and fault location) by discovering knowledge from the run-to-failure test and recording characteristic features from such datasets.

ACKNOWLEDGEMENT

The authors would like to thank GE Aviation for funding this project and Dr. Harvey from Southampton University for providing the test data.

REFERENCES

[1] Powrie H, McNicholas K. Gas path condition monitoring during accelerated mission testing of a demonstrator engine, AIAA 97, 1997, Paper 97-2904.

[2] Morris S, Wood RJK, Harvey TJ, Powrie H. Use of electrostatic charge monitoring for early detection of adhesive wear in oil lubricated contacts, ASME Journal of Tribology, 2002,124: 288-96.

[3] Harvey TJ, Wood, RJK, Powrie H. Electrostatic wear monitoring of rolling element bearings, Wear, 2007, 263: 1492-1501.

[4] Wang L, Hope AD. Bearing fault diagnosis using multi-layer neural networks. Insight: Non-Destructive Testing and Condition Monitoring 2004; 46(8): 451-55.

[5] Liu TI, Singonahalli JH, Iyer NR. Detection of roller bearing defects using expert system and fuzzy logic. Mechanical Systems and Signal Processing 1996; 10(5): 595-614.

[6] Zhang L, Jack LB, Nandi AK. Fault detection using genetic programming. Mechanical Systems and Signal processing 2005; 19: 271-89.

[7] Yang J, Zhang Y, Zhu Y. Intelligent fault diagnosis of rolling element bearing based on SVMs and fractal dimensions. Mechanical Systems and Signal Processing 2007; 21 (5): 2012-24.

[8] Sun W, Chen J, Li J. Decision tree and PCA-based fault diagnosis of rotating machinery. Mechanical Systems and Signal Processing 2007; 21 (3): 1300-17.

[9] Roberts SJ, Novelty detection using extreme value statistics, IEE Proc. on Vision, Image and Signal Processing. 1999; 146(3): 124-29.

[10] Banister PR, Tarassenko L. Learning jet engine vibration response for novelty detection. The 2nd World Congress on Engineering Assessment Management and the 4th International Conference on Condition Monitoring. 2007; 229-38.

[11] Kresta JV, MacGregor JF, Marlin TE. Multivariate statistical monitoring of process operating performance. Can. J. Chem. Eng. 1991; 69: 35–47.

[12] Kourti T, MacGregor JF, Process analysis, monitoring and diagnosis, using multivariate projection methods. Chemometrics Intell. Lab. Syst. 1995; 28: 3–21.

[13] Callan R, Larder B, Standiford J. An integrated approach to the development of an intelligent prognostic health management system. IEEE Aerospace Conference, Montana, USA, 2006.

[14] Dempster AP, Laird NM, Rubin DB, Maximum likelihood from incomplete data via the EM algorithm, Journal of the Royal Statistical Society 1977; B 39:1-38.

[15] Fraley C, Raftery AE. How many clusters? Which clustering method? Answer via model-based cluster analysis. The Computer Journal 1998; 41(8): 578-88.

[16] Callan R, Artificial Intelligence, Palgrave, UK, ISBN 0-333-80136-9, 2003.

[17] Anderson T. Am Introduction to Multivariate Statistical Analysis. New York: John Wiley and Sons, 2nd edn, 1984.

[18] Bishop CM, Neural Networks for Pattern Recognition. 1995, Oxford, UK, ISBN 0 19 853864 2 (pbk).

[19] Tandon N, Choudhury A. A review of vibration and acoustic measurement methods for the detection of defects in rolling element bearings. Tribol. Int. 1999; 32: 469-80.

CRACKS

Stochastic approach for crack identification in rotating shafts using Monte Carlo simulations and a hybrid mechanical model

T Szolc, P Tauzowski, R Stocki, J Knabel
Institute of Fundamental Technological Research,
Polish Academy of Sciences, Poland

1. INTRODUCTION

An effective identification and localization of cracks occurring in the rotor-shafts is a very important task for reliable and safe exploitation of modern, heavily affected rotating machines. Thus, recent advances in machinery dynamics, fracture mechanics and dynamic diagnostics have to be employed to build proper mechanical models and algorithms necessary for thorough investigations and monitoring of these defects. Vibration analyses of cracked rotors from the viewpoint of detection, localization and size identification of a crack have been carried out for many years by many authors. Some of them used to consider the crack influence on rotor-shaft bending vibrations only, e.g. in [1-3]. They focused their attention on the nonlinear dynamic behaviour of the rotor-shaft system due to crack breathing from the viewpoint of crack detection. In [4-6] there were studied coupling effects between shaft bending and torsional vibrations regarded as crack occurrence symptoms and caused by a local anisotropy of the cracked shaft cross-section. For this purpose, in [4] an analytical beam model is used and in [5,6] shaft discretization by means of beam finite elements is applied. The complete coupling effects due to anisotropy of the cracked shaft cross-section, i.e. between shaft bending, torsional and axial vibrations, are taken into consideration in [7-9]. The simulation results of coupled vibrations of the cracked rotor-shaft systems obtained in [7,8] by the use of the beam finite element model are qualitatively analysed in [7] and used for composing the cause-symptom relationships necessary for crack diagnostics in [8]. The analogous simulation results obtained in [9], using the hybrid model of the rotor-shaft system, have been applied for the fatigue life assessment of the cracked shaft.

In order to detect and localize a crack on the shaft, several methods of inverse mapping of the on-line measured response - investigated symptom relationships have been developed. Most of them use neural networks or other adaptive systems, see e.g. [8]. Since the fault identification routines employing the neural networks seem to be quite labour-consuming and not always sufficiently accurate, in the presented paper to identify crack depth and position on the rotor-shaft an approach based on searching for the most probable crack parameters in previously prepared database. This method consists of the two major steps. The first step is the database preparation, which is an intensive Monte-Carlo simulation of coupled dynamic responses of the cracked rotor-shaft system performed for various possible crack sizes and locations. For this purpose, a reliable and computationally efficient hybrid mechanical model of the rotor-shaft system is applied. The second step is the actual identification based on the data generated during the first phase. Several identification improvement strategies are also proposed and investigated. Their accuracy and efficiency are demonstrated using two numerical examples, the single-span rotor-shaft system and the rotor-shaft system of the typical 200 MW steam turbogenerator.

2. HYBRID DYNAMIC MODELING OF THE ROTOR-SHAFT SYSTEM

In order to obtain sufficiently reliable results of numerical simulations together with a reasonable computational efficiency, the vibrating rotor-shaft systems of rotor machines are usually modelled by means of the one-dimensional finite elements of the beam-type. Thus, in this paper dynamic investigations of the entire rotating system are performed by means of the one-dimensional model consisting of discrete oscillators and of flexurally, axially and torsionally deformable structural beam elements representing successive cylindrical segments of the stepped rotor-shaft. With a reasonable accuracy for practical purposes, the heavy rotors are represented here by rigid rings attached to the shafts using mass-less membranes enabling rotations of these rings around their diameters as well as their translational displacements in the shaft axial direction. Each journal bearing is represented by the use of a dynamic oscillator of two degrees of freedom, where apart from the oil-film interaction also visco-elastic properties of the bearing housing and foundation are taken into consideration. This bearing model makes it possible to represent, with relatively high accuracy, kinetostatic and dynamic anisotropic and anti-symmetric properties of the oil-film in the form of constant stiffness and damping coefficients, see [8]. An example of such a finite element model of the single-span rotor shaft system supported on two journal bearings is schematically presented in Fig. 1.

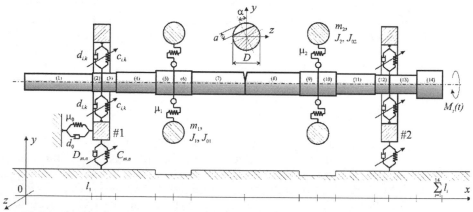

Figure 1 : Mechanical model of the two-bearing cracked rotor-shaft system

Nevertheless, finite element models commonly applied for large and complex real objects are characterized by a relatively high number of degrees of freedom in the range between hundreds and even thousands. A description of faults in the rotor-shaft systems usually makes motion equations of the finite element models non-linear and parametric, see e.g. [7,8]. Then, to obtain the system dynamic response, a direct integration of the motion equations is required. Thus, for such large finite element models proper algorithms reducing the number of degrees of freedom have to be employed in order to shorten computer simulation times. Moreover, for the Monte-Carlo simulation performed for numerous axial fault positions along the entire rotor-shaft line, the discretization mesh density of the finite element model must be appropriately modified in each case. Thus, the corresponding multiple reductions of degrees of freedom are troublesome and can lead to computational inaccuracies associated with unstable runs of simulation programs.

According to the above, in order to avoid the above-mentioned drawbacks of the classical finite element approach and to maintain the obvious advantages of this method, in the applied model of the rotor-shaft system, the inertial-visco-elastic properties of the beam finite elements representing successive cylindrical shaft segments have not been discretized, i.e. reduced to nodes using artificial shape functions, but in a natural way they remained continuously and uniformly distributed along the entire finite element length. Hence, such finite elements have been called, e.g. in [9,10], 'continuous visco-elastic macro-elements'. Then, the assumed dynamic models consisting of the continuous visco-elastic macro-elements and discrete oscillators are called 'hybrid models'.

In the hybrid model motion of cross-sections of each visco-elastic macro-element is governed by the partial differential equations derived using the Timoshenko and Rayleigh rotating beam theory for flexural motions and by the hyperbolic equations of the wave type, separately for torsional and axial motion. Similarly as in [9,10], mutual connections of the successive macro-elements that compose the stepped shaft as well as their interactions with the supports and rigid rings representing the heavy rotors, are described by equations of boundary conditions. These equations contain geometrical conditions of conformity for translational and rotational displacements of extreme cross-sections as well as linear, nonlinear and parametric equations of equilibrium for external forces and torques, static and dynamic unbalance forces and moments, inertial, elastic and external damping forces, support reactions and gyroscopic moments.

In the hybrid model in the selected rotor-shaft segment there is considered a transverse crack implemented in the form of a proper elastic connection of the respective adjacent left- and right-hand side parts of the faulty shaft segment. Additional flexural, torsional and axial flexibilities introduced by the crack into the shaft are represented here by means of mass-less springs connecting the adjacent beam macro-elements substituting the cracked shaft segment. Constant stiffness values of these springs are determined according to [4,7-9] using the fundamentals of fracture mechanics. In the inertial co-ordinate system these values become periodically variable because of shaft rotation. If in the rotor-shaft a transverse crack is assumed, its "breathing" process has been taken into consideration in the form of the well-known "hinge" crack model, as e.g. in [2,7-9]. Here, the system changes its configuration depending on whether the crack is temporarily "open" or "closed", which is indicated by the current difference value of inclination angles $\Delta\varphi(t)$ of the adjacent extreme cross-sections of the proper left- and right-hand side macro-elements representing the cracked shaft segment. Due to these configuration changes of the rotor-shaft structure and the above-mentioned stiffness periodic fluctuation of the springs representing the additional shaft flexibility introduced by the crack, the considered dynamic problem is formulated as a parametric-nonlinear one.

The solution for the forced vibration analysis has been obtained using the analytical-computational approach demonstrated in detail in [10]. By solving the differential eigenvalue problem for the linearized orthogonal system, two sets of bending, torsional and axial eigenfunctions are determined: for the un-cracked and cracked rotor-shaft, respectively. All parametric, anti-symmetric and gyroscopic terms omitted during system linearization have been regarded here as response- dependent external excitations. Generally, for linear continuous and discrete-continuous systems, an application of the Fourier solutions in the form of series in the orthogonal eigenfunctions leads to the infinite number of known separate ordinary differential equations in modal coordinates. But in the considered case, the above-mentioned response dependent external excitations couple these equations in the sense investigated e.g. in [11]. Thus, finally one obtains the following set of parametric ordinary differential equations in the modal co-ordinates contained in vector $\mathbf{r}(t)$:

$$\mathbf{M}(\Omega t)\ddot{\mathbf{r}}(t) + \mathbf{C}(\Omega,\Omega t)\dot{\mathbf{r}}(t) + \mathbf{K}(\Delta\varphi(t),\Omega t)\mathbf{r}(t) = \mathbf{F}(t,\Omega^2,\Omega t), \qquad (1)$$

where: $\mathbf{M}(\Omega t) = \mathbf{M}_0 + \mathbf{M}_u(\Omega t)$,

$\mathbf{C}(\Omega,\Omega t) = \mathbf{C}_0 + \mathbf{C}_g(\Omega) + \mathbf{C}_u(\Omega t), \quad \mathbf{K}(\Delta\varphi(t),\Omega t) = \mathbf{K}_0 + \mathbf{K}_b + \mathbf{K}_{cr}[\Delta\varphi_k(t),\Omega t].$

The symbols $\mathbf{M}_0$ and $\mathbf{K}_0$ denote, respectively, the constant diagonal modal mass and stiffness matrices, $\mathbf{C}_0$ is the constant symmetrical damping matrix and $\mathbf{C}_g(\Omega)$ denotes the skew -symmetrical matrix of gyroscopic effects. The terms of the unbalance effects are contained in the symmetrical matrix $\mathbf{M}_u(\Omega t)$ and in the non-symmetrical matrix $\mathbf{C}_u(\Omega t)$. Anti-symmetric elastic properties of the journal bearings are described by the skew-symmetrical matrix $\mathbf{K}_b$. Nonlinear and parametric properties of the breathing crack are described by the symmetrical matrix $\mathbf{K}_{cr}[\Delta\varphi_k(t),\Omega t]$ of periodically variable coefficients and the symbol $\mathbf{F}(t,\Omega^2,\Omega t)$ denotes the external excitation vector, e.g., due to the unbalance and gravitational forces. Theoretically, the number of Eqs. (1) is infinite, according to the Fourier solution. Nevertheless, a proved fast convergence of the Fourier solution enables us to reduce the appropriate number of the modal equations to be solved, in order to obtain a sufficient accuracy of results in the given range of frequency. Consequently, the number of Eqs. (1) corresponds to the number of eigenmodes taken into consideration, because the forced bending, torsional and axial vibrations of the rotor shaft are mutually coupled and thus, the total number of equations to be solved is a sum of all bending, torsional and axial eigenmodes of the rotor shaft model in the range of frequency of interest. Here, usually it is necessary to solve only 15÷40 coupled modal Equations (1), even in cases of great and complex mechanical systems. In this way, no degree-of- freedom reduction algorithms are required, contrary to the classical finite element formulation. Moreover, the hybrid modelling assures at least the same or even a better representation of real objects. In addition, its mathematical description is formally strict, demonstrates clearly the qualitative system properties and is much more convenient for a stable and efficient numerical simulation.

In order to obtain the system's dynamic response, Eqs. (1) are solved by means of direct integration. In the considered case of nonlinear-parametric problem, during simulation of forced vibrations, the angle difference $\Delta\varphi(t)$ determining the open/closed-stage of crack "breathing" has to be computationally predicted after each integration step by means of the explicit numerical method. Thus, there is indicated the actual base of eigenfunctions, in which the coupled dynamic response should be sought, i.e. the set of eigenfunctions corresponding respectively to the cracked or un-cracked rotor-shaft line. Then, at each "switch" from the open-crack-stage into the closed-crack-stage and vice-versa, the current system response becomes the initial condition for the "new" stage. Here, for a sufficiently small direct integration step using the Newmark method, almost no corrections of the time-step value were required in order to obtain sufficient accuracy of the computational routine.

3. SIMULATION PATTERNS FOR CRACK IDENTIFICATION IN THE ROTOR SHAFTS

The presented methodology of vibration analysis is demonstrated here using two computational examples. In the first one there is considered the single-span rotor-shaft system with two identical heavy rotors located on the shaft between two journal bearings, as shown in Fig. 1. The stepped-rotor shaft is characterized by the total length of 2.315 m and the bearing span equal to 1.6 m. The total mass of the considered rotating system equals ca. 590 kg. For this system the Monte-Carlo simulation (which is the basis of proposed crack identification methods) is performed for various crack depth ratios in the range $a/D=0.1\div0.4$ and for various axial positions on the entire shaft length. Since as it follows e.g. from [8,9], particularly for overweight shafts, also a circumferential crack position plays an important role in the system dynamic behaviour, the responses are obtained here for various circumferential crack positions on the shaft within $0\div360$ degrees at the constant rotational speed 6000 rpm.

In the second example there is investigated the rotor-shaft system of a typical 200 MW steam turbogenerator consisting of the single high- (HP), intermediate- (IP) and low-pressure (LP) turbines as well as of the generator-rotor (GEN). This rotor-shaft line of the total length of 25.9 m is supported by seven journal bearings, as shown in Fig. 2. For this system, the Monte-Carlo simulation of coupled forced vibrations is performed for various crack depth ratios in the range $a/D = 0.1\div0.3$ and for various axial positions on the entire length of the shaft. The system dynamic responses are obtained for the constant nominal rotational speed 3000 rpm and for various crack positions around the shaft circumference.

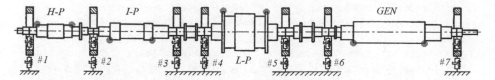

Figure 2: Mechanical model of the steam turbogenerator rotor-shaft system

In both examples, in addition to the static gravitational load, the only assumed sources of dynamic external excitation are static residual unbalances of the rotors. Thus, the torsional and axial vibrations can be regarded here as an output effect caused by bending vibrations of the rotor-shafts. The quantities that are of particular interest for the purpose of crack identification are the ones, which in regular exploitation conditions can be relatively easily measured on-line. These are lateral vibration displacements of the shaft at the bearing locations, fluctuation of shaft rotational speed due to torsional vibrations at the rotor locations as well as the longitudinal displacements of the shaft's both free ends and the shafts longitudinal displacement at the thrust bearing location. It should be emphasised that for the assumed sources of external excitations, the lateral responses of the considered uncracked rotor-shaft systems are harmonic in character with the only one synchronous frequency component 1X. Thus, shaft's lateral displacement orbits are elliptical. The very weak torsional-lateral coupling due to the residual unbalances results in negligible torsional response and lack of the lateral-axial coupling yields a zero axial response.

In Fig. 3a there are presented results of exemplary simulations obtained for the described above two-bearing rotor-shaft system with the crack of the depth ratio a/D = 0.4 localized at the bearing mid-span. In this case the system lateral response is not harmonic. The bearing journal displacement orbits are not elliptical and the corresponding amplitude spectra of displacement time-histories are affected also by the remarkable, even in the linear scale, frequency components 2X and 3X in addition to the fundamental synchronous component 1X. Analogous results obtained for the steam turbo-generator rotor-shaft with the crack of the depth ratio $a/D = 0.25$ localized at the generator rotor mid-span, i.e. between bearings #6 and #7, are presented in Fig. 3b. Here, the time history of the lateral displacement registered at bearing #6 is characterized by the predominant 3X frequency component, which emphasizes the nonlinear-parametric character of the dynamic response induced by this crack.

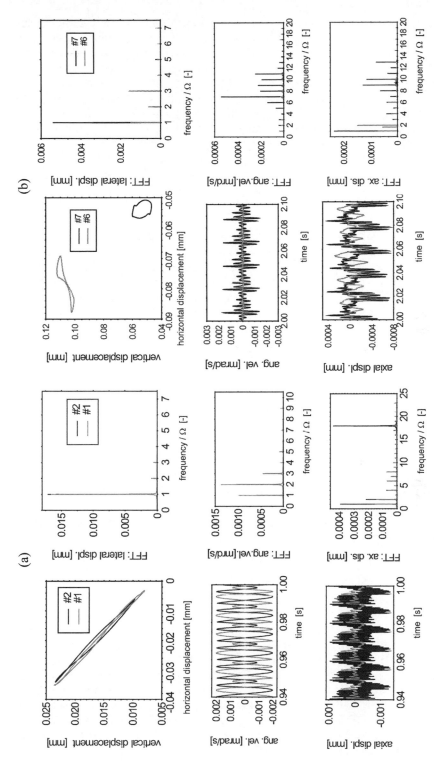

Figure 3: Coupled dynamic responses for the two-bearing rotor-shaft system (a) and for the turbo-generator rotor-shaft line (b).

For the investigated cracked rotor-shafts, the torsional responses are quite significant and the axial responses also become remarkable for measuring devices. It is noteworthy that the amplitude spectra of the torsional and axial responses, obtained for the mentioned crack depth ratios as well as those obtained for the cracks assumed at various locations on the shaft and for various other depths within the considered ranges are relatively "rich" and qualitatively similar, i.e. in general the same fluctuation components are excited with magnitudes rapidly increasing together with the crack a/D ratio. But mutual relations between the amplitude values of successive fluctuation components depend on particular cases of crack depths and locations. Similarly, the 2X and 3X frequency components of the system lateral responses also significantly increase together with the rise of a/D, whereas maximum values of the peak corresponding to the synchronous 1X component remain almost unchanged.

According to the observed facts, in both considered examples of the rotor machines, as diagnostic parameters necessary for crack identification in the shaft, the maximum amplitude values of the 2X and 3X frequency components of the system lateral responses as well as the resultant (global) amplitudes of the torsional and axial responses have been selected. The mentioned resultant (global) amplitudes of the considered torsional and axial time histories are regarded here as maximum fluctuation values with respect to their average values of the steady-state dynamic response. The values of response amplitudes listed above corresponding to randomly generated crack parameters will constitute the database of numerical experiments used in the identification process.

4. STOCHASTIC APPROACH IN IDENTIFICATION OF CRACK PARAMETERS

The objective of the proposed method is an efficient assessment of a possible damage of the vibrating rotor-shaft system when an information on system responses from a monitoring device is available. It can be described as solving the inverse problem at the point determined by the device readout. The approach consists of the two main steps: computation of the system responses corresponding to randomly sampled crack parameters and the actual identification based on the data generated during the sampling phase.

The first step is the random sampling performed in order to create a database of points (x,y) relating the crack parameters **x** to the responses y of the vibrating rotor-shaft system (see Sec. 3 for the discussion on the representative diagnostic responses). To generate a sample of crack parameters it is necessary to assume the probability distributions of the crack location x_c, the ratio a/D and the circumferential crack position α_p. The choice of probability density functions (PDFs) should guarantee a proper exploration of the entire parameter space as well as the space of responses. Given a large number of sample points, the uniform PDFs could serve this purpose. However, in order to assure a good "saturation" of the response space with experimental points in the regions corresponding to the most frequent crack occurrence scenarios, the probability distributions should account somehow for the dynamic behavior of the rotor-shaft systems.

Even though PDF for α_p can hardly be assumed other than uniform in the range 0 ÷ 360 degrees, the choice of PDFs for the remaining parameters is not obvious. Since it is much more likely that relatively shallow cracks are identified by monitoring devices (and some further actions are taken as a result of an inspection) rather than it is a very deep crack that is detected, the following half-normal PDF is employed for the crack depth a

$$f(a) = \frac{2\theta}{\pi} \exp\left[-\frac{(a - a_{min})^2 \theta^2}{\pi} \right], \qquad (2)$$

where a_{min} is the minimal crack depth that is considered and θ is the parameter selected such that the probability of crack depth greater than the maximal depth observable in practice is negligible. In the case of crack location x_c the corresponding PDF should account for the fundamental knowledge of fracture mechanics. It implies that the probability of having a crack at a particular point along the shaft is proportional to the maximal Huber - von Mises stress. Therefore, the PDF is taken to be proportional to the maximal reduced stress envelope obtained for the nominal uncracked rotor-shaft. The PDF corresponding to such stress envelope for the turbogenerator rotor-shaft system shown in Fig. 2 can be given as follows:

$$f(x_c) = \frac{|\sigma(x_c)|}{\int_{x_1}^{x_2} |\sigma(x)| \, dx}, \qquad x_c \in [x_1, x_2], \qquad (3)$$

where $\sigma(x)$ is the maximal reduced stress function and x_1 and x_2 are the axial coordinates of the outermost, i.e., respectively, of the first and the last bearings. The PDF (3) and the corresponding cumulative distribution function (CDF) are shown in Fig. 4. In the examples of the identification problem that are presented below, in the case of the crack depth and its location it was decided to generate 30% of the total number of sample points using the uniform PDF in the respective parameter domains, and 70% using the density function (2) for a and (3) for x_c.

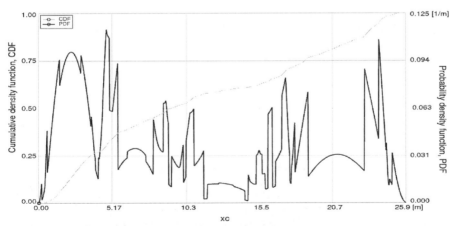

Figure 4: PDF and CDF for the crack location x_c along the rotor-shaft of the steam turbogenerator.

One of the major advantages of one-dimensional hybrid dynamical models of the rotor-shaft systems is their high computational efficiency. A single analysis of the parametric-nonlinear coupled vibration process of a single- or double-span rotor-shaft usually takes less than 10 seconds of CPU on modern PCs. Hence, accounting for inherent parallelism of random sampling in most of the cases it is affordable to perform hundreds of thousands or even millions of simulations. This allows for a thorough exploration of the response space facilitating the identification process.

The second step of the method, i.e. parameter identification, consists in localizing the point in the space of normalized responses that is closest to the readout of a monitoring device. Later in the text this approach is referred to as the nearest point (NP) method.

The other identification methods presented below aim at enhancing the database in the neighborhood of the NP estimation, which in consequence should lead to the identification accuracy improvement. It is assumed that the improvement phase cannot be too much CPU time-consuming in order to provide an additional information concerning the crack parameters in a relatively short time. Such an information facilitates a proper decision concerning the exploitation of the rotor machine. Moreover, the improvement methods permit to use smaller databases, reducing the costs of database preparation. The following three methods were investigated:

Orthogonal projection method (OP). The idea of orthogonal projection method is illustrated in Fig. 5. In the figure the readout is marked with the star ($\star$) and the diamond ($\blacklozenge$) markers denote sample points, i.e. they are the responses for randomly generated crack parameters.

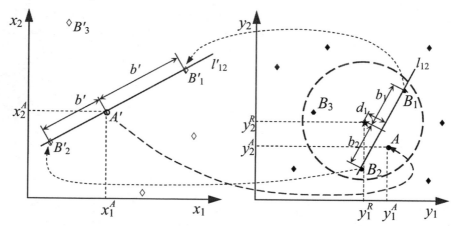

Figure 5: The idea of orthogonal projection method.

The OP algorithm starts from searching for the pair of points in the readout vicinity, which define the straight line that is closest to the readout. In Fig. 5 these are points B_1 and B_2 and the line is denoted by l_{12}. Next, the readout is projected on l_{12} determining two segments b_1 and b_2 connecting the projection point with B_1 and B_2, respectively. In the parameter space the points corresponding to B_1 and B_2 are B_1'

and B_2', respectively, and the straight line passing through these points is denoted by l_{12}'. Accounting for the proportion b_1/b_2 and the position of the projection point with respect to B_1 and B_2, point A' is established on l_{12}' as an approximation of the projection point mapping. Then, for the crack parameters determined by A', the dynamic responses of the rotor-shaft system are computed yielding point A with coordinates (y_1^A, y_2^A). This point is added to the database and the procedure restarts. If point A is localized in the vicinity of the readout, it will probably be used in the next iteration as one of the points constituting the closest straight line. This line, in turn, is likely to produce a new A point that is closer to the readout than the previous A point. If this is not the case, the procedure is stopped. Another stop criterion may be set on the number of iterations. The parameter values corresponding to the best A point (provided it is closer to the readout than any of the original database points) are taken to be the identification results.

The main advantage of the OP method is a small number of additional simulations of the rotor-shaft dynamic responses. However, due to a strongly nonlinear character of the considered relationship there is no guarantee that the methodology will always improve the NP identification results.

Nedler-Mead method (NM). This is the nonlinear simplex optimization method introduced by Nedler and Mead [12], which has been adopted to solve the considered identification problem. In our case the simplex is based on the set of $n +$ 1 points in the parameter space corresponding to the points closest to the readout in the response space, where n is the number of parameters. Each time, during the optimization process a new point (values of crack parameters $\mathbf{x}$) is established, the appropriate rotor-shaft responses $\mathbf{y}$ are computed and the new point is added to the database. The objective function that is to be minimized here is the distance in the response space between the readout and the closest database point. The procedure is stopped if the minimized distance is smaller than a certain ε tolerance or the number of calls to the rotor-shaft simulation program is exceeded.

Similarly to the OP method, the NM identification algorithm requires some additional simulations of the rotor-shaft system dynamic responses, which increase the identification time. However, this additional computational effort can be controlled in a quite flexible way depending on the user preferences. In terms of the distance from the readout the method always results in a solution, which is not worse than the NP-based identification.

Local Sampling method (LS). The LS method seems to be the simplest choice for improving the accuracy of the NP identification results. The method consists in finding the two points $\mathbf{y}_A$ and $\mathbf{y}_B$ that are closest to the readout $\mathbf{y}_R$ and computing the system responses for a certain number of new points sampled in the crack parameter space using PDFs determined in some way by $\mathbf{y}_A$ and $\mathbf{y}_B$. These new points are generated from the independent uniform distributions, where point $\mathbf{x}_A$ (related to the closest point $\mathbf{y}_A$) is the mean vector and the absolute differences between the respective parameter values of $\mathbf{x}_A$ and $\mathbf{x}_B$ are the standard deviations of the assumed distributions.

5. COMPUTATIONAL EXAMPLES

The accuracy of the NP identification method was tested on the two-bearing rotor-shaft system presented in Fig. 1. For the considered shaft there are nine diagnostic parameters described in Sec. 3. These are: 2×2 lateral responses, 2 torsional and 3 axial responses. The total number of sample points used in this study was equal to 500000. The points were generated using the PDFs introduced in Sec. 4. Preparation of such a database takes approximately 70 hours on a machine with 8 dual-core AMD Opteron processors.

The problem that should always be considered when generating the database, which is subsequently used for identification of crack parameters, is the choice of the number of numerical experiments that guarantees an acceptable identification accuracy. In order to assess the necessary database size, six subsets of the entire 500000 points database, consisting of 100, 1000, 10000, 50000, 100000, and 200000 points, were randomly selected. For each sub-database, the following test was performed 1000 times. First, the crack parameters were generated in a random way using the uniform distributions. Next, for such crack parameters the computed dynamic responses of interest were introduced to the identification algorithm as the readout. The identified parameters were then compared to the actual ones yielding the identification error. The graphs shown in Fig. 6 present the average identification error of the crack location x_c and depth a/D, and the 2-standard deviations confidence corridors as functions of the database size. By examining the figure, it seems that in the case of the considered two-bearing rotor-shaft system the database of 100 000 points provides the identification accuracy that seems to be sufficient in an engineering practice.

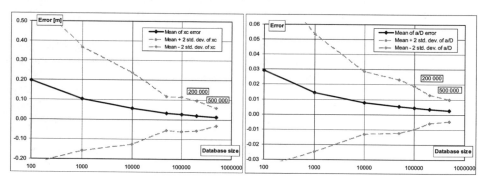

Figure 6: Identification error of x_c and a/D as functions of the database size for the two-bearing system.

The numerical efficiency of identification accuracy improvement methods presented in Sec. 4 was compared using the hybrid mechanical model of the steam turbogenerator rotor-shaft system shown in Fig. 2. In this case the response space is defined by twenty four diagnostic parameters described in Sec. 3. These are: 7×2 lateral responses, 7 torsional and 3 axial responses. The identification error was computed in the same way as in the case of the two-bearing rotor-shaft system.

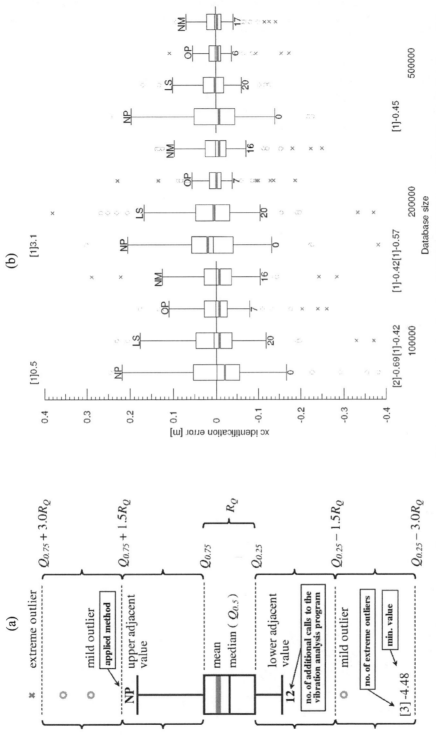

Figure 7: Box plot description (a) and xc identification errors obtained using 100000, 200000 and 500000 point databases (b).

Selected statistics for all the methods obtained for sub-databases of 100000, 200000 and 500000 points, respectively, as well as the average computational effort have been collected in one diagram, the so-called box plot, presented in Fig. 7.

The box plot consists of the following statistics, as shown in Fig. 7: the lower adjacent value (non-outlier simulation), the lower quartile $Q_{0.25}$, the median, the mean value, the upper quartile $Q_{0.75}$ and the upper adjacent value (non-outlier simulation). Moreover, the box plot depicts simulations considered to be unusual, called outliers, which are smaller than $Q_{0.25} - 1.5R_Q$ or greater than $Q_{0.75} + 1.5\ R_Q$, where $R_Q = Q_{0.75} - Q_{0.25}$ is the interquartile range. The outliers are divided into two groups: *mild* outliers (marked by the green circles) – simulations not smaller than $Q_{0.25} - 3.0R_Q$ or not greater than $Q_{0.75} + 3.0\ R_Q$, and *extreme* outliers (marked by the red crosses) – which are farther than the mild ones. For the sake of presentation quality, not always all the extreme outliers are shown. When this is the case, the number of hidden extreme outliers (in the brackets) and the maximal/minimal values are specified. Each of the box-plots presented below displays x_c identification statistics for 100 tests.

It is interesting to observe that for the three investigated databases, the scatter produced by the NP method is approximately the same. There is only one major identification error, registered in a test performed with 200000 points database. The dependence of the identification accuracy on the database size becomes more evident in the case of the improvement techniques OP, NM and LS. Unquestionably, the best results for all the examined methods are obtained by using the largest database; however, also with databases of 100000 and 200000 points one may obtain reliable and accurate results, especially when it is possible to use one of the improving algorithms. The fastest and the most efficient one seems to be the OP method, that needs only about 7 additional simulations.

6. SUMMARY

In the presented paper, non-linear and parametric components of coupled bending-torsional-axial vibrations have been used as a transverse crack occurrence symptom in rotor-shafts of the rotor machines. For this purpose, by means of the structural hybrid model of the real object, several databases containing results of dynamic responses corresponding to various possible crack depths and locations on the shaft were generated using the Monte-Carlo simulation. Special probability density functions of crack parameters were proposed in order to ensure good identification quality for most probable crack parameters. First, the nearest point (NP) method was tested on a simple two-bearing rotor-shaft system to check the algorithm performance. Then, more precise investigation has been carried out for the NP method and some identification improvement strategies, namely, the orthogonal projection method (OP), the Nadler-Mead method (NM) and the local sampling method (LS), to detect and localize the crack in the rotor-shaft system of large steam turbogenerator. Each of the methods was tested for their computational efficiency and the identification accuracy. From the numerical tests performed it follows that the random sampling approach for damage identification in vibrating rotor-shaft systems together with the proposed accuracy improvement methods seem to be an appropriate solution for industrial applications. The crucial advantage of the proposed procedure is the immediate identification process enabling to perform monitoring of rotor-shaft systems. The NP method is based on the previously prepared database of hundreds of thousands numerical experiments, that takes significant but still acceptable computational effort proportional to the demanded accuracy of the results. Additionally, the same database can be used to obtain in a relatively short time (5 to 20 minutes on a present-day PC) more precise estimation of the crack parameters. Based on the test results, the identification improvement methods which seem to be most efficient and accurate, are the OP and NM methods.

ACKNOWLEDGMENTS

The partial support from the Polish Committee for Scientific Research grants PBZ-KBN-105/T10/2003, No. N519 010 31/1601 and No. 3T11F00930.

REFERENCES

1. M. Zhao, Z. H. Luo, A convenient method for diagnosis of shafting crack, in: Proc. of the 12th Biennial ASME Conference on Mechanical Vibration and Noise. ASME, Rotating Machinery Dynamics, DE-Vol. 18-1, Montreal, Canada, 1989, pp. 29-34.
2. Y. Ishida, T. Yamamoto, K. Hirokawa, Vibrations of a rotating shaft containing a transverse crack (major critical speed of a horizontal shaft), in: Proc. of the 4th Int. IFToMM Conf. on Rotor Dynamics. Chicago, Sept. 1994, Vibration Institute Eds., Willowbrook, Ill., U.S.A., 1994, pp. 47–52.
3. P. Pennacchi, N. Bachschmid, A model-based identification method of transverse cracks in rotating shafts suitable for industrial machines, Mechanical Systems and Signal Processing, 20 (2006) 2112-2147.
4. C. A. Papadopoulos, A. D. Dimarogonas, Coupling of bending and torsional vibration of a cracked Timoshenko shaft, Ingenieur-Archiv 57 (1987) 257-266.
5. W. Ostachowicz, M. Krawczuk, Coupled torsional and bending vibrations of a rotor with an open crack, Archive of Applied Mechanics 62 (1992) 191-201.
6. N. Bachschmid, M. Mazza, E. Tanzi, Dynamic bending and torsion behaviour of a cracked pump shaft, in: Proc. of the 7th International Conference on Rotor Dynamics, organized by the IFToMM in Vienna, Sept. 2006, Austria, Paper ID-166, 2006.
7. A. Darpe, K. Gupta, A. Chawla, Coupled bending, longitudinal and torsional vibrations of a cracked rotor, Journal of Sound and Vibration 269 (2004) 33-60.
8. J. Kiciński, Rotor Dynamics, Eds. of the Institute of Fluid-Flow Machinery of the Polish Academy of Sciences, Gdańsk, 2006.
9. T. Szolc, T. Bednarek, I. Marczewska, A. Marczewski, W. Sosnowski, Fatigue analysis of the cracked rotor by means of the one- and three-dimensional dynamical model, in: Proc. of the 7th International Conference on Rotor Dynamics, org. by the IFToMM in Vienna, Sept. 2006, Austria, Paper ID-241, 2006.

10. T. Szolc, On the discrete-continuous modeling of rotor systems for the analysis of coupled lateral-torsional vibrations, Int. Journal of Rotating Machinery **6** (2) (2000) 135-149.

11. L. Meirovitch, Dynamics and Control of Structures, John Wiley & Sons, New York, 1990.

12. J. Nedler, R. Mead, Simplex method for function minimization, Computer Journal 7 (1965) 308-313.

A shaft crack identification technique based on vibration measurements

M Karthikeyan[+], R Tiwari[+], S Talukdar[†]

[+] Department of Mechanical Engineering,
Indian Institute of Technology Guwahati, India
[†] Department of Civil Engineering,
Indian Institute of Technology Guwahati, India

ABSTRACT

A technique has been experimentally investigated for the detection, localization, and sizing of a flaw (crack) in a shaft based on forced response measurements. The test rig consists of a circular shaft supported by rolling bearings at the ends. An artificial open flaw (transverse slit) was made on the shaft specimen. The specimen was excited by a sweep sine excitation through an exciter and the excitation force was measured by a force transducer. The corresponding transverse responses were measured by proximity sensors at the predetermined locations over the shaft span. The stored data were sampled and processed in frequency domain as required by the proposed flaw identification algorithm. The finite element system model was updated by taking the support flexibility of the shaft into consideration to overcome the discrepancies with intact shaft model. The estimated parameters (i.e., the flaw location, flaw flexibility coefficients, and flaw size) have been found in good agreement with the measured data.

Keywords: Experimental investigation; Shaft with a flaw; Forced responses; Model updating; Identification.

1. INTRODUCTION

The presence of various flaws (such as cracks, notches, slits, etc.) in any structures and machineries may lead to catastrophic failures. Therefore, it demands the detection and the diagnostic (i.e., its localization and sizing) of such flaws so that corrective action can be taken well before it grows critical. Such detection and diagnostic techniques should be practicable in terms of taking experimental measurements.

The importance of the topic on crack identification could be realized by looking into several review papers published on this topic. Wauer [1] presented a review of literatures in the field of dynamics of cracked rotors, including the modeling of the cracked part of structures; and determination of different detection procedures to diagnose fracture damages. Gasch [2] provided a survey of the stability behaviour of a rotating shaft with a crack, and on forced vibrations due to imbalances. Dimarogonas [3] reviewed the analytical, numerical and experimental investigations on the detection of structural flaws based on changes in dynamic characteristics. Salawu [4] reviewed the use of natural frequencies as a diagnostic parameter in the structural assessment procedures using the vibration monitoring. Various methods proposed for detecting damages by using natural frequencies were reviewed. Doebling *et al.* [5] provided an overview of methods to detect, locate, and characterize damages in the structural and mechanical systems by examining changes in measured vibration responses. The scope of this paper was limited to methods that use changes in modal properties. Recently, Sabnavis *et al.* [6] presented a review of literatures published since 1990 and some classical papers on the crack detection and severity estimation in shafts. The different type of crack model has been developed by various researchers. Dimarogonas and Paipetis [7] showed a beam with a transverse crack, in general, can be modeled in the vicinity of the crack by way of a local flexibility (compliance) matrix, connecting the longitudinal force, bending moment, and shear force; and corresponding displacements.

Most of the model based crack detection and diagnostics are based on the procedure that the experimental measurements from prototype structures are compared with predicted measurements from the corresponding finite element model. This is generally achieved by implementing the condensation scheme. Dharmaraju *et al.* [8] and Tiwari and Dharmaraju [9] developed algorithms for identifying the crack flexibility coefficients, and subsequently estimation of the equivalent crack depth based on the forced response information. They outlined the condensation scheme for eliminating the rotational degree of freedoms at crack element nodes based on the physical consideration of the problem, which was otherwise difficult to eliminate. However, the main practical limitation of the algorithm was that the location of the crack must be known a priori. Also the algorithm used the Euler-Bernoulli beam theory in the shaft model and without considering the damping in the system. Recently, Karthikeyan *et al.* [10] developed an algorithm for the crack localization and the sizing in a cracked shaft based on the forced response measurements by applying condensation schemes; however, it was limited to numerical illustrations.

In the present work, the identification method developed in [10] has been experimentally validated in order to detect, localize, and quantify a flaw in a shaft based on forced vibration measurements from a experimental test rig. An artificial open flaw (transverse slit) was made at arbitrary location on a shaft supported by two ball bearings. A harmonic force of known amplitude with sine-sweep frequency was used to dynamically excite the non-rotating shaft, over few (usually up to second) flexible modes, which was provided with the help of a modal exciter. Transverse linear responses were measured using proximity (i.e., eddy current based) sensors at various axial locations of the shaft and in both transverse planes. The FE model is updated by tuning the first resonance frequency obtained from the

620

experimental forced response of the shaft without flaw case. The developed identification algorithm uses the amplitude and phase information of the response, and it gives the location of the flaw, flaw flexibility coefficients, and subsequently the equivalent crack depth. The obtained flaw parameters are in good agreement with actual counterpart, and the estimated parameters have fairly good response reproduction capabilities.

2. SYSTEM MODELING

The Timoshenko beam theory is used for transverse vibrations; which incorporates the bending, rotary and shear deformations. The effect of proportional damping [10] is included in the formulation of system equations of motion. The FEM is used to develop the shaft model. Only single open crack has been considered in the shaft with its existence, location and size as unknown; for a shaft with flexible supports at ends. A harmonic force of known amplitude with a sine-sweep frequency is assumed to develop the identification algorithm in the frequency domain. Figure 1 shows a cracked shaft element subjected to a general loading. Let P_1 is the axial force, P_2 and P_3 are shearing forces, P_4 and P_5 are bending moments, and P_6 is the torque.

For transverse loadings of shafts, neglecting the axial and torsional loadings, the flexibility matrix of the crack can be expressed as [7]

$$[C_c]^{(e)} = \begin{bmatrix} C_{22} & 0 & 0 & 0 \\ 0 & C_{33} & 0 & 0 \\ 0 & 0 & C_{44} & C_{45} \\ 0 & 0 & C_{54} & C_{55} \end{bmatrix} \tag{1}$$

where C_{ij} is the crack flexibility coefficient.

2.1 System Equations of Motion and Forced Responses
The system equations of motion is given as [11]

$$[M]\{\ddot{q}(t)\} + [D]\{\dot{q}(t)\} + [K]\{q(t)\} = \{F(t)\} \tag{2}$$

$$\text{with } [D] = a_0[M] + a_1[K] \tag{3}$$

where $[M]$ is the assembled mass matrix, $[D]$ is the assembled damping matrix, $[K]$ is the assembled stiffness matrix, $\{F(t)\}$ is the assembled external force vector, and $\{q(t)\}$ is the assembled vector of nodal DOFs. Equation (2) has been obtained by assembling the equation of motion for shaft elements with and without flaws. For forced vibrations, by assuming the standard harmonic excitation, the equations of motion in frequency domain take the form

$$\left(-\omega^2[M] + j\omega[D] + [K]\right)\{Q\} = \{F\} \tag{4}$$

where ω is the forcing frequency, $\{F\}$ is the complex force vector, and $\{Q\}$ is the complex response vector (elements of which are complex quantities and contain both the amplitude and phase information). For given system properties (i.e.,$[M]$, $[D]$ and $[K]$, which includes the intact shaft and shaft with flaw element models, crack location, and size) the response $\{Q\}$ in frequency domain can be simulated from equation (4) corresponding to a given force $\{F\}$. The simulated response $\{Q\}$ corresponds to DOFs of all elements. Since only (few) transverse DOFs can be measured in practice, and it is difficult to measure transverse rotational DOFs accurately. Hence, all transverse rotational DOFs (and some of linear DOFs) are needed to be condensed from the response $\{Q\}$, which is called slave DOFs.

3. FLAW IDENTIFICATION WITH CONDENSATION SCHEMES

In the present section, the condensation has been performed in two stages[10] (i) the standard dynamic condensation for most of the linear DOFs and all angular DOFs, except at nodes of the shaft element with flaw; and (ii) the hybrid condensation for rotational DOFs at nodes of the shaft element with flaw. Estimation algorithm for flaw flexibility coefficients can be obtained from the condensed form of equation (4) by noting equation (1). The standard regression equation is given as [10]

$$[S^h]\{C\} = \{B_2^h\} \tag{5}$$

$$\text{with } \{C\} = \{C_{22} \quad C_{33} \quad C_{44} \quad C_{45} \quad C_{55}\}^T \tag{6}$$

In equation (5), the vector $\{C\}$ contains all unknown flaw flexibility coefficients as given by equation (6). The matrix $[S^h]$ and the vector $\{B_2^h\}$ contain all the known information, i.e., the intact shaft model, the flaw location, the force and corresponding response.

Theoretically, from the fracture mechanics approach, the flexibility coefficients of the compliance matrix are expressed as a function of the flaw depth ratio $\bar{a} = a/R$ [7]. The error function between the identified (superscript: id) and theoretical (superscript: th) flexibility coefficients can be defined as

$$\pi_{error} = \sum_{i=2}^{5} (C_{ii}^{th} - C_{ii}^{id})^2 + (C_{45}^{th} - C_{45}^{id})^2 \tag{7}$$

where C_{ij} are dimensionless flaw flexibility coefficients. Minimizing the error function with respect to the flaw depth ratio $\bar{a}$ in conjunction with the bi-section and Newton-Raphson methods, the equivalent crack depth ratio $(\bar{a}_e)$ could be obtained. In order to improve the conditioning of the matrix $[S^h]$, the Tikhonov regularization method [10] is incorporated in the estimation procedure.

4. THE FLAW DETECTION, LOCALIZATION, AND SIZING ALGORITHMS

The flaw location is important to be found out, as it required by the identification algorithm, which is outlined in Section 3. A crack localization procedure has been briefly outlined in this section based on the forced vibration data. Hence, it will be of more practical importance, if the position of the flaw can also be identified along with the flaw size. Experimentally measured transverse resonance frequencies, and responses (i.e., linear DOFs) will be used, iteratively, to obtain the flaw location and the size [10]. Various steps involved in proposed algorithm are outlined in Figure 2 as flow chart. It should be noted that as a by-product the present identification algorithm gives flexibility coefficients of the flaw and the damping matrix of the element with flaw. Extensive numerical simulations of the flaw identification algorithm have been carried out in [10].

5. EXPERIMENTAL SETUP AND MEASUREMENT PROCEDURES

The experimental setup consists of a shaft (with a flaw) test rig, and instrumentations for the signal and excitation generation, measurements, conditioning, capturing, and storing. The shaft test rig with flaw is shown in Figure 3 along with the measuring and analysis instrumentations. The physical and modeling parameters of the mild steel shaft of the circular cross section are given in Table 1. The shaft was supported by ball bearings at both ends, and with very high transverse stiffness. An artificial open flaw (transverse slit) was made by a saw on the shaft. Proximity sensors (eddy current based; Bentley Nevada) were used to sense vibrations (displacements) of the excited (non-rotating) shaft. The signal capturing unit, which consisted of the data acquisition system and the time capture module (Pulse System, B&K).

A slow-sine sweep excitation testing was adopted. Table 1 gives required information about shaft parameters, number of shaft elements, and flaw parameters (i.e., the location and the size). Proximity sensors observing the shaft displacement were adjusted to have 30 mil (0.000762 m) gap using a DC voltmeter. The sweep sinusoidal signal with the range of frequency as 5-200 Hz were generated, amplified, and sent to the modal exciter (50N, B&K). The excitation in time domain for the whole sine sweep period was measured, conditioned, and transferred to the DAQ (16-channel, B&K) for the digitization. Transverse plane linear displacements (four) were sensed by proximity sensors, amplified, and transferred to the DAQ for digitization. The digitized measurement data was recorded for the whole frequency sweep (with the sweeping rate of 1 Hz per second in the range 5-200 Hz). Similar measurements were taken by shifting the sensor stand at various axial locations of the shaft. The stored force and displacement data were resampled for a small time duration (i.e., equivalent to 1 Hz of frequency band) and the amplitude and phase information were extracted (i.e., the frequency response function). The procedure was repeated for other shafts with flaw, as well as for the intact shaft. The measured and processed data was used for the estimation of damping coefficients, updating of the FE model, and the identification of flaw parameters.

6. RESULTS AND DISCUSSIONS

The FE model of the test rig was updated by matching natural frequencies (up to second modes) from the theoretical model to that with experimental frequencies. The stiffness of rolling bearings was modeled by two transverse direction direct stiffness coefficients only, and obtained by the method given in [11] initially. For simplicity, both rolling bearings were assumed to be identical for a particular test case. For six shafts (e.g., Shafts 1, 2, ..., 6 given in Table 1), which were considered for the present test case, the updated direct stiffness coefficients (k_{xx} and k_{yy}) are given in Table 2. First, the intact shaft was mounted and measurements were taken, and then without dismounting the bearings a flaw was created in a form of a slit by a saw on the shaft to ensure almost same boundary conditions. Table 2 compares the first resonance frequency obtained experimentally with the finite element analysis by considering simply supported boundary conditions, and the updated finite element model by considering the flexibility of supports. Figure 4 shows the comparison of FRF (i.e., the receptance FRF [12]) obtained from experiments, FEM, and updated FE model for shaft 5. The updated FE model was used for the flaw detection, localization and sizing algorithm.

6.1 Flaw Detection, Localization, and Sizing

The proposed algorithm for the flaw detection, localization, and sizing was tested using the experimentally obtained responses of a shaft with flaw. Six test shafts were considered, which had a flaw at different locations and of dissimilar sizes, and their properties are given in Table 1. For shafts 1, 3 and 5 the flaw was made at the shaft mid span to study the effects of the flaw in the first mode of vibrations. This is due to the fact that effects is expected to be more in the first mode, and less (due to presence of a node) in the second mode of vibrations. Moreover, in order to study effects of the flaw in the second mode of vibrations, the flaw was made at around one-third of the span for shafts 2, 4, and 6.

6.1.1 Shaft 5 with flaw

For shaft 5, the first resonance frequency (experimental) of the intact shaft was found to be 35.50 Hz. For the present case the shaft had a flaw with the flaw depth ratio $\bar{a}$ = 0.75 and the flaw location ratio $\bar{x}$ = 0.5 (i.e., 14th element), and the first resonance frequency was found to reduce to 34.50 Hz. The change in resonance frequencies confirmed the presence of flaw, and was used in the flaw detection (see Table 2). Figure 5 shows the comparison of the measured and estimated responses, from the updated FE model (with the help of estimated flaw parameters) for shaft 5 with flaw, in the vertical plane. The comparison of the first resonance peaks is in very good agreement. The phase variation shows good comparisons of phase change of around 180^0 at resonances. However, it also shows the phase changes at the anti-resonance in the theoretical response. The experimental fundamental resonance frequency, and forced responses were fed to the identification algorithm in order to identify flaw parameters as discussed in Section 4.

6.1.2 Identification of flaw parameters

Figure 2 shows the procedure of finding the flaw location. For the flaw localization, the flaw depth ratio is assumed initially between 0.9 and 1.0 (based on the numerical simulation exercise [10]). For the initially assumed flaw depth ratio of $\bar{a}_{ig} = 0.9$, when the natural frequency is available experimentally, it gives the initial flaw location ratio to be 0.32 (i.e., 9th element). With this flaw location, flaw location flexibility coefficients are obtained from equation (5), and subsequently it gives equivalent flaw depth ratio from equation (7). The equivalent flaw depth ratio is used as a new flaw depth, and variations of resonance frequencies are obtained from equation (4) to get a new possible flaw location. The converged flaw location and size are 0.5 (i.e. 14th element) and 0.78 respectively. The convergence criteria adopted here is that the difference between flaw depth ratios of the two consecutive iterations should be less than 0.01.

It was found that two possible flaw locations it gives as subsequent guess value, however, finally all converges to a single value of the flaw location. Table 3 summarized the flaw parameter estimated for the above case in the fifth row. Apart from the flaw location and its size, it also shows the comparison of flaw flexibility coefficients. It can be seen that the prediction for the flaw location is good. The flexibility coefficients are compared with the theoretical ones corresponding to the estimated flaw size, and found to be very good in agreement. Table 3 also shows several case studies by using experimental measurements corresponding to different flaw locations and sizes of five more shafts with flaw. Since in actual case it is very difficult to make a thin flaw in slender beams (which has been considered in this work) the over estimation of the equivalent crack is always possible. It is observed that there is no appreciable variation in the estimation of flaw locations, flaw flexibility coefficients, and flaw sizes. This shows the consistency of the convergence of the present identification algorithm. Overall the number of iterations for the final convergence is very less (maximum is less than 10, however, 3 to 4 is very common).

7. CONCLUSIONS

The developed identification algorithm for the detection, localization and sizing of a flaw in shafts based forced vibration, was verified based on experimental forced vibration measurements. The details of experimental setup and procedure have been presented. Experimental measurements were correlated with numerically simulated measurements, and it was used to update the theoretical model accordingly. The comparison of the theoretical with identified flaw flexibility coefficients, and the comparison of the actual with identified flaw locations and sizes were provided. The identification for the flaw detection, localization and sizing were in good agreement with their actual counterpart. It would be interesting to explore the possibility of identifying multiple flaws.

REFERENCES

[1] J. Wauer, On the dynamics of cracked rotors: A literature survey, *Applied Mechanics Review* 1990; 43(1), 13–6.

[2] R. Gasch, A survey of the dynamic behaviour of a simple rotating shaft with a transverse crack, *Journal of Sound and Vibration* 1993; 160(2), 313-332.

[3] A. D. Dimarogonas, Vibration of cracked structures: A state of the art review, *Engineering Fracture Mechanics* 1996; 55(5), 831-857.

[4] O. S. Salawu, Detection of structural damage through changes infrequencies: A review, *Engineering Structures* 1997; 19(9), 718-723.

[5] S. W. Doebling, C. R. Farrar and M. B. Prime, A summary review of vibration based damage identification methods, *Shock and Vibration Digest*, 1998; 30(2), 91-105.

[6] G. Sabnavis, R. G. Kirk, M. Kasarda, and D. Quinn, Cracked shaft detection and diagnostics: A literature review, *Shock and Vibration Digest* 2004, 36(4), 287-296.

[7] A. D. Dimarogonas and S. A. Paipetis, *Analytical Methods in Rotor Dynamics*, London: Elsevier Applied science 1983.

[8] N. Dharmaraju, R. Tiwari and S. Talukdar, Development of a novel hybrid reduction scheme for identification of an open crack model in a beam, *Mechanical Systems and Signal Processing* 2005; 19, 633-657.

[9] R. Tiwari and N. Dharmaraju, Development of a condensation scheme for transverse rotational degrees of freedom elimination in identification of beam crack parameters, *Mechanical Systems and Signal Processing* 2006; 20(8), 2148-2170.

[10] M. Karthikeyan, R. Tiwari, and S. Talukdar, Development of a novel algorithm for a crack detection, localization and sizing in a beam based on forced response measurements, *Transactions of ASME, Journal of Vibration and Acoustics* 2008; 130(2), 021002.

[11] M. J. Goodwin, *Dynamics of rotor-bearing systems*, Unwin Hyman Publishers, London 1989.

[12] D. J. Ewins, *Modal Testing: Theory, Practice and Application*, 2nd edition, Research Studies Press Ltd., 2000.

[13] I. Green, and C. Casey, Crack detection in a rotor dynamic system by vibration monitoring - Part I: Analysis, *ASME Trans., Journal of Gas Turbines and Power* April 2005, 127(2), 425-436.

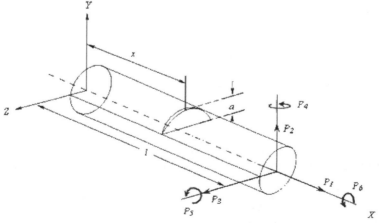

Figure 1 A cracked shaft in a general loading.

626

Start

Experimentation:

The fundamental natural frequency (ω_{nf}) and forced responses are measured from the experimental rig of the shaft with flaw.

Crack localization algorithm:

The fundamental natural frequency (ω_{nf}) is calculated by the numerical model for all possible flaw location $\overline{x}$ (i.e., for the shaft span) for initial guess of the flaw depth ratio ($\overline{a}_{ig}$). The variation of ω_{nf} is plotted against the assumed flaw locations over the shaft span and intersection with experimental ω_{nf} gives possible crack locations $\overline{x}_f$ which will be used in flaw-sizing algorithm.

Flaw sizing algorithm:

One of the obtained flaw locations $\overline{x}_f$ is selected to find out flaw flexibility coefficients and subsequently the equivalent flaw depth ratio ($\overline{a}_e$) are obtained using equations (5) & (7), respectively.

| Now update the value of $\overline{a}_{ig}$ equals to $\overline{a}_e$. | ← No — | Is $\left(\overline{a}_e - \overline{a}_{ig} \right) < 1 \times 10^{-2}$? |

Yes

Convergence of $\overline{a}_e$ and the corresponding $\overline{x}_f$ are achieved.

Stop

Figure 2 A flow chart for the flaw localization and sizing algorithm

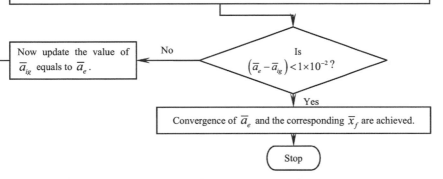

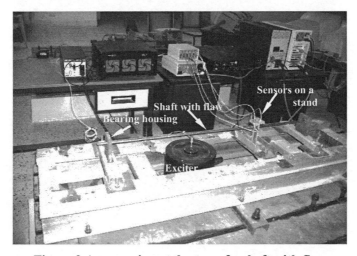

Figure 3 An experimental setup of a shaft with flaw

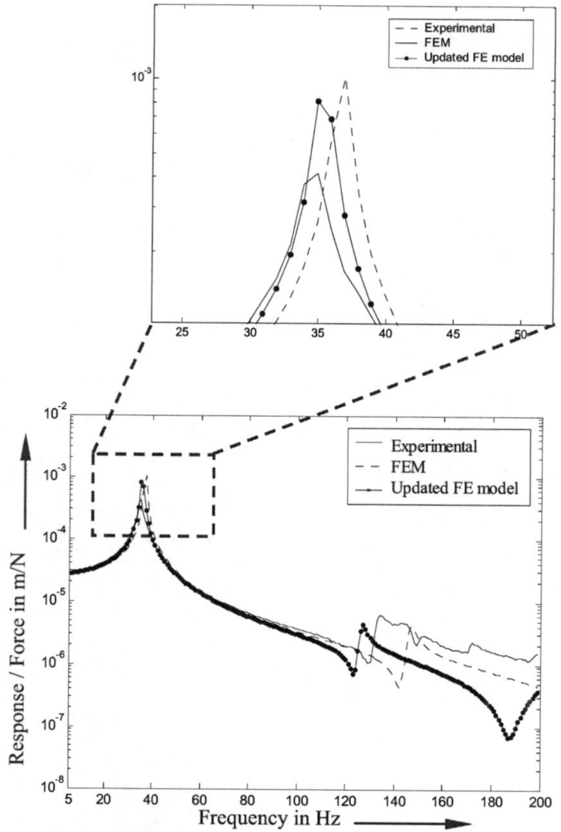

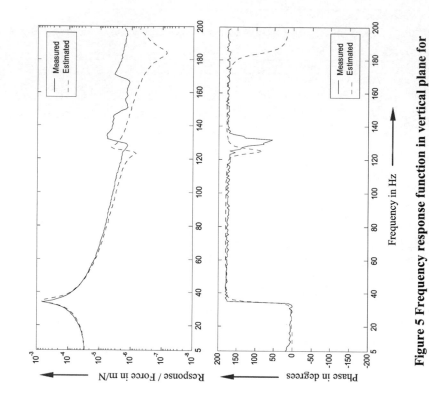

Figure 5 Frequency response function in vertical plane for shaft 5 with flaw

Figure 4 Comparison of FRFs for shaft 5 with flaw

Table 1 Parameters of the shaft for the experimental setup

Shaft no.	Parameters			
	Diameter D in m	Number of shaft elements, N	Shaft element no. with flaw	Location of the element with flaw and size of the flaw (x/L & a/R approximately)
1	0.0097	27	14	0.5 & 0.41
2	0.0097	27	9	0.33 & 0.72
3	0.0161	28	14	0.5 & 0.5
4	0.0161	28	7	0.25 & 0.5
5	0.0161	28	14	0.5 & 0.75
6	0.0161	28	7	0.25 & 0.75

Table 2 Comparison of the first resonance frequency

Shaft no.	Updated direct stiffness coefficients $k_{xx} = k_{yy}$ in N/m	Intact shaft resonance frequency (RF) in Hz			Shaft with flaw experimental RF in Hz (% of reduction in RF with intact shaft)
		Experimental	FEM with simply supports	Updated FE model	
1	75315	21.5	22.1	21.5	21.25 (1.16)
2	447215	22.0	22.1	22.0	21.625 (1.71)
3	584909	35.75	36.68	35.75	35.50 (0.70)
4	1049300	36.125	36.68	36.125	36.0 (0.35)
5	435559	35.375	36.68	35.375	34.5 (2.47)
6	2834016	36.25	36.68	36.25	36.0 (0.70)

Table 3 Actual and estimated flaw parameters for different shafts with flow

Shaft no.	Actual flaw parameters (approximately)		Estimated flaw parameters												
			Flaw flexibility coefficients (C^{id} and C^{th})†												
			C_{22} (mN^{-1})		C_{33} (mN^{-1})		C_{44} (mN)$^{-1}$		C_{45} (mN)$^{-1}$ (=C_{54})		C_{55} (mN)$^{-1}$			
	$\bar{x}$ $(n)^+$	$\bar{a}$	C_{22}^{id}	C_{22}^{th}	C_{33}^{id}	C_{33}^{th}	C_{44}^{id}	C_{44}^{th}	C_{45}^{id}	C_{45}^{th}	C_{55}^{id}	C_{55}^{th}	$\bar{x}$ (n)	$\bar{a}_e$
1	0.50 (14)	0.41	0.665	0.611	0.994	0.914	1.704	1.502	4.333	4.034	8.625	8.943	0.48 (13)	0.71
2	0.33 (9)	0.72	1.038	1.031	1.554	1.544	3.799	3.814	8.415	8.333	15.661	15.684	0.33 (9)	0.87
3	0.50 (14)	0.50	0.112	0.188	0.375	0.280	0.237	0.242	0.947	0.939	2.839	2.843	0.50 (14)	0.44
4	0.25 (7)	0.50	0.191	0.188	0.229	0.280	0.241	0.242	0.937	0.939	2.840	2.843	0.30 (8)	0.44
5	0.50 (14)	0.75	0.750	0.786	1.132	1.177	2.317	2.318	5.667	5.667	11.646	11.646	0.50 (14)	0.78
6	0.25 (7)	0.75	0.638	0.620	0.907	0.927	1.538	1.538	4.109	4.111	9.071	9.075	0.30 (8)	0.71

+ n is element number, and $\bar{a}_e$ is the equivalent flaw depth ratio; † Superscripts '*id*' and '*th*' indicate the estimated and theoretical, respectively (theoretically, flaw flexibility coefficients were obtained from linear fracture mechanics approach (Dimarogonas and Paipetis [7])).

Turbo-generator groups affected by transverse cracks: a sensitivity analysis of vibrations versus crack position and depth

N Bachschmid, E Tanzi, P Pennacchi
Dipartimento di Meccanica, Politecnico di Milano, Italy

ABSTRACT

The dynamic behaviour of heavy, horizontal axis, turbo-generator groups affected by transverse cracks can be analysed in the frequency domain by a quasi linear approach, using a simplified breathing crack model applied to a traditional finite element model of the shaft-line. This allows to perform a series of analyses with affordable efforts. The analysis of the modelling procedure allows to define an approximated approach for simulating the dynamical behaviour, which, combined to modal analysis, allows to predict the severity of the crack excited vibrations, resolving the old-age question on how deep a crack must be to be detected by means of vibration measurements.

The model of a 320 MW turbo-generator group has been used to perform a numerical sensitivity analysis, in which the vibrations of the shaft-line, and more in detail the vibrations of the shafts in correspondence to the bearings, have been calculated for all possible positions of the crack along the shaft-line. The calculated results confirm the predicted behaviour.

1. INTRODUCTION

The accurate modelling of the crack breathing behaviour in heavy horizontal shafts, which is the gradually crack opening and closing mechanism, and of the consequent cracked shaft dynamical behaviour, excited by the periodical, 1xrevolution shaft stiffness change, allows simulating the vibrations of a shaft affected by a transverse crack. More in detail the vibrations can be calculated in correspondence of the bearings of the machine, where they are measured in real machines, and its severity allows to predict the possibility of detecting the presence of the crack. Cracks in power station and industrial plant machinery, steam turbines, generators and pumps have been discovered in many European power plants as well as in the Far East and in the USA [1], [2]. They have generally been discovered, by analysing the monitored vibrations, and the machines have been stopped before the occurrence of a catastrophic failure, but in some cases the symptoms have not been recognized in time and the machines burst. Therefore the interest in determining the severity of the vibrations which could be measured in the bearings of a shaft-line as function of the position in which the crack has developed and of its depth.

2. MODEL OF THE BREATHING CRACK

The breathing mechanism is a result of the stress and strain distribution around the cracked area, which is due to static loads, like the weight and the bearing reaction forces, and to dynamical loads, like the unbalance and the vibration induced inertia force distribution. When the static loads overcome the dynamical ones, as it occurs in heavy horizontal rotating machines, the breathing is governed by the angular position of the shaft with respect to the stationary load direction, and the crack opens and closes again completely once each revolution. The transition from closed crack (full) stiffness to the open crack (weak) stiffness has been generally considered in literature abrupt or represented by a given cosine function, but can be calculated more accurately taking into account all static and dynamic loads. The stiffness in the different crack configurations (open, closed, or partially open) have been generally derived from fracture mechanics approach, which requires some limitations in partially open crack configurations, as shown e.g. in [3].

3D non linear finite element calculations allow the breathing mechanism to be predicted accurately, when the loads are known, but are extremely cumbersome, costly and time consuming (due to the need of a refined mesh in the crack region, and to the non-linear contact conditions) [4] and [5].

A simplified model, which assumes linear stress and strain distributions, for calculating the breathing behaviour, has been developed by the authors and proved to be very accurate [6]. The proposed model is as follows (see Figure 1): the initial position of the main inertia axes of the cracked cross section of the supposedly partially open crack is assumed as well as a linear axial stress distribution due to the bending load to which eventually also thermal stresses are superposed. Then the compressive and tensile stresses are defined: cracked surfaces where tensile stresses should appear are "open" areas, where compressive stresses appear are "closed" areas.

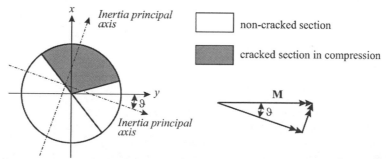

Fig 1 Main axis of inertia and the bending moment M decomposition in a generic position of the cracked section

Indeed open and closed areas define also the main inertia axes position. Therefore the procedure has to be repeated iteratively, until the position of the main axes remains stable.

With this simplified model accurate results have been found, despite the fact that the proposed approach assumes linear stress distribution where in reality the stress distribution is strongly non-linear, as it is well known from Fracture Mechanics and from 3D finite element non-linear analyses.

Once the breathing mechanism and the second moments of area have been defined for the different angular positions of the cracked section, the stiffness matrix of an *equivalent* cracked beam element of suitable length l_c can be calculated, assuming a Timoshenko beam with constant section and second moments of area along l_c, as shown in fig 2, for each different angular position. The neighbour beam elements are again circular section beams. The angular position dependent stiffness matrix $\mathbf{K}_c(\Omega t)$ is one per revolution periodical, has one constant term (the mean stiffness) and several harmonic components of which only the first three components are significant.

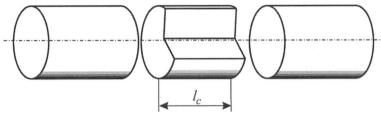

Fig 2 Equivalent cracked beam element: its configuration changes with its angular position

Note that the stiffness coefficients of $\mathbf{K}_c(\Omega t)$ are all proportional to the corresponding second moments J_x, J_y, J_{xy} of area (where x is the vertical and y the horizontal axis), which are functions of the depth and of the angular position Ωt since they depend on the one per revolution periodical breathing of the crack. The length l_c depends on the depth only and has been tuned by 3D finite element calculations, comparing the deflections, according to all considered degrees of freedom, of cracked beams with different depths loaded by unitary loads, and selecting the length l_c which gives the best fitting.

Figure 3 shows the dimensionless length l_c referred to the diameter as function of the dimensionless depth of the crack (also referred to diameter).

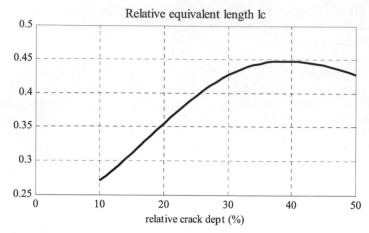

Fig 3 A-dimensional equivalent length of cracked element, referred to the diameter, as function of crack depth, referred to the diameter.

3. DYNAMICAL BEHAVIOUR OF CRACKED SHAFTS

When the equivalent cracked beam having a reduced cross section and a suitable length is inserted in the 1D standard rotor-dynamic FEM of the rotor, the global stiffness matrix $\mathbf{K}(\Omega t)$, which includes shaft, bearing and supporting structure stiffness, and which, due to the presence of $\mathbf{K}_c(\Omega t)$ of the cracked beam element, is one per revolution periodical, can be expanded in a Fourier series truncated at its third harmonic component according to eq.(1):

$$[\mathbf{K}(\Omega T)] = [\mathbf{K}_m] + [\Delta\mathbf{K}_1]e^{i\Omega t} + [\Delta\mathbf{K}_2]e^{i2\Omega t} + [\Delta\mathbf{K}_3]e^{i3\Omega t} \tag{1}$$

Each component depends on the crack depth only. The stiffness variation $\Delta\mathbf{K}_i$ is proportional to the variation of the second moments of area ΔJ_i over the equivalent length l_c. The equation of motion of the rotor can be expressed with eq. (2)

$$[\mathbf{M}]\ddot{\mathbf{x}} + ([\mathbf{R}] + [\mathbf{Ryr}]\Omega)\dot{\mathbf{x}} + \left([\mathbf{K}_m] + \sum_n [\Delta\mathbf{K}_n]e^{in\Omega t}\right)\mathbf{x} = \mathbf{F}e^{i\Omega t} + \mathbf{W} \qquad n = 1,2,3 \tag{2}$$

where $\mathbf{W}$ is the weight force vector and $\mathbf{F}$ the vector that contains the exciting forces acting on the rotor (e.g. unbalances). $\mathbf{M}$ is the mass matrix of the system (including the supporting structure), $\mathbf{R}$ and $\mathbf{Gyr}$ are respectively the damping and gyroscopic matrices. The rotor displacements $\mathbf{x}$ can be split in their static and dynamic components $\mathbf{x}_s$ and $\mathbf{x}_d$:

$$\mathbf{x} = \mathbf{x}_s + \mathbf{x}_d \tag{3}$$

where:

$$\mathbf{x}_s = [\mathbf{K}_m]^{-1}(\mathbf{F}_0 + \mathbf{W}) \tag{4}$$

$\mathbf{F}_0$ is a static force vector due to the presence of the crack. The values in $\mathbf{F}_0$ are extremely small with respect to the values in the weight force vector $\mathbf{W}$, and are neglected in the following.

Then, eq.(2) can be rewritten in the following form:

$$[\mathbf{M}]\ddot{\mathbf{x}}_d + \left([\mathbf{R}]+[\mathbf{Ryr}]\Omega\right)\dot{\mathbf{x}}_d + [\mathbf{K}_m]\mathbf{x}_d = \mathbf{F}e^{i\Omega t} - \sum_n [\Delta\mathbf{K}_n]e^{in\Omega t}(\mathbf{x}_s + \mathbf{x}_d) \qquad n = 1,2,3 \quad (5)$$

The last term of equation (5) represents the so-called *equivalent* crack forces which excite the vibrations of the shaft-line with stiffness $\mathbf{K}_m$, which is the mean stiffness of the system. The mean stiffness $\mathbf{K}_m$ (which accounts for the mean reduction due to the breathing crack) is only slightly different from the stiffness of the un-cracked shaft (the reduction of the mean stiffness is typically only few percent of the un-cracked stiffness even in case of deep crack, as can be derived from natural frequency measurements), because the crack affects the stiffness of only one element. Therefore one could say that the effect of the crack is given by the equivalent crack forces which excite the vibrations of the shaft-line as an external force system, applied to the nodes of the element in which the crack has developed, and which has a reduced stiffness.

The vibrations $\mathbf{x}_d$ in steady state conditions are periodical and can be expanded in a Fourier series:

$$\mathbf{x}_d = \mathbf{x}_1 e^{i\Omega t} + \mathbf{x}_2 e^{i2\Omega t} + \mathbf{x}_3 e^{i3\Omega t} + \ldots \qquad (6)$$

where $\mathbf{x}_1$, $\mathbf{x}_2$ and $\mathbf{x}_3$ are 1xrev, 2xrev and 3xrev harmonic vibration components.

The equivalent crack forces are calculated by substituting in eqs. (1) and (6) the forward and backward rotating components (instead of the only forward rotating components) and by introducing these expressions in eq. (5). Following expressions are found:

$$\mathbf{F}_1 e^{i\Omega t} = \left(\Delta\mathbf{K}_1\mathbf{x}_s + \tfrac{1}{2}\Delta\mathbf{K}_2\mathbf{x}_1^* + \tfrac{1}{2}\Delta\mathbf{K}_3\mathbf{x}_2^* + \tfrac{1}{2}\Delta\mathbf{K}_1^*\mathbf{x}_2 + \tfrac{1}{2}\Delta\mathbf{K}_2^*\mathbf{x}_3\right)e^{i\Omega t} \qquad (7)$$

$$\mathbf{F}_2 e^{i2\Omega t} = \left(\Delta\mathbf{K}_2\mathbf{x}_s + \tfrac{1}{2}\Delta\mathbf{K}_3\mathbf{x}_1^* + \tfrac{1}{2}\Delta\mathbf{K}_1\mathbf{x}_1 + \tfrac{1}{2}\Delta\mathbf{K}_1^*\mathbf{x}_2\right)e^{i2\Omega t} \qquad (8)$$

$$\mathbf{F}_3 e^{i3\Omega t} = \left(\Delta\mathbf{K}_3\mathbf{x}_s + \tfrac{1}{2}\Delta\mathbf{K}_1\mathbf{x}_2 + \tfrac{1}{2}\Delta\mathbf{K}_2\mathbf{x}_1\right)e^{i3\Omega t} \qquad (9)$$

where $\mathbf{x}_n{}^*$ and $\Delta\mathbf{K}_n{}^*$ are conjugate complex quantities of $\mathbf{x}_n$ and $\Delta\mathbf{K}_n$.

The static and dynamic response of the cracked rotor can be evaluated using eq. (5) that is solved with an iterative procedure for the three harmonic components, by considering the force vectors obtained with expressions (7-9).

Looking closer to expressions (7-9), and bearing in mind that a transverse crack affects mainly the shaft bending stiffness, and much less the shear stiffness (as shown in [7]) in correspondence of the crack, it results that the *equivalent* force vectors defined by above expressions are composed mainly by *equivalent* bending moments (given by bending stiffness $\Delta\mathbf{K}_n$ multiplied by angular deflections $\mathbf{x}_n$)

635

rather than by equivalent shear forces. The *equivalent* crack forces are represented by these rotating 1xrev., 2xrev. and 3xrev. bending moments, which amplitudes are all proportional to the static angular deflections in $\mathbf{x}_s$ of the nodes of the element in which the crack has developed (because all vibration components due to the crack are proportional to the static angular deflections, as can be seen developing the iterative procedure). These static angular deflections in turn are proportional to the value of the static bending moment M_b.

We could say that the *equivalent* crack forces (7-9) impose additional deflections to the shaft which in reality are generated by the periodical additional bending flexibility introduced by the breathing crack in the cracked element only. Therefore the *equivalent* crack force vector is composed by couples of bending moments equal and opposite in sign, applied to the end nodes of the cracked beam elements, which generate the additional bending deflection of the cracked element, without affecting all the other elements of the shaft.

The exciting crack forces are directly proportional to the static bending moment M_b. When the crack is located in regions along the shaft where the static bending moment is small (close to bearings) the excitation of vibrations is weak. When instead it is located where the static bending moment is large (at midspan between bearings) the excitation is much stronger. The complete modelling procedure is described in detail in [4]. Eq. (5) can be integrated in the frequency domain by means of an iterative procedure: this allows to calculate accurately the vibrations of a rotating shaft affected by a transverse crack. The 3 harmonic vibration components of a simple symmetrical shaft with a crack at mid-span, supported on two bearings, rotating at very low speed, have been evaluated in correspondence of the crack and referred to the mean static deflection in the same position, as a function of the crack depth. Figure 4 shows these ratios, which are independent of the used shaft model and which give an idea on how the vibration components increase as the crack propagates deeper (assuming a rectilinear tip).

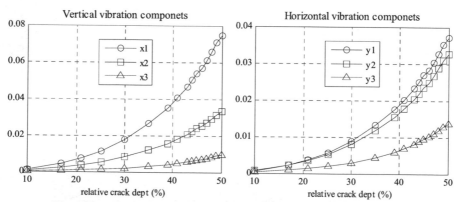

Fig 4 Crack induced dimensionless vibration components $\mathbf{x}_n/\mathbf{x}_s$ versus crack depth.

1x excitation is higher in vertical direction than in horizontal direction, 2xrev is almost equal in both directions, for any depth. The real value of the different vibration components (at very low speed when dynamic effects can be disregarded) can be obtained by the values given in Fig. 4 multiplied by the static deflection. In this simple shaft the static deflection is maximum at mid-span, where also the relative angular deflection of the cracked element is maximum because the static bending moment is maximum at mid-span. We have seen that in reality the vibration excitation is proportional to the angular deflection of the cracked element, and therefore to the static bending moment. So we can forecast the severity of vibration excitation from crack depth (Fig. 4) and from value of the static bending moment.

Let us now examine the static bending moment distribution along the shaft-line with reference to a typical turbo-group (which is described later). This analysis allows to understand where and when the excitation deriving from a crack can be either severe or weak.

4. STATIC BENDING MOMENTS EVALUATION

The actual static bending moment which is responsible for generating the crack forces and the related vibrations, depend on the distribution of masses along the shaft-line and on the bearing alignment conditions. Generally the first alignment in cold conditions is made by imposing null forces and moments in correspondence of the flanges of the different shafts. With respect to this situation some change may be introduced for increasing loads on bearings which, being lightly loaded, could operate close to its instability threshold. This occurs e.g. in the 320 MW turbo-group of Fig. 8 for the last bearing of the generator. At the operating speed, this alignment condition is slightly modified by the oil film thickness that builds up in the different bearings. In some machines also the thermal expansion of the supporting structure may modify the initial alignment conditions. In order to simulate accurately the effect of a crack the actual static bending moment distribution in operating conditions should be considered. Fig 5 shows a comparison of the distributions of the static bending moments in the vertical and in the horizontal plane at 0 speed and at the normal operating speed of 3000 rpm. The differences, recognizable only in horizontal direction, where at 0 speed the bending moments are null, when the alignment in this direction is according to a straight line (as it should theoretically be) are due to the oil film which builds up in the bearings and displaces the journals. It can be seen that the bending moments in horizontal direction are negligible with respect to the bending moments in vertical direction (which are represented in a different scale) in almost all positions along the shaft-line except in positions close to the bearings where the vertical bending moments become vanishing small.

Bearing in mind that the crack forces are proportional to $\mathbf{x}_s$ which is in each node proportional to above static bending moment, it results that cracks generate maximum excitation only when they are located close to the mid-span of the shafts.

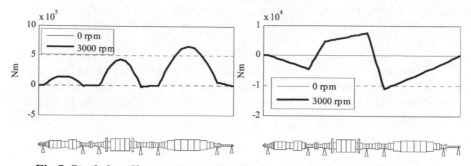

Fig 5 Static bending moments distributions along the shaft-line at rest and at normal operating condition.

5. CRACKED SHAFT-LINE DYNAMICAL BEHAVIOUR

The excitation of the different harmonic components of the vibrations depends on the depth of the crack (according to Fig. 4) and is proportional to the value of M_b static bending moment (which is function of the position of the crack as shown in Fig. 5). The resulting vibrations depend obviously not only on the intensity of excitation, but also on the position of the excitation and on the rotating speed: using the concepts of modal analysis we can forecast the severity of the resulting vibrations. This depends on the intensity of the generalized force component of the exciting equivalent bending moments which are function of the position of the cracked element with respect to the different vibration modes (eigenvectors), and on the dynamic amplification factor (q factor) function of the ratio frequency of excitation to the corresponding eigen-frequency and to the associated damping factor.

A crack located in a position where the considered eigenmode presents a high curvature, excites strongly this mode. If moreover the eigen-frequency of this mode is close to the 1xrev., or 2x rev. or 3 per rev. exciting frequency, then this mode is more strongly excited due to its closeness to resonance.

If the exciting forces are large but its generalized component is small in the position of the crack with respect to the vibration mode corresponding to the eigen-frequency which is closest to the $n\Omega$ exciting frequency, the resulting vibrations are small. If also the dynamic amplification factor is small (e.g. because the mode has a high damping factor or because the exciting frequency is far away from resonance), then the vibrations can hardly be measured, and the crack can hardly be detected.

Another problem is if these resulting vibrations are large enough to be measured in correspondence of the measuring stations (which are generally the bearings): this again depend on the amplitude of the excited eigenvectors in correspondence of the measuring station.

Only taking into account all these factors the measurable vibrations can be predicted.

Claiming that a crack of depth x% can be detected by vibration measurements, without defining all the above specified factors is nonsense.

Above considerations will be applied to the model of a turbo-group.

6. DESCRIPTION OF THE TURBO-GROUP

Figure 6 shows the FE model of the 320 MW turbo-group composed by a HP-IP steam turbine, a LP steam turbine and a generator. The group is equipped with 7 oil film bearings, which are represented by its linearised stiffness and damping coefficients, and with a supporting structure which is represented by simple pedestals (1 d.o.f. mass-spring-damper systems).

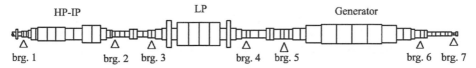

Fig 6 Finite element model of the turbo-group

7. UN-CRACKED SHAFT-LINE DYNAMICAL BEHAVIOUR

The natural frequencies (eigen-frequencies) and the associated mode shapes (eigenvectors) have been calculated at the normal running speed of 3000 rpm. The most significant mode shapes and associated frequencies and damping factors are shown in Table 1. Some highly damped modes have been disregarded. The first 6 modes correspond to classical modes of single shafts.

In the higher frequency range (corresponding e.g. to a 2xrev. excitation) the modes are rather complicated, involving mostly parts of the shafts in between bearings which are close together.

 The mode shapes should be taken into consideration when the generalized crack forces in the different mode shapes are estimated. Being the crack forces equivalent locally applied bending moments, their generalized forces are strong (which means that the energy introduced by the equivalent bending moments in that particular vibration mode is large), when the crack is located in a position where the mode shape presents a high curvature. Cracks at mid-span of single shafts excite strongly its first bending modes when the shaft is rotating at the corresponding frequency (critical speed), but excite very weakly its second bending modes. These are more strongly excited when the crack is located closer to its main bearings.

Table 1 Main natural frequencies, damping ratios and associated modes

Mode	frequency (Hz)	Damping ratio	Description
1	18.98	0.0026	1.st bending mode of generator
2	21.22	0.018	1.st bending mode of LP turbine
3	24.34	0.010	1.st bending mode of HP-IP turbine
4	33.19	0.110	Other 1.st bending mode of HP-IP turbine
5	43.00	0.125	2.nd bending mode of LP turbine
6	48.01	0.251	2.nd bending mode of generator
7	57.42	0.160	Other 2.nd bending mode of generator

8. NUMERICAL SENSITIVITY ANALYSIS: VIBRATIONS IN BEARINGS AT RATED SPEED DUE TO A CRACK WHICH HAS DEVELOPED IN ANY POSITIONS ALONG THE TURBOGROUP

A numerical analysis has been performed calculating the vibration behaviour of the above said turbo-group affected by a real crack with rectilinear tip and with a depth of 25% of the diameter.

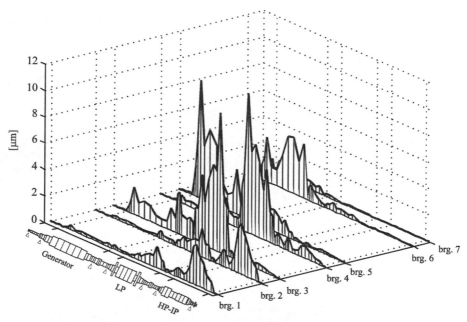

Fig 7 Waterfall plot of 1xrev. vibration amplitudes in correspondence of the bearings (at normal running speed of 3000 rpm) as a function of the position of a 25% deep crack

Figures 7 and 8 show the waterfall plots respectively of the 1xrev. and 2xrev. vibrations in the bearings in vertical direction as function of the crack position. 1xrev. vibration excitation is larger in vertical than in horizontal direction (which are not shown for brevity), 2xrev vibration excitation is instead almost equal in horizontal and vertical direction. Both results were predicted in Figure 4. Figure 7 shows that the maximum 1xrev vibration amplitudes in the bearings are less than 4 μm when crack is in between the HP and the IP sections of the turbine (which shows the effect of the severity of excitation), around 10μm when the crack is in the LP turbine close to one or the other bearing, and something less than 10 μm when the crack is on the generator, close to bearing 6. This situation is due to the effect of the dynamical behaviour of the group at the frequency of 50 Hz.

Figure 8 shows that the maximum 2xrev vibration amplitudes in the bearings are less than 2 μm when crack is in between the HP and the IP sections of the turbine, less than 3 μm when the crack is in the LP turbine close to one or the other bearing, and something less than 5 μm when the crack is on the generator, close to bearing 6. Both last effects are due to the dynamical behaviour of the group at the frequency of 100 Hz. It results that at rated speed the 1xrev crack forces introduce maximum energy in the corresponding vibration mode when the crack is located close to bearing 6.

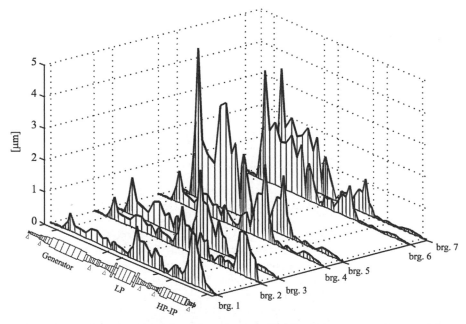

Fig 8 Waterfall plot of 2xrev. vibration amplitudes in correspondence of the bearings (at normal running speed of 3000 rpm) as a function of the position of a 25% deep crack.

The 2xrev excitation is roughly half the 1xrev excitation, as already shown in Fig 4, but dynamic effects are different at 50 and 100 Hz. The 2xrev. component, which is the most reliable symptom related to a developing crack, is rather small for the 25%

deep crack and could hardly be detected by measurements taken in the bearings at rated speed. This holds for any position of the crack along the shaft-line, with the exception of the few above specified positions.

From the above graphs it can be seen that the severity of the crack excited vibrations in the bearings, at normal operating speed, depend strongly not only on excitation severity (static bending moment and crack depth) but also on the dynamic behaviour of the shaft-line in correspondence of the crack position and in correspondence of the bearing which is the measuring station.

As the crack propagates deeper the different vibration amplitudes increase according to the trend shown in Fig. 4.

Much more information can be gathered if a run down speed transient is analysed, in which resonances can amplify the 2xrev. symptoms.

As an example in Fig 9 the frequency response curves for the 3 harmonic components which could be measured in bearing 3 of the LP turbine, excited by a 25% crack located at mid-span, are represented.

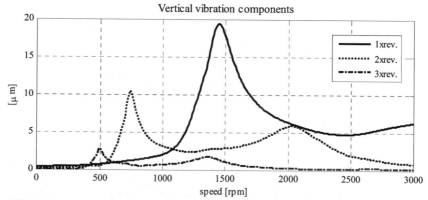

Fig 9 Frequency response curves in bearing 3 for the 25% crack located at mid-span of the LP turbine.

At mid-span the bending moment is maximum (strong crack force excitation), and the generalized crack force components is relevant in the first mode (at first bending critical speed) and is small in the second mode (at its second bending critical speed), which is not at all excited. Regarding the 2xrev component, we have resonances at half the critical speed values for the 1.st and 3.rd critical speed. Finally also a peak at 1/3.rd of the first bending critical speed is clearly recognizable in the 3xrev component. These symptoms help strongly in identifying the presence of a crack.

It should be always reminded that the 1xrev. vibration component excitation can be due to different causes (crack included) whereas the 2xrev. and 3xrev vibration excitation are due to the stiffness asymmetry associated to the crack: the 2xrev vibration (and the 3xrev vibration where measurable) are the most reliable symptom of the presence of a transverse crack in the rotating shaft.

9. CONCLUSIONS

It can be concluded from this analysis that a manifold and complex dynamic response can be expected as function of the position of the crack. The shaft vibrations in correspondence of the bearings are in general more sensitive to cracks located close to mid-span, where the crack forces are relevant, especially in correspondence of the first bending 1x, 2x or 3x critical speeds, when the generalized crack forces in the first bending mode reach maximum values and when the dynamic amplification factor is maximum. At rated speed the situation is strongly influenced by the dynamic behaviour of the shaft-line and maximum crack response is found for cracks in positions where the crack forces introduce much energy in the vibration mode.

Summarizing it results from this sensitivity analysis that small cracks located close to mid-span could generate measurable vibrations, at suitable rotating speeds (typically during coast down transients) and that deep cracks located in other parts of the shaft-line could sometimes hardly be detected by vibration measurements in the bearings at rated speed during normal operating conditions of the machine.

REFERENCES

[1] Allianz Berichte, nr. 24 Nov. 1987, ISSN 0569-0692.
[2] Ishida Y. "Cracked rotors: industrial machine case histories and non linear effects shown by simple Jeffcott rotor" MSSP 22 (2008), pp 805-817
[3] Papadopoulos C. A. "The strain energy release approach for modelling cracks in rotors: a state of art review" MSSP 22 (2008), pp 763-789.
[4] Stoisser C.M., Audebert S. "A comprehensive theoretical, numerical and experimental approach for crack detection in power plant rotating machinery" MSSP 22 (2008), pp 818-848
[5] Bachschmid N., Pennacchi P. and Tanzi E."Some remarks on breathing mechanism, on non-linear effects and on slant and helicoidal cracks" MSSP 22 (2008), pp 879-904
[6] Bachschmid N., Pennacchi P. and Tanzi E.(2007) "Rotating shafts affected by transverse cracks: Experimental behaviour and modelling techniques" *Int. J. Materials and Structural Integrity*, Vol. 1, Nos 1/2/3, pp. 71-116
[7] Bachschmid N., Tanzi E.(2006) "Effect of shear forces on cracked beam deflections" Crack Paths 2006 Conference, Parma (Italy) Sept. 2006

Detecting cracked rotors using an active magnetic actuator

M I Friswell[1], Y Y He[2], J E T Penny[3], J T Sawicki[4]

[1] Department of Aerospace Engineering, Bristol University, UK
[2] Department of Precision Instruments, Tsinghua University, PR China
[3] School of Engineering, Aston University, UK
[4] Department of Mechanical Engineering, Cleveland State University, USA

ABSTRACT

Cracked rotors are not only important from a practical and economic viewpoint, they also exhibit interesting dynamics. This paper will look the modelling and analysis of machines with breathing cracks, which open and close due to the self-weight of the rotor, producing a parametric excitation. After considering the modelling of cracked rotors, the paper considers a novel method for crack detection, mainly focusing on the use of active magnetic bearings to apply force to the shaft of a machine. Consideration is given of some issues to enable this approach to become a robust condition monitoring technique for cracked shafts.

1.   INTRODUCTION

The idea that changes in a rotor's dynamic behaviour could be used for general fault detection and monitoring was first proposed in the 1970s. Of all machine faults, probably cracks in the rotor pose the greatest danger and research in crack detection has been ongoing for the past 30 years. Current methods examine the response of the machine to unbalance excitation during run-up, run down or during normal operation. In principle, a crack in the rotor will change the dynamic behaviour of the system but in practice it has been found that small or medium size cracks make such a small change to the dynamics of machine system that they are virtually undetectable by this means. Only if the crack grows to a potentially dangerous size can they be readily detected. It is a race between detection and destruction!

Detecting faults in rotating machines present certain problems that do not occur in fixed structures. First of all, in addition to cracks there are several other types of faults that can be present in a machine, such as a seal rub or a misaligned bearing. Ideally we wish to detect a crack in a machine rotor at an early stage in its development and estimate its size and location. However, compared to static structures, access to a rotor can be limited either in order to contain the working fluid or for safety reasons. When two or more machines are coupled together it is generally sufficient to be able to determine which machine rotor is cracked so that the correct stator is removed to gain access to the rotor. On the other hand, compared to a large static structure, the rotor of a machine is easily excited by the residual out-of-balance that always exists or by the use of active magnetic bearings.

Crack detection methods fall into two groups, model updating and pattern recognition. In the former method, the dynamic behaviour of the rotor is used to update a model of the rotor and in the process determine both the severity and location of any crack. Clearly the crack model used must be adequate for the task. Bachschmid *et al.* [1] used a least squares method in the frequency domain to identify and locate machine faults (including a cracked rotor) whereas Markert *et al.* [2] used a least squares method in the time domain to identify and locate machine faults. If the pattern recognition approach is used, then whilst a crack model is not directly required, it is desirable to have some idea of the dynamic behaviour that will result from a cracked rotor in order that it can be recognised in the pattern of behaviour. Again a model is required.

If the vibration due to any out-of-balance forces acting on a rotor is greater than the static deflection of the rotor due to gravity, then the crack will remain either opened or closed depending on the size and location of the unbalance masses. In the case of the permanently opened crack, the rotor is then asymmetric and this condition can lead to stability problems. If the vibration due to any out-of-balance forces acting on a rotor is less than the static deflection of the rotor due to gravity the crack will open and close (or breathe) as the rotor turns. This is the situation that exists in the case of large horizontal machines and this paper will concentrate on this condition.

2.   EQUATIONS OF MOTION OF CRACKED ROTOR

The analysis of a rotor-bearing system may be performed in fixed or rotating co-ordinates. If neither the bearings and foundations nor the rotor is axi-symmetric then the resulting differential equations, whether described in fixed or rotating coordinates, will be linear equations with harmonic coefficients. Typically foundations of a large machine will be stiffer vertically than horizontally and in this case the cracked rotor will not be axi-symmetric when the crack is open. Thus there is no compelling reason to used fixed or rotating coordinates for the analysis. To determine the stiffness of the rotor as the crack opens and closes it is easier to work in co-ordinates that fixed to the rotor and rotate with it. The reduction in stiffness due to a crack is then calculated in directions perpendicular to and parallel to the crack face, and these directions will rotate with the rotor. Having determined the rotor stiffness in rotating coordinates we transform the stiffness matrix to fixed coordinates and join the stiffness to the system inertia to obtain the equation of motion in fixed coordinates.

Let the stiffness matrix in rotating co-ordinates for the un-cracked rotor be $\tilde{\mathbf{K}}_0$ and the reduction in stiffness due to a crack be $\tilde{\mathbf{K}}_c(\theta)$, where θ is the angle between the crack axis and the rotor response at the crack location and determines the extent to which the crack is open. Thus the stiffness of the cracked rotor is

$$\tilde{\mathbf{K}}_{cr} = \tilde{\mathbf{K}}_0 - \tilde{\mathbf{K}}_c(\theta). \tag{1}$$

This stiffness matrix is transformed from rotating to fixed co-ordinates using the transformation matrix $\mathbf{T}(\Omega t)$, to give, assuming the un-cracked rotor is axi-symmetric,

$$\mathbf{K}_{cr} = \mathbf{T}^T\tilde{\mathbf{K}}_0\mathbf{T} - \mathbf{T}^T\tilde{\mathbf{K}}_c(\theta)\mathbf{T} = \mathbf{K}_0 - \mathbf{K}_c(\theta,t). \tag{2}$$

Let the deflection of the system be $\mathbf{q} = \mathbf{q}_{st} + \mathbf{q}_{dy}$ where $\mathbf{q}_{st}$ is the static deflection of the un-cracked rotor due to gravity, and $\mathbf{q}_{dy}$ is the dynamic deflection due to the rotating out of balance and the effects of the crack. Thus, $\dot{\mathbf{q}} = \dot{\mathbf{q}}_{dy}$ and $\ddot{\mathbf{q}} = \ddot{\mathbf{q}}_{dy}$, and the equation of motion for the rotor in fixed co-ordinates is

$$\mathbf{M}\ddot{\mathbf{q}}_{dy} + (\mathbf{D}+\mathbf{G})\dot{\mathbf{q}}_{dy} + (\mathbf{K}_0 - \mathbf{K}_c(\theta,t))(\mathbf{q}_{st}+\mathbf{q}_{dy}) = \mathbf{Q}_u + \mathbf{W} \tag{3}$$

where $\mathbf{Q}_u$ and $\mathbf{W}$ are the out of balance forces, and the gravitational force respectively. Damping and gyroscopic effects have been included as a symmetric positive semi-definite matrix $\mathbf{D}$ and a skew-symmetric matrix $\mathbf{G}$, although they have little direct bearing on the analysis. If there is axi-symmetric damping in the rotor then there will also be a skew-symmetric contribution to the undamaged stiffness matrix, $\mathbf{K}_0$. We refer to Equation (3) as the "full equations".

The steady state deflection of the rotor varies over each revolution of the rotor since $\mathbf{K}_c$ varies. However, the stiffness reduction due to the crack is usually small, and we may make the reasonable assumption that $\|\mathbf{K}_0\| \gg \|\mathbf{K}_c(\theta,t)\|$. With this assumption the steady state deflection is effectively constant and equal to the static deflection, $\mathbf{q}_{st}$. The second approximation commonly used in the analysis of cracked rotors is weight dominance. If the system is weight dominated it means that the static deflection of the rotor is much greater than the response due to the unbalance or rotating asymmetry, that is $|\mathbf{q}_{st}| \gg |\mathbf{q}_{dy}|$. For example, for a large turbine rotor the static deflection might be of the order of 1 mm whereas at running speed the amplitude of vibration is typically 50 μm. Even at a critical speed the allowable level of vibration will only be 250 μm. In this situation that the crack opening and closing is dependent only on the static deflection and thus $\theta = \Omega t + \theta_0$, where Ω is the rotor speed and θ_0 is the initial angle. With these two assumptions, Equation (3) becomes

$$\mathbf{M}\ddot{\mathbf{q}}_{dy} + (\mathbf{D}+\mathbf{G})\dot{\mathbf{q}}_{dy} + (\mathbf{K}_0 - \mathbf{K}_c(t))\mathbf{q}_{dy} = \mathbf{Q}_u \tag{4}$$

where $\mathbf{K}_c$ is now independent of θ.

3. MODELS OF BREATHING CRACKS

The breathing crack was initially studied by Gasch [3] who modelled the crack as a *hinge*. In this model the crack is open for one half and closed for the other half a revolution of the rotor, and the transition from open to closed (and vice-versa) occurs abruptly as the rotor turns. Mayes and Davies [4, 5] developed a similar model except that the transition from fully open to fully closed is described by a cosine function, that will be used in this paper. Penny and Friswell [6-8] compared the response due to different crack models and considered the effect on the dynamic response of the rotor.

In this paper the fully open crack is modelled by reducing the element stiffness in orthogonal directions (parallel and perpendicular to the crack face). The stiffness matrix of the machine when the crack is open, in rotating co-ordinates, is then $\tilde{\mathbf{K}}_1$.

If weight dominance is assumed, then the opening and closing of the crack is periodic at the rotor spin speed. In the Mayes model the time dependent stiffness matrix in rotating coordinates is

$$\tilde{\mathbf{K}}_c(t) = 0.5 \times \left(1 - \cos(\Omega t + \theta_1)\right)\left[\tilde{\mathbf{K}}_0 - \tilde{\mathbf{K}}_1\right] \tag{5}$$

where θ_1 depends on the crack orientation and the initial angle of the rotor. When $\cos(\Omega t + \theta_1) = 1$ the crack is fully closed and $\tilde{\mathbf{K}}_{cr}(t) = \tilde{\mathbf{K}}_0$, the un-cracked rotor stiffness, where $\tilde{\mathbf{K}}_{cr}$ is defined in Equation (1). Thus the rotor is axi-symmetric when the crack is closed. When $\cos(\Omega t + \theta_1) = -1$ the crack is fully open so that $\tilde{\mathbf{K}}_{cr}(t) = \tilde{\mathbf{K}}_1$. Note that when the crack is open the rotor is asymmetric.

Converting from rotating to fixed co-ordinates is performed using the transformation given in Equation (2). The stiffness matrix in stationary co-ordinates, $\mathbf{K}_c(t)$, is a periodic function of time only and the full non-linear Equation (3) becomes a linear parametrically excited equation. Penny and Friswell [6] showed that the model generates a constant term plus 1X, 2X and 3X rotor angular velocity components in the stiffness matrix.

4. CONDITION MONITORING USING ACTIVE MAGNETIC ACTUATORS

Active magnetic bearings (AMB) have been used in high-speed applications or where oil contamination must be prevented, although their low load capacity restricts the scope of applications. Recently a number of authors have considered the use of AMBs as an actuator that is able to apply force to the shaft of a machine [9-11]. If the applied force is periodic, then the presence of the crack generates responses containing frequencies at combinations of the rotor spin speed and applied forcing frequency. The excitation by unbalance and AMB forces produces combination resonances between critical speed of the shaft, the rotor spin speed and the frequency of the AMB excitation. The key is to determine the correct excitation frequency to induce a combination resonance that can be used to identify the magnitude of the time-dependent stiffness arising from the breathing mode of the rotor crack.

The force applied on the rotor by the AMBs must be included in the equations of motion. Thus Equation (4) becomes

$$\mathbf{M}\ddot{\mathbf{q}}_{dy} + (\mathbf{D} + \mathbf{G})\dot{\mathbf{q}}_{dy} + (\mathbf{K}_0 - \mathbf{K}_c(t))\mathbf{q}_{dy} = \mathbf{Q}_u + \mathbf{Q}_{\text{AMB}} \tag{6}$$

where $\mathbf{Q}_{\text{AMB}}$ is the external forces applied to the rotor by the active magnetic bearing. This force will probably be chosen to be harmonic, either in one or two directions. Other waveforms would be possible if they were perceived to offer some advantage.

The key aspect of the analysis is that the system has three different frequencies, namely the natural frequency (or critical speed), the rotor spin speed and the forcing frequency from the AMB. The parametric terms in the equations of motion (or non-linear terms in the full equations) cause combinational resonances in the response of the machine. Mani *et al.* [9, 11] and Quinn *et al.* [10] used a multiple scales analysis to determine the conditions required for a combinational resonance, which occurs when

$$\Omega_2 = |n\Omega - \omega_i|, \qquad \text{for} \quad n = \pm 1, \pm 2, \pm 3 \tag{7}$$

where Ω is the rotor spin speed, Ω_2 is the frequency of the AMB force, and ω_i is a natural frequency of the system. This analysis was based on a two degree of freedom Jeffcott rotor model with weight dominance, equivalent to that described in this paper. Mani *et al.* [11] also considered the effect of detuning, that is when the excitation is close to this exact excitation frequency for resonance, and investigated the effect on the magnitude of the primary resonance close to the natural frequency of the machine. In the examples the running speed of the machine was five times higher than the natural frequency. This ratio is not practical since there is likely to be a second un-modelled resonance below the running speed. Indeed the fact that higher resonances are not modelled is a serious omission, particularly as the combinational resonances are likely to excite any higher frequency resonances.

5.   THE TIME SIMULATION

The analysis thus far has indicated the combinational resonances that are likely to occur in a machine with a breathing crack, excited by a magnetic actuator. Most of the analysis in the literature has been performed on simple two degree of freedom models of the machine, with simplifying assumptions concerning the crack model, gyroscopic effects, higher modes and so on. In order to check the robustness of the frequency content of the machine response a time simulation will be performed on a detailed model of the machine. This will allow realistic features of the real machine to be easily incorporated. To ensure the transient response decays within a reasonable time, damping is added to the bearings and/or disks. The equations of motion are integrated using ode45 in MATLAB. However the number of degrees of freedom of a detailed finite model is likely to be large, requiring a long computational time to simulate the response. Thus the equations of motion in the rotating frame are reduced using the lower mode shapes of the undamped and undamaged machine, neglecting gyroscopic effects. A sufficient number of modes should be included to simulate the range of excitation frequencies, and also any significant combinational resonances. This reduction has two beneficial effects; not only are the number of degrees of freedom reduced, leading to a lower computational cost per time step, but also the higher frequencies are removed, thus allowing a larger time step.

The reduction procedure is to calculate the eigenvectors of the undamped and undamaged machine as

$$\left[-\omega_{0i}^2 \mathbf{M} + \mathbf{K}_0\right]\boldsymbol{\phi}_i = \mathbf{0} \tag{8}$$

where ω_{0i} and $\boldsymbol{\phi}_i$ are the i th natural frequency and mode shape. If the lower r modes are retained then the reduction transformation is

$$\mathbf{T}_r = \left[\boldsymbol{\phi}_1 \ \boldsymbol{\phi}_2 \cdots \boldsymbol{\phi}_r\right]. \tag{9}$$

The reduced equations of motion, assuming the mode shapes are mass normalized, are then

$$\ddot{\mathbf{q}}_r + \left(\mathbf{D}_r + \mathbf{G}_r\right)\dot{\mathbf{q}}_r + \left(\boldsymbol{\Lambda}_0 - \mathbf{K}_{rc}\left(t\right)\right)\mathbf{q}_r = \mathbf{Q}_{ru} + \mathbf{Q}_{rAMB} \tag{10}$$

where

$$\mathbf{q}_{dy} = \mathbf{T}_r \mathbf{q}_r, \qquad \mathbf{D}_r = \mathbf{T}_r^T \mathbf{D}_r \mathbf{T}_r, \qquad \mathbf{G}_r = \mathbf{T}_r^T \mathbf{G}_r \mathbf{T}_r, \qquad \mathbf{K}_{rc} = \mathbf{T}_r^T \mathbf{K}_c \mathbf{T}_r,$$
$$\mathbf{Q}_{ru} = \mathbf{T}_r^T \mathbf{Q}_u, \qquad \mathbf{Q}_{rAMB} = \mathbf{T}_r^T \mathbf{Q}_{AMB}, \tag{11}$$

and $\boldsymbol{\Lambda}_0 = \mathrm{diag}\left(\omega_{01}^2, \omega_{02}^2, \ldots, \omega_{0r}^2\right)$ is the diagonal matrix of undamped and undamaged eigenvalues.

The equations are integrated until a steady state has been established and then the FFT is calculated. The steady state response should only contain the excitation and rotor spin frequencies, and the combinational resonances, and therefore the spectrum of the response should only contain discrete frequencies. However, leakage is likely to occur because of the difficulties in choosing a sample period so that every sinusoidal component in the response has an integer number of cycles in the sample. The effect of leakage may be reduced by using time window functions. Furthermore the sample period must be sufficiently long to ensure that the frequency increment is small enough to distinguish the individual frequency components.

6.   CASE STUDY

The rotor model used for the following simulations is shown in Fig. 1. The shaft is 750 mm long and 20 mm diameter. Both disks are 25 mm thick, and disk 1 has a diameter of 250 mm, and disk 2 has a diameter of 150 mm. The rotor is supported on isotropic bearings and foundations of stiffness 1 MN/m and damping 1 kNs/m. A damper of value 100 Ns/m is added at the location of disk 2; this damper is used to increase damping in the first mode, and hence reduce the time required for the transient response to decay. The shaft is modelled using 6 elements, and the length of the cracked element was taken as 20 mm. The element length is small, relative to the wavelengths of the lower modes, and hence the results from using such an equivalent model for the crack will not be sensitive to the length of the crack element over a reasonable range. For all simulations the rotor speed was 1500 rev/min (25 Hz). For the undamaged rotor at this spin speed, the first and second damped natural frequencies for the forward rotating modes are 21.7 and 84.3 Hz. An out of balance with magnitude 10^{-3} Nm was added at disk 2 and, when applied, the active magnetic bearing (AMB) applied a peak harmonic force of 100 N. Fig. 2 shows the Campbell diagram for the rotor system and Fig. 3 shows the first six mode shapes at a rotor spin speed of 1500 rev/min.

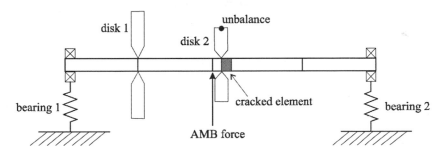

Fig. 1. The example rotor system.

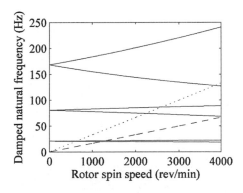

Fig. 2. Campbell diagram for the example system.

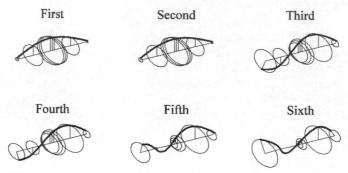

First   Second   Third

Fourth   Fifth   Sixth

Fig. 3. The first six mode shapes for the example system at 1500 rev/min.

Fig. 4 shows the response of the rotor at disk 2 when the rotor was undamaged, and where there AMB force is not present; clearly only the synchronous response due to the unbalance is visible. The integration is performed over 300 cycles of the rotor spin frequency, and the first 100 cycles are discarded to allow the transients to decay. Fig. 5 shows the response of the undamaged rotor when the AMB force with frequency 28.5 Hz is present; the response only occurs at the rotor spin frequency and the AMB forcing frequency. Note that the number of cycles has been chosen so that there is no leakage effects at the two significant frequencies at which the response is non-zero.

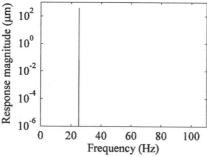

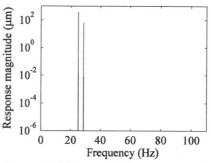

Fig. 4. Response for the uncracked rotor with no AMB excitation

Fig. 5. Response for the uncracked rotor excited by the AMB (Ω_2=28.5Hz)

Suppose a crack is now added to the shaft, modelled by assuming the stiffness in one direction for the element is reduced by 20%. The crack opening and closing is assumed to be weight dominated, and hence the equations of motion are linear and parametrically excited. Fig. 6 shows the response at disk 2, for zero force from the AMB. In this case the only response is at the rotor spin speed and its harmonics. Suppose the rotor is now excited by the AMB with a frequency of 28.5 Hz, based on an assumed natural frequency of 21.5 Hz (which is slightly less than the true natural frequency of the uncracked rotor), with $n = 2$ in Equation (7). Thus for the cracked rotor we clearly see clusters of frequencies that should be readily detected.

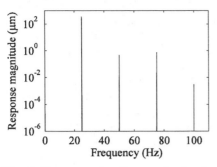

Fig. 6 Response for the cracked rotor with no AMB excitation

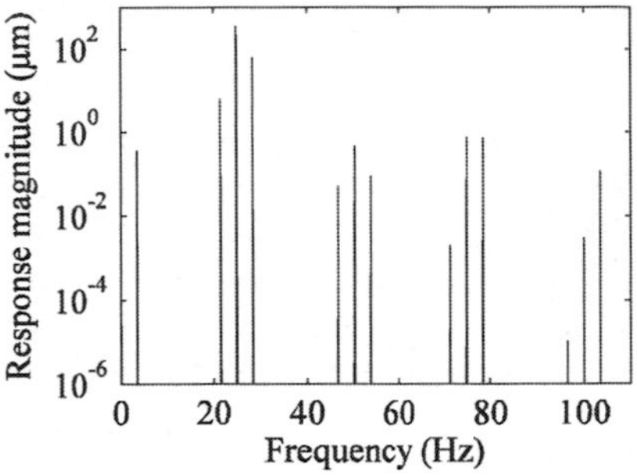

Fig. 7 Response for the cracked rotor excited by the AMB (Ω_2=28.5Hz)

Although the frequency lines in Figs. 4 to 7 are very distinct, it should be emphasised that this is a very idealised case. A number of issues that arise in practice will now be discussed. A damper was added to disk 2 to increase the damping in the first mode. The only other damping arises in the bearings, and since the displacement in the first mode at the bearings is small, without this damper at the disk the first mode is very lightly damped. Fig. 8 shows the response (equivalent to Fig. 7) when this damper is removed. Clearly the first modes (both forward and backward whirling) are now excited and transients have not had sufficient time to decay. This means that the responses at the discrete frequencies are not so clearly identified. Fig. 9 shows the effect when 201 cycles of the rotor spin frequency are used for the measurement. This means that the time period does not contain an integer number of cycles of the AMB forcing, and Fig. 9 clearing shows leakage effects around the AMB frequency of 28.5 Hz (again compare to Fig. 7). It is possible to overcome the leakage effects since we know very accurately the rotor spin speed (assuming it is constant) and the AMB frequency; this allows us to choose the sample period, or alternative use an estimator for the magnitude of the two frequency terms.

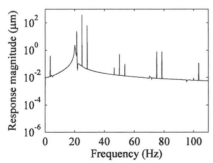

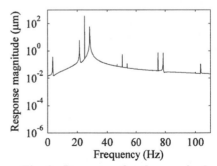

Fig. 8. Response for the cracked rotor with no damper at the disk

Fig. 9. Response for the cracked rotor with leakage effects

Next consider random noise. Here a uniform random noise with a maximum amplitude of 1% of the response amplitude has been added to the time data used to produce Fig. 7. The resulting spectrum is shown in Fig. 10. Although the frequencies around the rotor spin speed are clearly above the noise floor, many of the other response at discrete frequencies are now hidden. Given that we know which combinational frequencies we are looking for there is the possibility to average to help to emphasise the response at frequencies of interest. As a final practical issue consider the accuracy of the estimated natural frequency. The first natural frequency of the example machine is 21.7 Hz and this was estimated as 21.5 Hz to give the AMB frequency from Equation (7) using $n = 2$ that was used to generate Figs. 7 to 10. Fig. 11 shows the effect of estimating the natural frequency as 20.0 Hz, giving an AMB frequency of 30.0 Hz. Clearly the result is very similar to that in Fig. 7, and highlights that the natural frequencies do not have to be estimated very accurately.

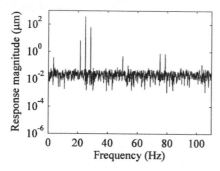

Fig. 10. Response for the cracked rotor with random noise

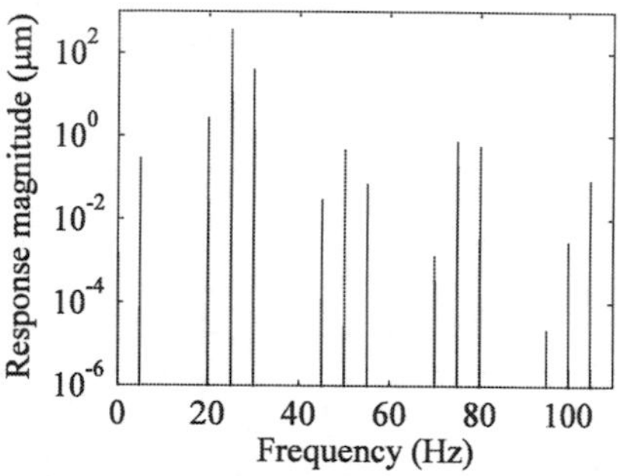

Fig. 11. Response for the cracked rotor excited by the AMB (Ω_2=30.0Hz)

Thus far the AMB frequency has been calculated using $n = 2$ and an estimate of the first natural frequency in Equation (7). Suppose that $n = -1$ with an estimated first natural frequency of 21.5 Hz, giving an AMB frequency of 46.5 Hz. Fig. 12 shows the spectrum in this case, which is very similar to Fig. 7, with the same combination frequencies, although different response amplitudes. Fig. 13 shows the effect of using the second natural frequency (estimated as 84.0 Hz) and $n = 4$ in Equation (7), giving an AMB frequency of 16 Hz. Clearly combination frequencies are excited, although the character of the spectrum is different because of the relative frequencies of the rotor spin speed and the AMB force.

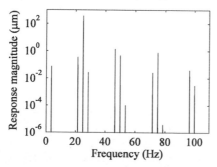

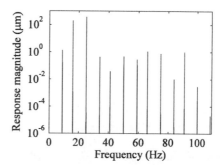

Fig. 12. Response for the cracked rotor excited by the AMB (Ω_2=46.5Hz)

Fig. 13. Response for the cracked rotor excited by the AMB (Ω_2=16.0Hz)

7. CONCLUSIONS

A variety of approaches have been proposed to try to make use of the changes in the dynamics of a rotor to identify and possibly locate crack (and other faults) in a rotor at an early stage in their development. The simulations shown suggest that the use of an auxiliary active magnetic bearing (AMB) to help identify crack in the rotor has some merit, but further work is needed to produce a robust condition monitoring technique. Applying a sinusoidal force from the AMB produces combination frequencies based on the AMB frequency, the rotational speed, and the natural frequencies, that could be used to detect cracks in the rotor. However a robust method is needed to determine the presence, location and severity of a crack from the combination frequencies in the response. Furthermore the effect of adding an extra force to the system might encourage a faster crack growth, which is obviously a disadvantage.

8. REFERENCES

1. Bachschmid, N., Pennacchi, P., Tanzi, E. and Vania, A., Identification of transverse crack position and depth in rotor systems, *Meccanica*, **35**, 563-582, 2000.
2. Markert, R., Platz, R. and Seidler, M., Model based fault identification in rotor systems by least squares fitting, *International Journal of Rotating Machinery*, **7**(5), 311-321, 2001.
3. Gasch, R.A., Survey of the dynamic behavior of a simple rotating shaft with a transverse crack, *Journal of Sound and Vibration*, **160**, 313-332, 1993.
4. Mayes, I.W. and Davies, W.G.R., A method of calculating the vibrational behaviour of coupled rotating shafts containing a transverse crack, *IMechE International Conference on Vibrations in Rotating Machinery*, Paper C254/80, 17-27, 1980.
5. Mayes, I. W. and Davies, W. G. R., Analysis of the response of a multi-rotor-bearing system containing a transverse crack in a rotor, *ASME Journal of Vibration, Acoustics, Stress, and Reliability in Design*, **106**, 139-145, 1984.
6. Penny, J.E.T. and Friswell, M.I., Simplified modelling of rotor cracks, *ISMA 27*, Leuven, Belgium, 607-615, 2002.
7. Penny, J.E.T, Friswell, M.I. and Zhou, C., Condition monitoring of rotating machinery using active magnetic bearings, *ISMA 2006*, Leuven, Belgium, 3497-3506, 2006.
8. Penny, J.E.T. and Friswell, M.I., The dynamics of cracked rotors, *IMAC XXV*, Orlando, Florida, paper 198, 2007.
9. Mani, G., Quinn, D.D., Kasarda, M.E.F., Inman, D.J. and Kirk, R.G., Health monitoring of rotating machinery through external forcing, *ISCORMA-3*, Cleveland, Ohio, USA, 2005.
10. Quinn, D.D., Mani, G., Kasarda, M.E.F., Bash, T.J. Inman, D.J. and Kirk, R.G., Damage detection of a rotating cracked shaft using an active magnetic bearing as a force actuator - analysis and experimental verification, *IEEE/ASME Transactions on Mechatronics*, **10**(6), 640-647, 2005.

11. Mani, G., Quinn, D.D., and Kasarda, M., 2005, Active health monitoring in a rotating cracked shaft using active magnetic bearings as force actuators, *Journal of Sound and Vibration*, **294**, 454-465.

Flexibility coefficients of a shaft with an elliptical front crack

L Rubio, B Muñoz-Abella
Mechanical Engineering Department, University Carlos III of Madrid, Spain

ABSTRACT

Cracks in mechanical components produce changes in their behaviour like increases of displacements or decreases of frequencies due to the flexibility increase. Most of the works related with the analysis of the behaviour of cracked shafts consider the front of the transversal fatigue cracks to be straight, but experience says that the front of these kind of cracks is approximately elliptical. Many expressions have been given for the flexibility of a cracked shaft with a straight front. It is not the same for the case of elliptical front for which one example of approximate expressions for the flexibility has been developed. In this work, flexibility expressions for cracked shafts having elliptical front cracks have been obtained, based on the polinomial fitting of the stress intensity factors, taking into account the size and shape of the elliptical cracks.

1. INTRODUCTION

It is very well known that the presence of a crack in a mechanical element produces an increase in its flexibility which is accompanied by an increase in cross displacements and a decrease in vibration frequencies or modifications of the transversal section orbits, among others. Those changes could be used to detect the presence of a crack and also, to identify its position and size. So, it is of great interest to establish a relationship between the flexibility of a cracked mechanical element and the properties of the crack (shape, position and size).

In rotating shafts, cracks grow up following transversal planes to their longitudinal axis, due to the fatigue produced by the cyclic loads to which they are submitted. In order to detect the presence of the cracks and to identify their position and size, some studies have been carried out since the first works of Gasch [1].In many of those works, a relationship is established between the mechanical behaviour of the element and the presence of a crack with a straight front [2, 3]. However, the study of the flexibility of a shaft with a crack with an elliptical front has not been so exhaustive, having found an example of approximate expressions [4], whose results are not as promising as they should be.

The study of the behaviour of cracked shafts submitted to bending moments and axial loads or the combination of both has become a matter of interest in mechanical engineering, because the catastrophic failures of those cracked elements should be produced by those efforts on static shafts, as well as, in rotating shafts, in which the breathing of the crack takes place.

In this work, expressions for the flexibility coefficients of a shaft with an elliptical front crack have been obtained. The model has been derived considering the formulation of Fracture Mechanics, establishing the relationship between the Energy Release Rate (as a function of the flexibility) and the Stress Intensity Factor (SIF). For that, the polynomial fittings of the stress intensity factors, proposed previously by Shin et al. [5], have been taken into account. Different values of flexibility have been obtained for the variables involved: length of the crack, position of the crack and shape of the crack front (from straight to semicircular front passing through the elliptical front). The flexibility coefficients have also been fitted with polynomials of the characteristic variables of the problem such as the length, the shape and the position of the elliptical crack with very promising results. The work includes the comparison of the same results obtained considering other flexibility expressions derived by other authors [4].

2. PROBLEM FORMULATION

Changes in the mechanical behaviour of a cracked element are produced as a result of a rigidity lost in the cracked section or of an increase of flexibility.

Fig 1 Cracked shaft submitted to bending moments and axial loads

The presence of a crack in an element submitted to an axial load, N, and to a bending moment, M, (see figure1), can be modelled considering discontinuities in longitudinal displacements, Δu, and in slope, $\Delta\theta$, in the cracked section that can be written in terms of the transmitted efforts [6]:

$$\Delta u = \lambda_{nn} N + \lambda_{mn} M \tag{1}$$

$$\Delta\theta = \lambda_{mm} M + \lambda_{mn} N \tag{2}$$

where λ_{ij} are the flexibility coefficients, and where

$$\Delta u = u_2 - u_1 \tag{3}$$

$$\Delta\theta = \theta_2 - \theta_1 \tag{4}$$

u_1 and θ_1 being the displacement and slope at the left of side of the crack and u_2 and θ_2, being the displacement and slope at the right side of the crack.

The flexibility coefficients of a cracked mechanical element can be obtained considering the expression of the Energy Release Rate, G, in terms of the Stress Intensity Factor, K_I, that in plane strain conditions is:

$$G_N = \frac{1-\nu^2}{E} K_{I,N}^2 = \frac{N^2}{2} \frac{d\lambda_{nn}}{dA} \tag{5}$$

$$G_M = \frac{1-\nu^2}{E} K_{I,M}^2 = \frac{M^2}{2} \frac{d\lambda_{mm}}{dA} \tag{6}$$

where E and ν are the Modulus of Elasticity and the Poisson's ratio of the material, respectively, and where dA is the elemental crack area.

By integration, one can obtain the flexibility coefficients for bending and axial efforts and the combination of both:

$$\lambda_{nn} = \frac{2(1-\nu^2)}{E} \int_A \left(\frac{K_{I,N}}{N}\right)^2 dA \tag{7}$$

$$\lambda_{mm} = \frac{2(1-\nu^2)}{E} \int_A \left(\frac{K_{I,M}}{M}\right)^2 dA \tag{8}$$

$$\lambda_{mn} = \frac{2(1-\nu^2)}{E} \int_A \left(\frac{K_{I,N}}{N}\right) \left(\frac{K_{I,M}}{M}\right) dA \tag{9}$$

In those expressions, $K_{I,N}$ and $K_{I,M}$ are the Stress Intensity Factors in mode I for the considered efforts.

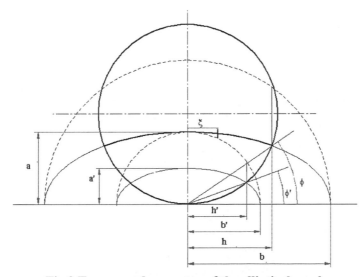

Fig 2 Transversal geometry of the elliptical crack

In order to study of the behaviour of cracked shafts and to calculate the flexibility in a simple way, many authors (i.e. Carpinteri [7]) assume that fatigue cracks in a shaft present a straight front. Although this hypothesis is valid and leads to conservative results, it does not represent the reality of the cracked shafts, in which the crack is approximately elliptical (figure 2) as experience says.

The characteristic parameters of an elliptical crack are the following:

- $\alpha = \dfrac{a}{D}$, characteristic length of the crack

- $\beta = \dfrac{a}{b}$, shape factor of the crack ($\beta = 1$ corresponds to a semicircular crack and

  $\beta = 0$ corresponds to a straight crack)

- $\gamma = \dfrac{\xi}{h}$, relative position in the front of the crack

Taking into account those geometrical considerations $dA = 2Sda$, where S is the length of the elliptical arc between points A y B of the elliptical front, see figure 2, and takes the form:

$$S = \int_{ar\cos\bar{h}}^{\frac{\pi}{2}} a\sqrt{\left(\frac{b}{a}\right)^2 \sin^2\phi + \cos^2\phi}\; d\phi \tag{10}$$

where $\bar{h}$ represents the relation $\dfrac{h}{b}$.

The Stress Intensity Factors can be expressed also considering the same variables that characterise the geometry of the elliptical front:

- $K_{I,N} = K_{I,N}(\alpha,\beta,\gamma)$
- $K_{I,M} = K_{I,M}(\alpha,\beta,\gamma)$

Some authors have studied the same problem given expressions for the stress intensity factor (SIF) as can be seen in the review of Álvarez et al. [8]. Couroneau [9] and Astiz [10], among others, obtained the SIF by interpolation from one point in the front crack. Shih y Chen [4] have published expressions for $K_{I,N}$ and $K_{I,M}$ particularized in points A y B (see figure 2) and have approximated the values of the rest of the elliptical front by a mean value. In a latter work the same authors [11] have obtained close-form expressions for the SIF. More recently Shin y Cai [5] have obtained polinomial expressions of the SIF that improve the results:

$$K_I = F_I \sigma \sqrt{\pi a} \tag{11}$$

$$F_{I,N} = \sum_{i=0}^{2}\sum_{j=0}^{7}\sum_{k=0}^{2} M_{ijk}(\beta)^i (\alpha)^j (\gamma)^k \tag{12}$$

$$F_{I,M} = \sum_{i=0}^{2}\sum_{j=0}^{6}\sum_{k=0}^{2} N_{ijk}(\beta)^i (\alpha)^j (\gamma)^k \tag{13}$$

The values of coefficients M_{ijk} y N_{ijk} of equations (12) y (13) can be found in [5].

3. FLEXIBILITY COEFFICIENTS

With the knowledge of the SIF expressions considering the three parameters that characterise the crack, equations (11)-(13), the flexibility functions can be calculated as follows, taking as an example the calculation of the flexibility for bending, λ_{mm}.

The maximum stress at the circular section of the shaft due to the bending can be written as:

$$\sigma = \frac{M\dfrac{D}{2}}{\dfrac{1}{4}\pi\left(\dfrac{D}{4}\right)^4} = \frac{32M}{\pi D^3} \tag{14}$$

substituting (14) in (11) and latter in (8), the flexibility for bending takes the form:

$$\lambda_{mm} = \frac{2(1-\nu^2)}{E}\int_A \frac{32^2\,\pi a}{\pi^2 D^6} F_{I,M}^2(\alpha,\beta,\gamma)dA \tag{15}$$

where,

$$dA = dS\,da = D^2 d\bar{s}\,d\alpha \tag{16}$$

being $\bar{s} = \dfrac{s}{D}$.

Now, taking into account equation (10), the expression for the elemental elliptical arc, $d\bar{s}$, will be:

$$d\bar{s} = \alpha\sqrt{\frac{1}{\beta^2}\sin^2\phi + \cos^2\phi}\;d\phi \tag{17}$$

and substituting in equation (15), the bending flexibility appears like:

$$\lambda_{mm} = \frac{4096(1-\nu^2)}{\pi E D^3}\overline{\lambda}_{mm} \tag{18}$$

where,

$$\overline{\lambda}_{mm} = \int_0^{\alpha_f}\int_{ar\cos\overline{h}}^{\frac{\pi}{2}} F_{I,M}^2(\alpha,\beta,\gamma)\alpha d\bar{s}d\alpha \tag{19}$$

is the nondimensional flexibility coefficient for bending, corresponding α_f to the final coefficient $\dfrac{a}{D}$.

In order to integrate this last expression it is needed to set a geometrical relation between γ, β and $\overline{h}$. To do this $\overline{h} = \dfrac{h}{b}$ has to be obtained, where h, see figure 2, represents the intersection of the external circular contour of the shaft given by equation:

$$x^2 + \left(y - \frac{D}{2}\right)^2 = \left(\frac{D}{2}\right)^2 \tag{20}$$

with the elliptical shape of the crack,

$$\left(\frac{x}{b}\right)^2 + \left(\frac{y}{a}\right)^2 = 1 \tag{21}$$

The intersection of both curves allows to obtain the expression for $\overline{h}$,

$$\overline{h} = \beta \left[\frac{\sqrt{\beta^4 - 4\alpha^2(\beta^2 - 1)} + 2\alpha^2(\beta^2 - 1) - \beta^2}{2\alpha^2(\beta^2 - 1)^2} \right]^{\frac{1}{2}} \tag{22}$$

In this last expression the relation between $\overline{h}$ and γ is a function of the slope ϕ, see figure 2, and is given by:

$$\gamma = \frac{1}{h}\cos\phi \tag{23}$$

So, the equation (19) can be written,

$$\overline{\lambda}_{mm} = \int_0^{\alpha_f} \int_{ar\cos\overline{h}}^{\frac{\pi}{2}} H_{I,M}(\alpha,\beta,\gamma)\,d\overline{s}d\alpha \tag{24}$$

where,

$$H_{I,M}(\alpha,\beta,\gamma) = \alpha^2 F_{I,M}^2(\alpha,\beta,\gamma) \left[\frac{1}{\beta^2}\sin^2\phi + \cos^2\phi \right]^{\frac{1}{2}} \tag{25}$$

The values of $\overline{\lambda}_{mm}$, that are shown in figure 3, have been obtained by the numerical integration of expression (24) using the very well known trapezium rule.

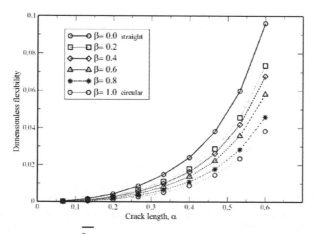

Fig 3 Variation of $\overline{\lambda}_{mm}$ as function of α for different values of β

The same methodology is applied for the calculation of the flexibility functions for axial loads, λ_{nn}, and for the combination of bending moment and axial loads, λ_{mn}. The obtained results are resumed in figures 4 and 5.

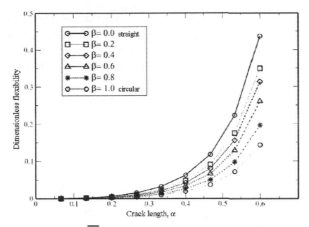

Fig 4 Variation of $\overline{\lambda}_{nn}$ as function of α for different values of β

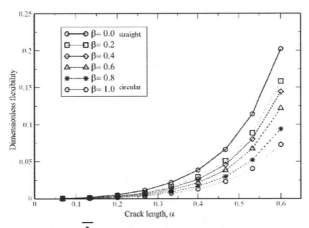

Fig 5 Variation of $\overline{\lambda}_{mn}$ as function of α for different values of β

As it was expected, and according to the literature results, the flexibility, in all the cases under consideration, increases with the crack length and with straight shapes of the crack front. Having the values of the flexibility for different shapes and sizes of cracks, a polynomial fitting for the flexibility coefficients can be done for the considered loads.

$$\overline{\lambda}_{mm} = \sum_{i=0}^{p}\sum_{j=0}^{q} C_{ij}^{mm}\alpha^i\beta^j \tag{26}$$

$$\overline{\lambda}_{nn} = \sum_{i=0}^{p}\sum_{j=0}^{q} C_{ij}^{nn}\alpha^i\beta^j \tag{27}$$

$$\overline{\lambda}_{mn} = \sum_{i=0}^{p}\sum_{j=0}^{q} C_{ij}^{mn}\alpha^i\beta^j \qquad (28)$$

where, C_{ij}^{mm}, C_{ij}^{nn} and C_{ij}^{mn} are the coefficients of the fittings for the nondimensional flexibility in bending, axial loads and the combination, respectively, and where p is the polynomial grade in α (length of the crack), and q is the polynomial grade in β (shape factor of the crack). In order to obtain the coefficients of the fittings, multiple regression technique has been used.

The fittings were done considering cracks of small length, up to 1/3 of the diameter. Those cracks are the most interesting for detecting and identifying. On the other hand, there have been taken into account shapes of the front crack varying from straight front to circular front. For those conditions, the best fittings were reached with grade 3 polynomials in α and with grade 7 in β for bending with a coefficient $R^2 = 0.99994$; with grade 4 in α and grade 6 in β for axial with $R^2 = 0.99981$; and with grade 4 in α and and grade 6 in β for the combination of bending and axial with $R^2 = 0.99976$. The coefficients of these fittings are shown in tables 1, 2 y 3.

Table 1: Fitting coefficients for $\overline{\lambda}_{mm}$

C_{ij}^{mm}	0	1	2	3	4	5	6	7
0	0	0.0012	-0.0085	0.02784	-0.0328	-0.0056	0.0350	-0.0171
1	0.0014	-0.0313	0.2073	-0.7672	1.2435	-0.6587	-0.2349	0.2397
2	0.0557	-0.0843	0.7227	-3.8517	11.3679	-18.9255	15.9576	-5.2378
3	0.2236	-1.4211	11.1389	-41.8703	82.4508	-86.3086	44.4635	-8.5512

Table 2: Fitting coefficients for $\overline{\lambda}_{nn}$

C_{ij}^{nn}	0	1	2	3	4	5	6
0	0.0000	-0.0058	0.0491	-0.1546	0.2517	-0.2152	0.07528
1	-0.0099	0.1900	-1.4878	4.4782	-6.8399	5.4485	-1.7944
2	0.2588	-2.3364	18.0395	-59.3744	98.7133	-83.1750	28.10945
3	-0.882	7.4109	-58.1128	190.8939	-315.5058	264.3339	-88.8941
4	3.1810	-16.5881	125.2311	-431.0101	743.2852	-638.7608	216.7398

Table 3: Fitting coefficients for $\overline{\lambda}_{mn}$

C_{ij}^{mn}	0	1	2	3	4	5	6
0	-0.0000	-0.0012	0.0039	0.0162	-0.0684	0.0794	-0.0299
1	-0.0032	0.0519	-0.2694	0.1116	0.9942	-1.5143	0.6287
2	0.1289	-0.8434	4.6697	-9.2186	5.3858	1.9437	-2.0465
3	-0.11975	1.69850	-9.34935	13.130223	6.5709	-23.1852	11.2768
4	1.0419	-5.5034	34.3777	-91.5844	116.052	-71.702	17.6543

In the same direction, the errors in the estimation of the flexibility have been calculated as:

$$\varepsilon = \frac{\overline{\lambda}_{fit} - \overline{\lambda}_{calc}}{\overline{\lambda}_{calc}} \qquad (29)$$

$\overline{\lambda}_{calc}$ being the values of flexibility obtained through the integration of the stress intensity factor, and $\overline{\lambda}_{fit}$ the values obtained from the polynomial expressions for the flexibility (equations 26, 27 and 28).

These errors are plotted in figures 6, 7 and 8.

From the analysis of the errors one can conclude that the fitting polynomials obtained estimate very well, with errors minor than 2% in almost all the cases, the flexibility for the three considered type of efforts for any crack front shape (from straight to circular). This is true for cracks with small lengths, that is, for incipient cracks whose determination and identification has become quite an interesting subject in engineering.

Lastly, a comparison between the values obtained with the flexibility function for bending proposed in this work, and the one proposed by Shih et al. [4] has been made and the results are shown in figure 9.

Considering the results given in figure 9 the conclusion is that the flexibility function proposed in this work improves very much other flexibility functions proposed before. In particular, the one by Shih [4], in which flexibility took negative values for most of the shape fronts and sizes of small cracks.

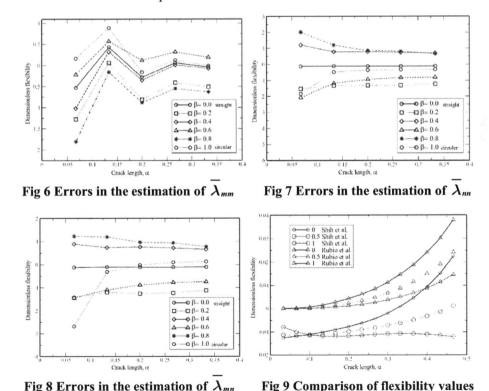

Fig 6 Errors in the estimation of $\overline{\lambda}_{mm}$ Fig 7 Errors in the estimation of $\overline{\lambda}_{nn}$

Fig 8 Errors in the estimation of $\overline{\lambda}_{mn}$ Fig 9 Comparison of flexibility values

4. CONCLUSIONS

In this work, polynomial expressions for the flexibility functions of a cracked shaft submitted to axial loads, bending moments and the combination of both, have been obtained. The main contribution is that, in this study, the transversal crack has been supposed to present an elliptical front, which is a more real configuration for a fatigue crack in a shaft.

To get the results, polynomial expressions of the stress intensity factor previously proposed in literature, have been used. Those expressions have been integrated numerically along the elliptical front characterized by the shape and length parameters of the crack. The results indicate that flexibility increases with straight fronts, so that the use of this model configuration for a cracked shaft is conservative. Translating this to the estimation of shaft life, the consideration of straight crack fronts is worse than the consideration of elliptical fronts.

The fitting of the flexibility functions from the integration flexibility values has been made by multiple regression techniques. The error analysis confirms that the polynomial expressions obtained for the flexibility coefficients give a good approximation for the cracked shaft flexibility.

Comparing the flexibility results obtained using the expressions proposed here and previous ones taken from literature, one can conclude that the present work improve the estimation of flexibility of cracked shafts presenting small cracks.

ACKNOWLEDGEMENT

The authors would like to thank the Spanish Ministerio de Educacion y Ciencia for the support for this work through the project DPI2006-09906.

REFERENCES

[1] R. Gasch, Dynamic behaviour of a simple rotor with a cross-sectional crack, *Vibrations in rotating machinery, ImechE Conference paper, C178/76*, 1976.

[2] A.D. Dimarogonas, C. Papadopoulos, Vibrations of cracked shafts in bending, *Journal of Sound and Vibration 91*, pp. 583–593, 1983.

[3] C.A. Papadopoulos, A.D. Dimarogonas, Coupled longitudinal and bending vibrations of a rotating shaft with an open crack, *Journal of Sound and Vibration 117 (1)*, pp. 81–93, 1987.

[4] Y-S. Shih, J-J. Chen, Analysis of fatigue crack growth on a cracked shaft, *International Journal of Fatigue 19*, pp. 477–485, 1997.

[5] C.S. Shin, C.Q. Cai, Experimental and finite element analyses on stress intensity factors of an elliptical surface crack in a circular shaft under tension and bending, *International Journal of Fracture 129*, pp. 239–264, 2004.

[6] T.M. Tharp, A finite element for edge-cracked beam columns, *International Journal for Numerical Methods in Engineering 24*, pp. 1941–1950, 1987.

[7]   A. Carpinteri, Stress intensity factors for straight-fronted edge cracks in round bars, *Engineering Fracture Mechanics 42*, pp. 1035–1940, 1992.

[8]   N. Álvarez, B. González, J.C. Matos, J. Toribio, Factores de intensidad de tensiones y propagación de fisuras por fatiga en geometrías cilíndricas, *Anales de mecánica de la fractura (23)*, pp. 333–338,

[9]   N. Couroneau, J. Royer, Simplified model for the fatigue growth analysis of surface cracks in round bars undeer mode I, *International Journal of Fatigue 20*, pp. 711–718, 1998.

[10]   M.A. Astiz, An incompatible singular elastic element of two- and three-dimensional crack problems, *International Journal of Fracture 31*, pp. 105–124, 1986.

[11]   Y-S. Shih, J-J. Chen, The stress intensity factor study of an elliptical cracked shaft, *Nuclear Engineering and Design 214*, pp. 137–145, 2002.

TURBOCHARGERS

Transient modal analysis of the nonlinear dynamics of a turbocharger

P Bonello
School of MACE, University of Manchester, UK

ABSTRACT

The prediction of nonlinear vibration phenomena introduced by high-speed oil film bearings is a major preoccupation of turbocharger designers. The purpose of this paper is to present the computational basis for the efficient solution of such a problem and to use the technique to study the self-excited phenomena in a typical turbocharger running on floating ring bearings. Both fully-floating and semi-floating rings are considered and the findings of the simulations indicate that the latter configuration may lead to significant reduction of self-excited vibration.

1.   INTRODUCTION

Floating ring bearings (FRBs) are a highly cost-effective means of supporting ultra-high speed, high temperature rotating machinery such as turbochargers. However, such bearings are well-known to be highly nonlinear elements that introduce self-excitation [1]. This results in vibration that contains sub-synchronous frequency components over the entire operating speed range, in addition to the usual synchronous frequency component excited by the rotor unbalance. Such oscillations are undesirable since they introduce noise and can be of large amplitude, potentially causing rotor-stator rub and affecting the fatigue integrity of the rotor if not properly accounted for at the design stage.

The prediction of the nonlinear vibration is a major preoccupation of turbocharger designers. Despite this fact, there exists relatively little published research on limit cycle analysis of realistic turbochargers models. For example, the research in [1] was restricted to a simple turbocharger model running on linearised FRBs. The analysis in [2] considered nonlinear behaviour of a turbocharger flexible rotor-bearing system but was highly restricted by the limited number of physical degrees of freedom that the model could accommodate. More recent analysis [3] considered a more sophisticated rotor model but was restricted in other aspects e.g. requiring the user to input an assumed ring spin speed to rotor speed ratio and the apparent lack of a facility to accommodate the dynamics of the support structure.

The aims of this paper are two-fold: (i) to present the computational basis of a program designed to overcome some of the restrictions of commercial rotordynamic software packages currently used to solve the nonlinear turbocharger problem; (ii) to present a study of the nonlinear dynamics of a turbocharger. The solution is achieved by time marching using a robust Matlab® solver with adaptive time-step control suitable for the numerically stiff problem encountered in limit cycle prediction. The program uses the undamped modes of the free-free rotor under non-rotational conditions, which are pre-computed separately to any desired accuracy by finite element or other techniques, without compromising the nonlinear routine's efficiency. The dynamics of the support structure can be included if its modal parameters are known. This latter is an important facility given that turbochargers are tested up to speeds of 180000 rpm (3 kHz). Gyroscopic effects are considered as "right-hand" excitations in the nonlinear computation, along with the bearing forces, unbalance forces and static loads. The rings of the bearings could be either fully floating (spinning and orbiting) or semi-floating (i.e. not spinning, but orbiting). In the former case, the instantaneous ring speed is calculated by the program, a facility that does not appear to be available in commercial software used in UK industry. The bearing model currently used in the program is the short bearing solution, with allowance for supply pressure (if oil supply grooves are used), and a choice of cavitation (oil film rupture) pressure. For turbocharger work, the cavitation model for the oil film has hitherto been restricted to the assumption of oil film rupture at atmospheric pressure (e.g. the short pi-film used in [2]) despite research indicating that oil film rupture at absolute zero pressure is more likely [4].

2.   THEORY

The following theory is with reference to Figure 1(a), where there are two FRBs, resulting in four oil films (OFs). It is noted that OF 1 and OF 3 are respectively the inner and outer films of FRB 1. OF 2 and OF 4 similarly belong to FRB 2. OF i ($i = 1...4$) will have an inner member ("journal") and outer member ("bore") with centres J_i, B_i respectively (see Figure 1(b)). For $i = 1, 2$: J_i is the shaft centre and B_i the ring centre at the FRB containing OF i; for $i = 3, 4$: J_i is the ring centre and B_i the bearing housing centre at the FRB containing OF i. Note that J_3 and B_1 define the same point, as do J_4 and B_2.

In line with previous works e.g. [1-3], the following theory neglects the inertias of the oil films.

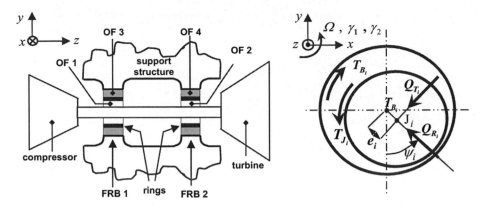

(a) Turbocharger schematic   **(b)** Forces and torques due to generic oil film

Figure 1: Turbocharger assembly with floating ring bearings (FRBs)

Let $\boldsymbol{\gamma} = \begin{bmatrix} \gamma_1 & \gamma_2 \end{bmatrix}^{\mathrm{T}}$ be the vector of instantaneous rotations of the rings. Let $\hat{\boldsymbol{\gamma}} = \begin{bmatrix} \hat{\gamma}_1 & \dots & \hat{\gamma}_4 \end{bmatrix}^{\mathrm{T}}$ where $\hat{\gamma}_i$ is the ring rotation associated with OF i. Hence

$$\hat{\boldsymbol{\gamma}} = \mathbf{A}\boldsymbol{\gamma} \, , \quad \mathbf{A} = \begin{bmatrix} 1 & 0 & 1 & 0 \\ 0 & 1 & 0 & 1 \end{bmatrix}^{\mathrm{T}} \tag{1}$$

The Cartesian fluid forces on the journal of OF i, $\mathbf{p}_i = \begin{bmatrix} Q_{x_i} & Q_{y_i} \end{bmatrix}^{\mathrm{T}}$, are functions of the rotor's rotational speed Ω and the vector $\boldsymbol{\lambda}_i = \begin{bmatrix} x_i & y_i & \dot{x}_i & \dot{y}_i & \dot{\hat{\gamma}}_i \end{bmatrix}^{\mathrm{T}}$ where x_i, y_i are the displacements of J_i relative to B_i. If e_i, ψ_i are the corresponding polar coordinates (Figure 1(b)), it follows from the Reynolds Equation [5] that the radial and tangential forces Q_{R_i}, Q_{T_i} of the oil film can be calculated by considering it as a pure (i.e. non-rotating) squeeze-film damper with the same e_i, $\dot{e}_i$ values but with a modified rate of change of attitude angle $\dot{\psi}_{eq_i}$ where:

$$\dot{\psi}_{eq_i} = \dot{\psi}_i - 0.5 \left(\dot{\hat{\gamma}}_i + \delta_i \Omega \right) \tag{2}$$

… δ_i being 1 for an inner film and 0 for an outer film. Hence,

$$\mathbf{p}_i \left(\Omega, \boldsymbol{\lambda}_i \right) = \begin{bmatrix} -\sin\psi_i & -\cos\psi_i \\ \cos\psi_i & -\sin\psi_i \end{bmatrix} \begin{bmatrix} Q_{R_i} \left(e_i, \dot{e}_i, \dot{\psi}_{eq_i} \right) \\ Q_{T_i} \left(e_i, \dot{e}_i, \dot{\psi}_{eq_i} \right) \end{bmatrix} = \begin{bmatrix} -\sin\psi_i & -\cos\psi_i \\ \cos\psi_i & -\sin\psi_i \end{bmatrix} \mathbf{v}_i \left(\boldsymbol{\sigma}_i \right) \tag{3}$$

where $\mathbf{v}_i = \begin{bmatrix} Q_{R_i} & Q_{T_i} \end{bmatrix}^{\mathrm{T}}$ is purely a function of $\boldsymbol{\sigma}_i = [e_i \quad \dot{e}_i \quad \dot{\psi}_{eq_i}]^{\mathrm{T}}$.

If T_{J_i} and T_{B_i} are the oil film torques on the journal and bore of OF i (Figure 1b) then [5]:

$$T_{J_i}, T_{B_i} = \pm 0.5 e_i Q_{T_i} - E_i \left(\Delta \Omega_i \right) \left(1 - e_i^2/c_i^2 \right)^{-0.5} \qquad (4a,b)$$

...where the "+" sign applies for T_{J_i} and the "-" sign applies for T_{B_i}. $E_i = 2\pi\mu_i R_i^3 L_i/c_i$ where μ_i, R_i, c_i, L_i are respectively the oil viscosity, bore radius, radial clearance and total effective oil film length. $\Delta\Omega_i$ is $\Omega - \dot{\gamma}_i$ for an inner film and $\dot{\gamma}_i$ for an outer film.

We define the "linear part" of the assembly as the system that remains in Figure 1 when the oil films are replaced by gaps. Let $\mathbf{q}$ be the $H \times 1$ vector of modal coordinates of the linear part and Λ the diagonal matrix of the squares of the natural frequencies. The equations of motion are written as:

$$\ddot{\mathbf{q}} + \Lambda\mathbf{q} = \mathbf{H}_u^T\mathbf{u} + \mathbf{H}_w^T\mathbf{w} + \mathbf{H}_g^T\mathbf{P}\mathbf{H}_\theta\dot{\mathbf{q}} + \sum_{i=1}^{4} \mathbf{H}_{P_i}^T \mathbf{p}_i \qquad (5a)$$

$$\ddot{\gamma} = \Gamma^{-1}\left\{ \begin{bmatrix} 0 & F \end{bmatrix}\mathbf{t}_J - \begin{bmatrix} F & 0 \end{bmatrix}\mathbf{t}_B \right\} \qquad (5b)$$

where:

$$\mathbf{H}_{P_i} = \mathbf{H}_{J_i} - \mathbf{H}_{B_i} = \begin{bmatrix} \boldsymbol{\psi}_{J_i}^{(1)} & \cdots & \boldsymbol{\psi}_{J_i}^{(H)} \end{bmatrix} - \begin{bmatrix} \boldsymbol{\psi}_{B_i}^{(1)} & \cdots & \boldsymbol{\psi}_{B_i}^{(H)} \end{bmatrix} \qquad (6)$$

$\boldsymbol{\psi}_{J_i}^{(r)}$, $r = 1 \ldots H$, is the mass-normalised 2×1 eigenvector whose rows respectively define the x and y displacements of J_i in mode no. r. Similarly for $\boldsymbol{\psi}_{B_i}^{(r)}$.

$\mathbf{H}_u$ and $\mathbf{H}_w$ are the matrices whose columns are the mass-normalised eigenvectors $\boldsymbol{\psi}_u^{(r)}$, $\boldsymbol{\psi}_w^{(r)}$ evaluated at the degrees of freedom corresponding to the directions and locations of the elements of the vector of unbalance forces $\mathbf{u}$ and vector of static loads $\mathbf{w}$.

The modes in eq. (5a) pertain to the linear part at zero rotor speed. The right-hand term $\mathbf{H}_g^T\mathbf{P}\mathbf{H}_\theta\dot{\mathbf{q}}$ accounts for the gyroscopic effect on the rotating nonlinear assembly. This effect is assumed to be concentrated at G points on the rotor. $\mathbf{P}$ is the diagonal matrix: $\mathbf{P} = \Omega \, \mathbf{diag}\{-I_1, I_1, \ldots, -I_G, I_G\}$, I_p being the polar moment of inertia at rotor location p. $\mathbf{H}_g$ and $\mathbf{H}_\theta$ are the matrices whose columns are the mass-normalised eigenvectors $\boldsymbol{\psi}_g^{(r)}$, $\boldsymbol{\psi}_\theta^{(r)}$ taken at the degrees of freedom corresponding to the directions and locations of the elements of the gyroscopic moment and flexural rotation vectors $\mathbf{g}$ and $\boldsymbol{\theta}$ defined in another paper [6].

The modes in eq. (5a) exclude the rigid body rotation of the rings, which is accounted for in eqs. (5b). In this equation, $\Gamma = \mathbf{diag}\{\Gamma_1, \Gamma_2\}$, where Γ_1, Γ_2 are the polar moments of inertia of the rings, $\mathbf{t_J} = \begin{bmatrix} T_{J_1} & \ldots & T_{J_4} \end{bmatrix}^T$, $\mathbf{t_B} = \begin{bmatrix} T_{B_1} & \ldots & T_{B_4} \end{bmatrix}^T$. The diagonal matrix $\mathbf{F}$ is used to control the rotational constraint on the individual rings. If both rings are free to spin: $\mathbf{F} = \mathbf{diag}\{1,1\}$, if the ring of FRB 2 only is free to spin: $\mathbf{F} = \mathbf{diag}\{0,1\}$, …etc.

The numerical time domain integration of eqs. (5a,b) by most available solvers (e.g. [7]) requires transformation into a system of $2H + 2$ first-order equations $\dot{\mathbf{s}} = \mathbf{h}(t,\mathbf{s})$ where $\mathbf{s} = \begin{bmatrix} \mathbf{q}^T & \dot{\mathbf{q}}^T & \dot{\gamma}^T \end{bmatrix}^T$ and $\mathbf{h}$ is a vector function in terms of $\mathbf{s}$ and (if rotational unbalance is considered) t. For given $\mathbf{s}$ the value of λ_i required to compute the forces of OF i is given by:

$$\lambda_i = \begin{bmatrix} \begin{bmatrix} x_{0_i} & y_{0_i} \end{bmatrix}^T \\ 0 \end{bmatrix} + \hat{\mathbf{H}}_{\mathbf{P}_i} \mathbf{s} \tag{7}$$

x_{0_i}, y_{0_i} are arbitrarily assigned x, y offsets and $\hat{\mathbf{H}}_{\mathbf{P}_i}$ is the block diagonal matrix:

$$\hat{\mathbf{H}}_{\mathbf{P}_i} = \mathbf{blkdiag}\{\mathbf{H}_{\mathbf{P}_i}, \mathbf{H}_{\mathbf{P}_i}, \mathbf{A}_i\} \tag{8}$$

… $\mathbf{A}_i$ being the ith row of matrix $\mathbf{A}$ in eq. (1).

The implementation of a robust implicit solver such as the Matlab® routine *ode23s*© [7] used in the present work requires the computation of the Jacobian matrix $\partial \mathbf{h}/\partial \mathbf{s}$ at each time step. The solution process is considerably accelerated if an expression for the Jacobian is supplied to the solver [7]. The full expression for the Jacobian is not presented here for reasons of space, but follows from eqs. (5). The key to rapid computation is the consideration that:

$$\frac{\partial}{\partial \mathbf{s}} \left\{ \sum_{i=1}^{4} \mathbf{H}_{\mathbf{P}_i}^T \mathbf{p}_i \right\} = \sum_{i=1}^{4} \mathbf{H}_{\mathbf{P}_i}^T \frac{\partial \mathbf{p}_i}{\partial \lambda_i} \hat{\mathbf{H}}_{\mathbf{P}_i} \tag{9}$$

where, from eq. (3):

$$\frac{\partial \mathbf{p}_i}{\partial \lambda_i} = \begin{bmatrix} -\sin\psi_i & -\cos\psi_i \\ \cos\psi_i & -\sin\psi_i \end{bmatrix} \frac{\partial \mathbf{v}_i}{\partial \sigma_i} \frac{\partial \sigma_i}{\partial \lambda_i} + \begin{bmatrix} -Q_{R_i} & -Q_{T_i} \\ -Q_{T_i} & Q_{R_i} \end{bmatrix} \begin{bmatrix} \cos\psi_i \\ -\sin\psi_i \end{bmatrix} \frac{\partial \psi_i}{\partial \lambda_i} \tag{10}$$

Since analytical expressions can be derived for the matrices $\partial \sigma_i/\partial \lambda_i$, $\partial \psi_i/\partial \lambda_i$, the only partial derivatives that require numerical evaluation are those in the matrices:

$$\frac{\partial \mathbf{v}_i}{\partial \sigma_i} = \begin{bmatrix} \partial Q_{R_i}/\partial e_i & \partial Q_{R_i}/\partial \dot{e}_i & \partial Q_{R_i}/\partial \dot{\psi}_{eq_i} \\ \partial Q_{T_i}/\partial e_i & \partial Q_{T_i}/\partial \dot{e}_i & \partial Q_{T_i}/\partial \dot{\psi}_{eq_i} \end{bmatrix} \tag{11}$$

Moreover, from eqs. (4a,b), the Jacobian term $\partial \mathbf{t}_{\mathbf{J,B}}/\partial \mathbf{s}$ is given by:

$$\frac{\partial t_{J,B}}{\partial s} = \left[\left\{ \left(\pm 0.5 Q_{T_1} \frac{\partial e_1}{\partial \sigma_1} \pm 0.5 e_1 \frac{\partial Q_{T_1}}{\partial \sigma_1} - E_1 \left(\Delta \Omega_1 \right) \frac{\partial}{\partial \sigma_1} \left(1 - \frac{e_1^2}{c_1^2} \right)^{-0.5} \right) \frac{\partial \sigma_1}{\partial \lambda_1} - E_1 \left(1 - \frac{e_1^2}{c_1^2} \right)^{-0.5} \frac{\partial \Delta \Omega_1}{\partial \lambda_1} \right\} \hat{H}_{P_1} \\ \vdots \right]$$

(12)

The only non-analytical partial derivatives in eq. (12) are the terms $\partial Q_{T_i} / \partial \sigma_i$ and these will have already been found since they are the lower row of the matrices defined by eq. (11).

3.   SIMULATIONS

The simulations presented are for a turbocharger with a simplified flexible rotor described in Figure 2 running on FRBs described in Table 1. The details in Figure 2 were based on an actual turbocharger rotor running on the same FRBs. The support structure was taken to be rigid.

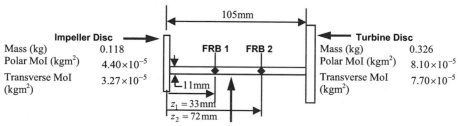

Figure 2 labels:

105mm

Impeller Disc
Mass (kg)	0.118
Polar MoI (kgm²)	4.40×10^{-5}
Transverse MoI (kgm²)	3.27×10^{-5}

FRB 1   FRB 2

Turbine Disc
Mass (kg)	0.326
Polar MoI (kgm²)	8.10×10^{-5}
Transverse MoI (kgm²)	7.70×10^{-5}

$-$11mm
$z_1 = 33$mm
$z_2 = 72$mm

Shaft of density 7860kg/m³, Young Modulus 200GN/m²

Figure 2: Details of rotor used for simulation

Table 1: FRB details for rotor in Figure 2

	viscosity (centipoise)		radial clearance (10^{-6}m)		bore radius (mm)		length (mm)		rings	
									mass (grams)	polar MoI (kgm²)
	inner	outer	inner	outer	inner	outer	inner	outer		
FRB 1	6.4	6.4	34	74	5.534	8.000	6.5	9.0	7.2	1.15×10^{-7}
FRB 2	4.9	4.9	34	74	5.534	8.000	6.5	9.0	7.2	1.15×10^{-7}

Prior to performing the nonlinear simulation, a modal analysis of the linear part of the assembly had to be performed. A total of 14 assembly modes were used for the nonlinear simulation. The modes $r = 1 \ldots 10$ defined the free-free rotor modes, alternately in the xz and yz planes. Of these, the first four were rigid body modes (two per plane) and the rest were flexural modes (three per plane). These modes included the effect of the distributed shaft inertia and shear deformation. Figure 3 compares the "exact" frequency response functions (FRFs) for the vibration of the free-free rotor in one plane computed by the dynamic stiffness technique [4] with approximate FRFs computed from a modal series truncated beyond the fifth mode. It is clear that the number of rotor modes used more than adequately described the rotor dynamics up 3 kHz. The remaining assembly modes $r = 11...14$ referred to the rigid body translation of the two rings in the xz and yz planes (one per plane per ring).

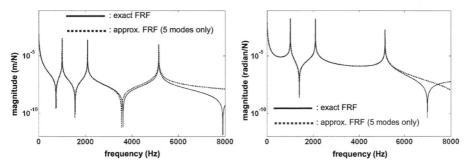

(a) Displacement per unit force at z_1 (Figure 2) (b) Rotation at impeller disc per unit force at z_1

Figure 3: Frequency response functions (FRFs) of free-free rotor vibration in yz or xz plane

The results of the above modal analysis were used to assemble the modal matrices Λ, $\mathbf{H_{P}}_i$, $\mathbf{H_w}$, $\mathbf{H_g}$, $\mathbf{H_\theta}$,...etc. in eqs. (5). It is to be noted that the use of a more refined rotor model and/or the inclusion of the support structure dynamics would merely involve the use of different and/or additional entries into these matrices, in accordance with the dynamics of the linear part.

The instantaneous pressure distribution in each oil film was represented by the pressure distribution of an equivalent short squeeze-film [4] with the modified rate of change of attitude angle $\dot{\psi}_{eq_i}$ given by eq. (2) to account for the shaft and ring rotations. Since the FRBs used had no oil distribution grooves, the supply pressure term was omitted from this pressure expression. Each distribution was truncated below a cavitation pressure of absolute zero ($-101.325\,\text{kPa}$) and numerically integrated to yield the associated FRB film forces.

Gyroscopic effects were included in the nonlinear simulations but rotor unbalance was not included since the aim was to investigate vibration that was purely of a self-excited nature.

The final conditions for the simulation at a given rotor speed were used as initial conditions for the subsequent speed. The results below relate to steady-state conditions, achieved by removing the first several hundred shaft revolutions of the vibration time history at each speed. The time step size was automatically chosen and varied by the solver to efficiently maintain a pre-set accuracy [7]: as an example, the step-sizes for the steady-state part of the vibration at speeds of 92,000 rpm and 180,000 rpm typically corresponded respectively to around 1/69 and 1/55 of a shaft revolution.

4.   DISCUSSION OF RESULTS

In the results presented (Figs. 4-8), "S", "R" and "H" respectively refer to the shaft centre, ring centre and housing centre at an FRB.   All results at an FRB are normalised with respect to c_{in}, c_{ou} and c_t where c_{in}, c_{ou} are the radial clearances of the inner and outer films respectively and $c_t = c_{in} + c_{ou}$.

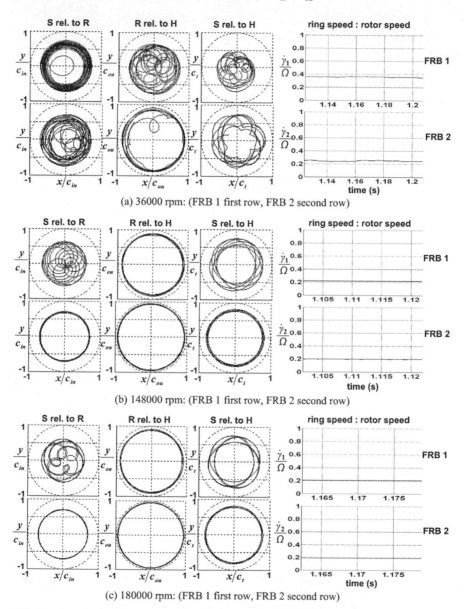

(a) 36000 rpm: (FRB 1 first row, FRB 2 second row)

(b) 148000 rpm: (FRB 1 first row, FRB 2 second row)

(c) 180000 rpm: (FRB 1 first row, FRB 2 second row)

Figure 4: Limit cycle orbits for FRBs with fully-floating rings over 50 shaft revs.

The results in Figures 4 and 5 show the predicted vibration for FRBs with fully floating rings. For a given speed, the spin speeds of the rings are virtually constant. At the lower speeds there is significant difference between the ring speeds of FRB 1 and FRB 2. As the rotational speed is increased both ring speeds equalise to around 20% of the shaft speed. Hence, the assumption in [3] of a ring speed to rotor speed ratio of 20% is consistent with these results. The waterfall diagrams of Figure 5 indicate that, overall there are three sub-synchronous frequency trains. There is a dominant train that reaches around 325 Hz at 3000 rev/s and a lesser one that is significant for FRB 1 and reaches 765 Hz at 3000 rev/s. There is also a very faint low-frequency train that appears to branch out from the dominant one at 1500 rev/s.

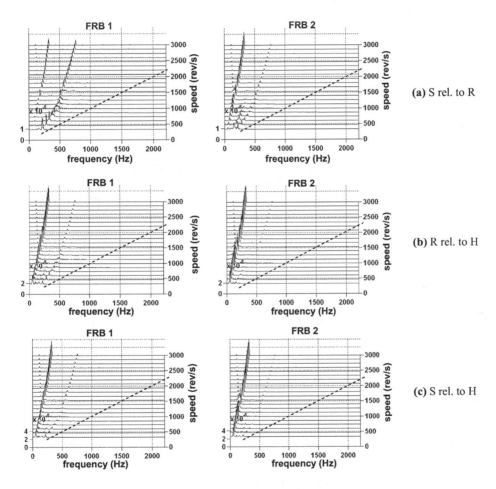

Figure 5: Waterfall diagrams of *y* vibration spectra for FRBs with fully floating rings (for reference purposes, dashed lines defining synchronous conditions have been included)

679

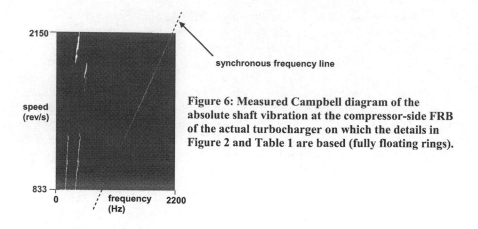

Figure 6: Measured Campbell diagram of the absolute shaft vibration at the compressor-side FRB of the actual turbocharger on which the details in Figure 2 and Table 1 are based (fully floating rings).

As in the previous works e.g. [1, 3], the direct comparison of the actual values of predicted and measured sub-synchronous frequencies is outside the scope of this preliminary study since it would require comprehensive experimental results and a more refined input into the program. However, the manufacturer's measurement in Figure 7 does provide evidence of strong sub-synchronous self-excited vibration, as predicted in the simulations. This diagram includes a relatively faint synchronous frequency train due to the inherent rotational unbalance. It is noted however that, unlike the simulations, the measurement shows a speed range for which the sub-synchronous frequencies are very faint. Other experimental waterfall diagrams [1] show uninterruptedly-strong sub-synchronous frequency trains over the entire speed range. It is also interesting to note that certain waterfall diagrams [1] exhibit one dominant sub-synchronous frequency train which then switches abruptly to a higher sub-synchronous train as the speed is increased. Such a feature is not apparent in either the simulations of the present study or the measurement in Figure 7.

Discussions with industrial specialists have revealed an interest in trying out FRBs with semi-floating (non-rotating) rings. Figures 7 and 8 show the predicted vibration for such FRBs. In this case, the outer film is acting as a pure squeeze-film damper. Hence, just like the unsupported squeeze-film damper used in aero-engine bearings [4], the outer film will require dynamic loading in order to support the gravity loading on the rotor and rings. In the absence of rotor unbalance (as in the present case), this dynamic loading is provided by the relative vibration between the shaft and ring caused by the self-excitation produced by the inner rotating film. In Figure 7 it is observed that there is a difference in mean position of the rings of the two bearings. This is due to the fact that the centre of gravity of the rotor is located merely 1.4 mm to the right of FRB 2 (see Figure 2(b)). The waterfall diagrams indicate that at the higher end of the rotor speed range the vibration is dominated by one or the other of two distinct trains of frequencies typically approaching 0.26EO ("Engine Order") and 0.35EO respectively. More importantly, comparison of Figures 7 and 8 with Figures 4 and 5 clearly indicate that, at least for the case studied, the transformation of the outer oil films into pure squeeze-films tended to significantly suppress limit cycle activity.

5.   CONCLUSIONS

The computational basis for the prediction of the nonlinear vibration of a turbocharger on floating ring bearings using a modal-based approach has been presented.   The technique was used to study the self-excited phenomena in a turbocharger with a simplified rotor based on an actual turbocharger rotor running on the same bearings.   For the case of fully floating bearings the simulations exhibited the extensive and strong sub-synchronous activity routinely observed in industrial measurements.   Other simulations revealed that the use of a rotational constraint on the rings may significantly reduce the self-excited vibration.   Current efforts are directed at re-running the program with a more refined structural input and the deployment of more advanced bearing models in the program.

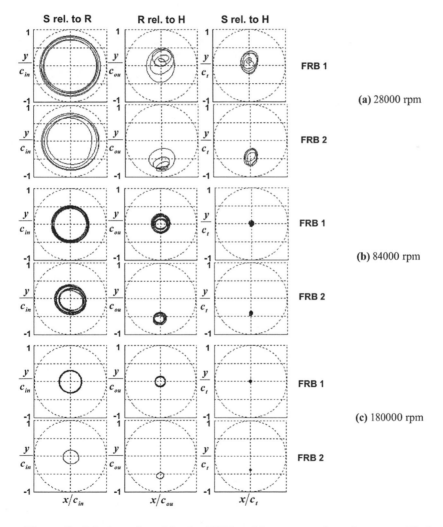

Figure 7: Limit cycle orbits for FRBs with non-rotating rings over 50 shaft revs.

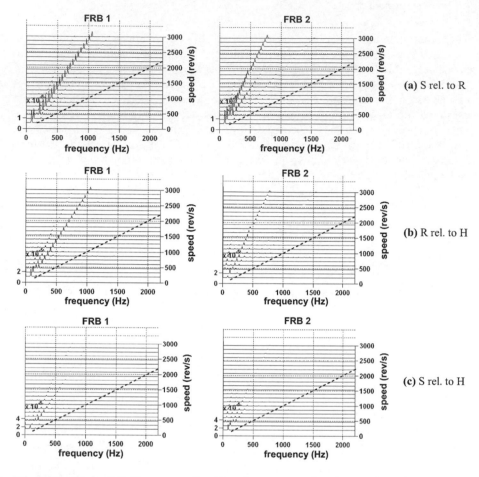

Figure 8: Waterfall diagrams of *y* vibration spectra for FRBs with non-rotating rings (for reference purposes, dashed lines defining synchronous conditions have been included)

ACKNOWLEDGEMENTS

The author would like to acknowledge Cummins Turbo Technologies Ltd for the provision of data.

REFERENCES

[1] R. Holmes, M. J. Brennan and B. Gottrand. "Vibration of an automotive turbocharger - a case study", *IMechE Conference Transactions, 8th International Conference on Vibrations in Rotating Machinery, Swansea, UK, 7-9 Sept. 2004, 445-455 (Professional Engineering Publishing)*.

[2] C. H. Li. "Dynamics of rotor bearing system supported by floating ring bearings", *Trans. ASME Journal of Lubrication Technology*, **104**, 469-477, 1982.

[3] E. J. Gunter and W. J. Chen. "Dynamic analysis of a turbocharger in floating bushing bearings", *ISCORMA-3, Cleveland, Ohio, 19-23 Sept. 2005*.

[4] P. Bonello, M. J. Brennan and R. Holmes. "Non-linear modelling of rotor dynamic systems with squeeze film dampers – an efficient integrated approach", *Journal of Sound and Vibration*, **249**(4), 743-773, 2002.

[5] A. Cameron. *Basic Lubrication Theory*. Longman, 1971.

[6] P. M. Hai and P. Bonello. "The nonlinear computational analysis of the vibration of a whole-engine model with squeeze-film bearings – time domain approach", *submitted to 9th International Conference on Vibrations in Rotating Machinery, Exeter, UK, Sept. 2008, manuscript C663/004/2008*.

[7] L. F. Shampine and M. W. Reichelt. "The Matlab ODE suite", *SIAM J. Sci. Comput.*, **18**(1), 1-22, 1997.

Experimental study of high speed turbocharger dynamic stability

R G Kirk
Virginia Polytechnic Institute and State University, USA

J C Nicholas
Rotating Machinery Technology, Inc., USA

ABSTRACT

The experimental documentation of high speed turbocharger instability for on-engine testing has been the focus of research at Virginia Tech for the past three years. Diesel engine turbochargers are known to have operation in the self-excited unstable region. This unstable operation with a nonlinear limit cycle has been tolerated on most applications to date. The need for quieter and smoother operation and in addition reduced emission levels has prompted new evaluations of the rotor bearing design for these systems. In previous research a commercial rotor-bearing dynamics computer program was used to evaluate the stability and transient response of a high speed diesel engine turbocharger. The results from unloaded and loaded engine testing have produced experimental results that are similar to the predicted analytical results. Testing of two different stock bearings and one alternate custom design has provided some insight into how difficult the total resolution of this problem will be. The summary results of the on-engine testing over the past year are documented in this paper.

1.   INTRODUCTION

Turbochargers are intended to increase the power of internal combustion engines. The first turbocharger was invented in the early twentieth century by the Swiss engineer Alfred Buchi who introduced a prototype to increase the power of a diesel engine. Turbocharging has now become standard for most diesel engines [1] and is also used in many gasoline engines. Engineers and other researchers are still searching for ways to improve turbocharger designs for better performance and lower manufacturing cost.

Since vibration-induced stresses and bearing performance are major failure factors, rotordynamic analysis should be an important part of the turbocharger design process. However, a thorough rotordynamic investigation was very difficult and relatively few detailed studies have been published.

Advances in rotordynamic analysis using modern computation techniques have made the dynamics of the turbocharger rotor-bearing system much easier and manufacturers can now use these tools to develop more dynamically stable turbochargers. Design improvements cannot depend on computational studies alone and on-engine test data are needed to validate the analytical predictions.

The research at Virginia Tech has studied the dynamic stability of a diesel engine turbocharger rotor-bearing system using both analysis and on-engine testing. A previous investigation [2] used a commercial finite element analysis (FEA) computer program [3] to model the dynamics of the turbocharger. That investigation demonstrated how linear and non-linear analysis can be beneficial for understanding the dynamic performance of the turbocharger system; the current paper summarizes a small portion of the experimental results obtained during on-engine testing over the past two years.

2.   BACKGROUND

A turbocharger consists basically of a compressor and a turbine coupled on a common shaft. The turbocharger increases the power output of an engine by compressing excess air into the engine cylinder, which increases the amount of oxygen available for combustion. Since the output of reciprocating internal combustion engines is limited by the oxygen intake, this increases engine power [4]. Since the turbine is driven using energy from the exhaust, turbocharging has little effect on engine efficiency. By contrast, a supercharger using power from the engine shaft to drive a compressor also increases power, but with an efficiency penalty.

An important factor in the design of an automotive turbocharger is the initial cost. The same power increase provided by the turbocharger can be provided by simply building a larger engine. Since engine weight is not a major part of overall weight for a diesel truck, the turbocharger is only competitive if it is less expensive than increasing engine size. For passenger cars the turbocharged diesel must compete with lighter and less expensive gasoline engines. To keep costs down while maintaining reliability, the designs of automotive turbochargers are usually as simple as possible. A common design assembly consists of a radial outflow compressor and a radial inflow turbine on a single shaft. Bearings are mounted inboard, with the compressor and turbine overhung as shown in Figure 1. The turbine rotor in most common automotive turbochargers is connected to its shaft by a friction or electron beam welding method. The compressor wheel, or impeller, is usually a clearance or very light interference fit on the other end of the shaft. A locknut is used to hold the impeller against a shoulder on the shaft. Friction from the interference fit and/or nut clamping pressure is generally sufficient to transmit the torque, so no splines or keys are needed. Figure 2 shows the elements of a typical automotive turbocharger.

Most automotive-size turbochargers incorporate floating bushing journal bearings. These bearings are designed for fully hydrodynamic lubrication at normal operating

Figure 1. A typical assembled turbocharger rotor

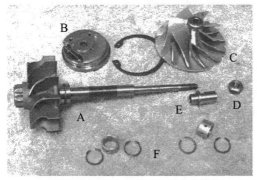

Figure 2. Common turbocharger components

A - turbine and shaft   B – thrust bearing assembly
C- compressor impeller   D – stock shaft nut
E – test nut and shaft probe target
F – floating ring bearing with retaining rings

speeds. For low cost and simple maintenance, turbochargers use the engine oil system for lubrication instead of having a separate system.

The primary consideration in the rotordynamic design of high-speed machinery is to control and minimize vibration. Large-amplitude vibration is undesirable in that it generates noise and can have large amplitudes that cause rotor-stator rub. In most rotating machinery, the dominant vibration is a forced response to rotor imbalance. There exists, however, another class of vibration termed rotordynamic instability or self-excited vibration. Vibration of this type requires a different design approach. Almost all rotors of automotive turbochargers exhibit both forced vibrations and self-excited vibrations [4, 5].

Forced vibrations from imbalance are harmonic and occur at the turbo shaft speed. They are generally driven by either mass eccentricity in the rotor or shaft bow. Mass eccentricity is a result of manufacturing tolerances, while shaft bow can be due to manufacturing tolerances or thermal effects. Unbalance vibrations can be minimized by designing the rotating element so that no natural frequencies are close to the desired operating speed range.

Self-excited vibrations usually occur at frequencies that are a fraction, rather than a multiple, of shaft speed. The sub-synchronous vibrations do not require a driving imbalance in the rotating element, but are due to the interaction between the inertia and elasticity of the rotating elements, the aerodynamic forces on the rotor and the hydrodynamic forces in the bearings.

Rotordynamic design of turbochargers has been based on both linear and nonlinear vibration analysis [6, 7, 8]. It was found that floating bushing bearings were more resistant to self-excited vibration than plain journal bearings, and these became widely used. Even with floating bushing bearings many turbochargers show high levels of sub-synchronous vibration [2, 6, 9].

In recent years developments in computational methods have made the analysis of self-excited vibration easier, faster, and more reliable [10-14]. Such analysis is becoming a fairly common part of the turbocharger rotordynamic design process. There has been only limited experimental data for verification of modeling results [6, 13, 15]. The goal of the work presented here is to provide additional experimental data from on-engine testing for comparison with future modeling results.

3. DESCRIPTION OF THE ENGINE TEST STAND

The test turbocharger was installed on a 3.9 liter 130 HP 4 cylinder diesel engine. Design and setup of the test stand was performed by Mechanical Engineering senior students as an undergraduate design project [16]. The engine was installed, using its stock mounts, on a heavy cast-iron test base. It was coupled to a chassis dynamometer with a flywheel adapter plate and a shaft floating between two universal joints. The purpose of the dynamometer was to control turbocharger speed by controlling engine speed and load, not to measure engine performance. Fuel, coolant, exhaust, and control connections were made as needed. Figure 3 shows the engine on the Virginia Tech IC Engine Laboratory test stand.

The basic engine instrumentation included engine rpm from an optical speed sensor monitoring the main shaft, coolant temperature gauge and a turbo boost gauge. Additional engine parameters can be monitored in future testing if desired. A second optical speed sensor and non-contact displacement probes were added to the turbocharger. In addition, a velocity sensor was mounted on the side of the engine to document the basic engine vibration. The use of available instrumentation was helpful but special order equipment was also required to monitor this high frequency and small diameter shafting. In addition, the lighting for the optical speed sensor target is critical for proper speed detection.

Turbocharger shaft deflection was measured by two special order eddy current proximity sensors measuring the horizontal and vertical displacement of the custom small diameter impeller shaft target nut.

Initially, data collection was limited to turbocharger speeds below 60 000 r/min, since that was the maximum data collection speed of the PC based data acquisition system available for the project. Initially the team was investigating ways to reduce the once per turn signal, but a commercial Keyphasor® conditioner was found that could give the desired speed signal reduction. Changing the Keyphasor® to $1/n$ caused the test speed monitor to respond to every nth timing mark, so that the actual maximum test speed was increased to 60 000 x n r/min.

Figure 3. Diesel engine on test stand at Virginia Tech showing compressor inlet with optical speed sensor

The instrumentation used for the project is further described in Table I. Future testing can make use of the available 16 channel data acquisition system when additional temperature and pressure sensors are added to the engine test stand.

Table I Instrumentation for experimental work

2 - 8 ch PC based data acquisition	2 – optical speed sensors
4 - non-contact shaft probes	2 – velocity seismic probes
2 – amplitude, phase and speed vector filters	1 – boost pressure gauge
2 – normal probe drivers	1 – Keyphasor® conditioner
2 - x-y Oscilloscopes	1 – temperature gauge
2 – special probe drivers	2 – power supplies
1 - 400 line FFT Analyzer	1 – engine dynamometer

4. RESULTS AND DISCUSSION

The test results have been documented using a standard commercial PC based data acquisition and reduction system that was available for the students use. The limitation on the reference shaft frequency was much lower than required for this high speed turbocharger. This made it necessary to use a Keyphasor® conditioner to allow the actual shaft speed to be reduced by a factor of three (3), thereby permitting a shaft speed of up to 180 000 r/min to be documented for spectrum content. All full speed and loaded test runs use a Keyphasor® factor of 1/3. For no load full speed, it was possible to use a factor of 1/2, since the top turbocharger speed is less than 120 000 r/min as shown below. The vibration probes are mounted to monitor the special target nut at the compressor inlet end of the turbocharger shaft.

4.1 Test Results for No Load Engine Full Speed Operation with Stock Bearings
A Keyphasor® conditioner was available to allow the altered speed signal for the data acquisition system and the engine was tested at full speed by the year one team, but still without the dynamometer connected, to see the maximum turbo speed for this condition. It was very interesting to see the turbo was able to go to near 100 000 r/min, without any load on the engine. This fact could be very useful for future proof tests of bearing stability, since the dynamometer is not required to achieve these high speeds. However, the engine operation at such high speed and no load is not generally a good idea for long duration operation. Figure 4 shows the waterfall plot for the first run to top engine speed and no load, which occurred on April 18, 2006. The turbo is noted to be operating at 1620 Hz (97 200 r/min) for the max point recorded. A new, higher frequency instability of 565 Hz (33 900 cpm) is noted to appear above a turbocharger speed of 1330 Hz (80 000 r/min). The lower instability at 300 Hz (18 000 cpm) seems to drop out at the same time the new one occurs. The reverse actions occur on deceleration. The vibration remained essentially forward whirl for all frequency components. The largest instability was noted to have a double amplitude of 4.35 mil-pp at a frequency of 590 Hz (35 400 cpm) when the turbo was at max speed of 162 Hz (97 200 r/min).

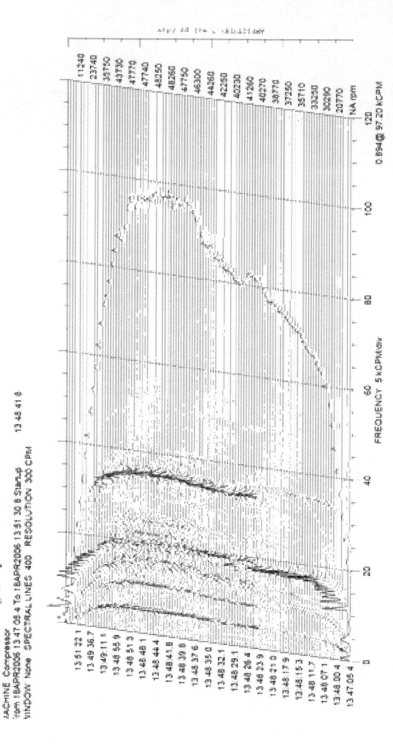

Figure 4. Waterfall of vertical probe for first full speed run of engine April 18, 2006. Turbocharger top speed of 162 Hz (97 200 r/min) and dominant instability at 580 Hz (35 400 cpm). Keyphasor® at ½ so turbocharger speeds are 50% of actual frequency.

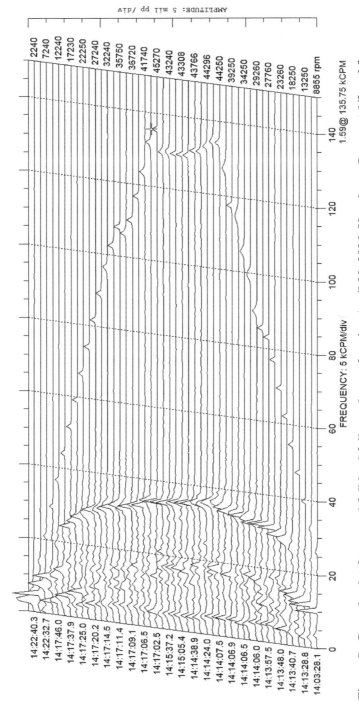

AMPLITUDE: 5 mil pp /div

FREQUENCY: 5 kCPM/div

1.59 @ 135.75 kCPM

Figure 5. Cascade plot for second full load full speed run of engine April 20, 2006, Keyphasor® now at 1/3 and frequency range adequate to show actual 1x speed when operation at full speed. Results here show the same higher dominant instability at 590 Hz (35 400 cpm).

4.2 Test Results for Full Load Engine Full Speed Operation with Original Stock Bearings

The engine was coupled to the dynamometer and on April 19, 2006 the first run to full speed and full load was documented using a Keyphasor® of 1/3, meaning every 3rd timing mark was counted. Hence, the actual speed is 3 times the documented turbocharger speed for each spectrum. Figure 5 shows the waterfall plot presentation where the lower instability at 250 Hz (15 000 cpm), drops out while the new, higher instability, now occurs at a turbocharger speed of 900 Hz (54 000 r/min) with an amplitude of 2.96 mil-pp and a frequency of 455 Hz (27 300 cpm). At the top turbocharger speed of 2280 Hz (137 000 r/min), the instability is 3.79 mil-pp at a frequency of 590 Hz (35 400 cpm).

4.3 Additional Test Results for Full Load Engine Full Speed Operation with Stock Bearings

The project team for year two repeated the stock bearing runs from the year one team, shown in Figs 4 and 5. With confirmed repeatability, the turbocharger was rebuilt several times with different stock bearings [17]. The testing covered the minimum and maximum tolerance for both stock bearing designs: the no-groove design without an outer groove and the grooved design that includes an outer groove. The results of these runs in the form of observed whirl frequency ratio, are presented in Figs 6 and 7. The mode two instability appears at shaft speeds from 30 000 r/min to 50 000 r/min and remains for the entire speed range for the minimum clearance no-groove design and the grooved maximum clearance designs. The second mode drops out at 120 000 r/min for the maximum clearance no-groove design and drops out at 90 000 r/min for the grooved minimum clearance design. Note that "max G" refers to the grooved design at maximum clearance, etc., while "NG" refers to no groove.

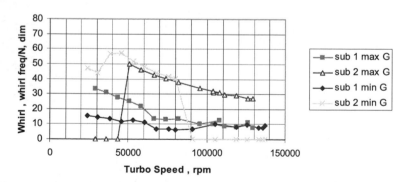

Stock With Outer Groove - % whirl comparison

Figure 6 Whirl frequency ratio for stock bearings with outer groove (G). Data from spring 2007 test runs. Note: 'sub 1' indicates mode 1 instability and 'sub 2' indicates mode 2 instability.

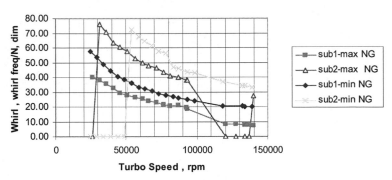

Figure 7 Whirl frequency ratio for stock bearings with no outer groove (NG). Data from spring 2007 test runs. Note: 'sub 1' indicates mode 1 instability and 'sub 2' indicates mode 2 instability.

4.4 Test Results for Full Load Engine Full Speed Operation with Custom Bearings

In spring 2007 a custom floating bush bearing having additional internal axial multi-groove design for the inner bearing surface was tested. The initial no load testing looked promising and the loaded test results were clear of the second mode to speeds just over 100 000 r/min (see Fig 8 with "MG" referring to the multi-groove design). This was the highest speed without the excitation of the second mode that has been observed in all the testing to date. This is the good news. The bad news was that when the instability did appear, the amplitudes were at the highest amplitudes recorded to date. The turbocharger was rebuilt with the original stock bearings and the results repeated the original builds, regarding vibration levels and instability onset speeds. The results of this testing has prompted a redesign of the custom bearing for testing by the year three project team.

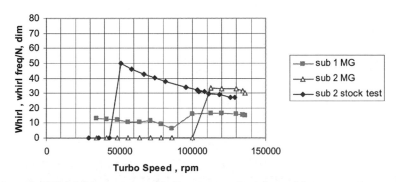

Figure 8 Whirl frequency ratio for custom bearing with outer groove and multiple inner grooves (MG) compared to the stock outer groove mode 2 result. Data from spring 2007 test runs. Note: 'sub 1' indicates mode 1 instability and 'sub 2' indicates mode 2 instability.

693

5. CONCLUSIONS

The testing to date has provided increased confidence that the on-engine testing can be used to evaluate different bearing designs by rebuilding the turbocharger utilizing a consistent procedure for comparison to previous runs. The three different bearing designs tested to date, and discussed in this paper, are shown in Fig 9. The no groove and grooved outer surface design have a circular bore, while the multi-groove design has both axial grooves in the inner bearing surface plus the circumferential outer land groove. The testing has shown varying levels of instability for changes in outer grooving and clearance conditions. The results of the custom design bearings are promising and a modified custom design is scheduled for testing in spring 2008. The results will be of great interest and will provide even further insight into the design modifications required to eliminate the higher frequency mode-two instability. This is the major goal of the continuing on-engine testing.

In addition, a bench test stand for off-engine test comparison of the custom bearings is currently under design by the current eleven member year three project team. This capability will give much needed information to better understand the excitation sources for the turbocharger rotor bearing system.

**Figure 9   Comparison of stock bearings a) without (NG) and b) with
outer groove (G) plus c) the custom bearing with outer groove
plus the multiple inner groove design (MG).**

6. ACKNOWLEDGMENTS

The authors would like to thank the Virginia Tech Rotordynamics Lab Industry Affiliates Group for their support of this research project. The project results would not have been possible without the additional help and support provided by Bently Nevada, LTD and Rotating Machinery Technology, Inc. Special thanks are also given to the 28 student members of the Mechanical Engineering Senior Capstone Turbocharger Design Projects for both 2005-06 [16] and 2006-07 [17].

694

7. REFERENCES

1. Watson, N. and Janota, M. S., *Turbocharging the Internal Combustion Engine*, Wiley, New York, 1982.
2. Alsaeed, A. A., "Dynamic Stability Evaluation of an Automotive Turbocharger Rotor-Bearing System," M.S. Thesis, Virginia Tech Libraries, Blacksburg, VA, 2005.
3. Gunter, E. J. and Chen, W. J., "Dynamic Analysis of a Turbocharger in Floating Bushing Bearings," *Proc. 3rd International Symposium on Stability Control of Rotating Machinery*, Cleveland, OH., 2005.
4. Ward D. et al., US Patent 6,709,160, "Turbo bearing lubrication system," 2004.
5. Choudhury Pranabesh De, "Rotordynamic stability Case Studies," *Internatiomal Journal of Rotating Machinery*, 2004, 203-211.
6. Holmes, R., Brennan, M. J. and Gottrand, B. ,"Vibration of an Automotive Turbocharger – A Case Study," *Proc. 8th International Conference on Vibrations in Rotating Machinery,* Swansea, UK, 2004, 445-450.
7. Li, C. H. and Rohde, S. M., "On the Steady State and Dynamic Performance Characteristics of Floating Ring Bearings," *Trans. ASME Journal of Lubrication Technology*, **103**, 1981, 389-397.
8. Shaw, M. C. and Nussdorfer, T. J., "An Analysis of the Full-Floating Journal Bearing," Report No. 866, National Advisory Committee for Aeronautics (NACA), 1947.
9. Tanaka, M., Hatakenaka, K. and Suzuki, K., "A Theoretical Analysis of Floating Bush Journal Bearing with Axial Oil Film Rupture Being Considered," *Trans. ASME Journal of Tribology*, **124**, 2002, 494-505.
10. Gunter, E. J. and Chen, W. J., *DyRoBeS© - Dynamics of Rotor Bearing Systems User's Manual*, RODYN Vibration Analysis, Inc., Charlottesville, VA, 2000.
11. Kirk, R.G., "Stability and Damped Critical Speeds: How to Calculate and Interpret the Results," *Compressed Air and Gas Institute Technical Digest*, **12**(2), 1980, 1-14.
12. Swanson, E., "Fixed-Geometry, Hydrodynamic Bearing with Enhanced Stability Characteristics," *STLE Tribology Transactions*, **48**(1), 2005, 82-92.
13. Born, H. R., "Analytical & Experimental Investigation of the Stability of the Rotor-Bearing System of New Small Turbocharger," *Proc. Gas Turbine Conference and Exhibition*, Anaheim, CA, 1987, 10.
14. Kirk, R.G., Alsaeed, A.A. and Gunter, E.J., 2007, "Stability Analysis of a High-Speed Automotive Turbocharger," *Tribology Transactions*, **50**(3), pp 427-434.
15. Andres, L. and Kerth, J.,"Thermal Effects on the Performance of Floating Ring Bearings for Turbochargers," *Proc. Instn Mech Engrs J Engineering Tribology*, **218**(J), 2004, 437-450.
16. Kirk R.G., Alsaeed A, Liptrap J, Lindsey C, Sutherland D, Dillon B, Saunders E, Chappell M, Nawshin S, Christian E, Ellis A, Mondschein B, Oliver J and Sterling J, "Experimental Test Results for Vibration of a High Speed Diesel Engine Turbocharger," Proceedings of 2007 Annual Meeting, Philadelphia, PA, May 7-10, 2007.

17. Kirk, R.G., J. Sterling, R. Utara, G. Biggins, D. Hodge, S. Johnson, J. Dean, B. Mastropieri, L. Fernandez, R. Rumeet, T. Miles, F. Chen, A. Johnson, H. Ko, A. Riggs, M. Price, and A. Cerni, "Diesel Engine Turbocharger Rebuild and Experimental Testing," ASME/STLE International Joint Tribology Conference, paper IJTC2007-44417, San Diego, CA, Oct 22-24, 2007.

An investigation of gyroscopic effects on the dynamic behaviour of a turbocharger

P Kamesh, M J Brennan
Institute of Sound and Vibration Research, University of Southampton, UK
R Holmes
School of Engineering Sciences, University of Southampton, UK

ABSTRACT

Automotive turbochargers exhibit strong sub-harmonic vibrations due to oil whirl instability, which are principally in the form of two modes: a conical mode and an in-phase whirl mode. These units operate at very high speeds, exceeding 180,000 rpm. The gyroscopic effect caused by the rotating motion and the polar moment of inertia of the rotating body can have an important dynamic effect in a rotor system. This paper describes an investigation into the effect of gyroscopic moment on the conical whirl mode instability of a turbocharger induced by tilting motion of the rotor. It is shown that for certain geometries the gyroscopic moment has the effect of suppressing the conical whirl mode instability.

NOTATION

a, a_{rs}, a_{sr} stiffness coefficient $\left(= \mu l^3 R\pi\omega/2C^3\right) = A\omega$

A $= \left(\mu l_b^{\,3} R\pi/2C^3\right)$

b, b_{rr}, b_{ss} damping coefficient $\left(= \mu l^3 R\pi/C^3\right) = 2A$

C bearing radial clearance

D differential operator $= \left(d/dt\right)$

F total external load per bearing,

F_x, F_y oil film forces along x and y axes

I transverse moment of inertia

J polar moment of inertia

j $\sqrt{-1}$

l distance between the bearing centres

l_b bearing length

m half symmetrical rotor mass

M_x, M_y moments about x and y axes

r, s co-ordinates of bearing centre

r_1, s_1 co-ordinates of journal centre

R journal radius

$\overline{s}$ $= D/\omega_k$

β $= J/I$

μ lubricant viscosity

θ, ϕ angular coordinates of the rotor centre of gravity

ω rotor speed, radians/sec

ω_k $= Al^2/I$

Ω $= \omega/\omega_k$

C_B bearing centre

C_{JS} static journal centre

C_{JD} dynamic journal centre

O stationary casing centre

1   INTRODUCTION

Although the gyroscopic effect is well known in rotor dynamics, its effect on the stability in rotor - journal bearing systems has not been fully investigated. The journal bearings, which support the vast majority of automotive turbocharger rotors, are of the oil-film variety. With all bearing designs, waterfall diagrams reveal strong sub-harmonic vibrations, which exist over most of the speed range, typically from 30,000 to 180,000 rpm. These are in addition to the more usual vibrations due to unbalance. These oil whirl frequencies are usually found to vary from about 10% to 50% of the rotor speed and cause a low frequency noise (rumble). Holmes [1] proposed a simple model for the symmetric mode of vibration of a turbocharger with floating ring and press-fit bearings using a rigid rotor. His model showed qualitative agreement with waterfall plots from commercial turbochargers. Holmes *et al.* [2] extended the investigation into an asymmetric model of a turbocharger with floating ring bearings. The investigation revealed two sub-synchronous frequencies corresponding to an in-phase whirl mode and a conical mode. The aim of this paper is to use a similar model to that developed in [2] to investigate the effect of the gyroscopic moment on the stability of the conical mode of a turbocharger rotor.

2   EQUATIONS OF MOTION

Figure 1 shows a schematic diagram of a turbocharger with a *symmetric* rotor and two identical bearings for *tilt* motion (For clarity, the discs representing the turbine and the compressor at each end of the shaft are not shown). Tilt motion only is considered here. The angular co-ordinates about the y and x axes are θ and ϕ

698

respectively, and l is the distance between the bearing centres. It is assumed that the rotor is rigid and symmetrical with transverse moment of inertia I and mounted in two identical bearings of $360°$ oil film (i.e. no cavitation or oil film rupture [3]) with rigid bearing housings as shown in **Figure 2**. For zero static eccentricity [3], the equations of motion for the symmetric mode of rotor *tilt motion* including gyroscopic moments are given by

$$I\ddot\theta - F_x l + J\omega\dot\phi = 0 \tag{1a}$$

$$I\ddot\phi - F_y l - J\omega\dot\theta = 0 \tag{1b}$$

where J is the polar moment of inertia of the rotor assembly, and F_x and F_y are the oil film forces given by [2]

$$F_x = 2A\dot{x}_1 + A\omega\dot{y}_1 \tag{2a}$$

$$F_y = 2A\dot{y}_1 - A\omega x_1 \tag{2b}$$

where $A = \mu l_b^3 R\pi/2C^3$, with μ_b being the viscosity of the oil, l_b the length of each bearing, R the journal radius, C the radial clearance in the bearings, and ω the rotational speed of the rotor assembly. Substituting for F_x and F_y from equations (2a,b), into equations (1a,b), and writing the linear co-ordinates x_1, y_1 in terms of the angular co-ordinates θ, ϕ for small displacements, gives

$$I\ddot\theta + Al^2\dot\theta + \frac{Al^2\omega}{2}\phi + J\omega\dot\phi = 0 \tag{3a}$$

$$I\ddot\phi + Al^2\dot\phi - \frac{Al^2\omega}{2}\theta - J\omega\dot\theta = 0 \tag{3b}$$

Dividing by $I\omega_k^2$, where $\omega_k = Al^2/I$, which is the speed (frequency) at which the inertial moment is equal to the damping moment if the stiffness and the gyroscopic moments are set to zero, and assuming that $\theta = \Theta e^{st}$ and $\phi = \Phi e^{st}$, equations (3a,b) can be written in matrix form as

$$\begin{bmatrix} \bar{s}^2 + \bar{s} & \Omega(1/2 + \beta\bar{s}) \\ -\Omega(1/2 + \beta\bar{s}) & \bar{s}^2 + \bar{s} \end{bmatrix} \begin{Bmatrix} \Theta \\ \Phi \end{Bmatrix} = 0 \tag{4}$$

where $\bar{s} = s/\omega_k$, $\Omega = \omega/\omega_k$ and $\beta = J/I$ is the ratio of polar to the diametral moment of inertia.

3   STABILITY ANALYSIS

In this section the effects of the gyroscopic moments on stability are investigated. For comparison the behaviour of the system without gyroscopic effects is first considered. The matrix equation (4) can be simplified by noting that the equations can be summed as $\theta + j\phi$, because θ and ϕ are 90° out of phase and the model is symmetric. The characteristic equation of tilt motion for $\beta = 0$ is thus

$$\bar{s}^2 + \bar{s} - j(\Omega/2) = 0 \tag{5}$$

which is a quadratic equation with three non-dimensional moments, $\bar{s}^2$- inertial moment, $\bar{s}$- damping moment and $-j(\Omega/2)$- cross-coupled stiffness moment. Solving equation (5) for $\bar{s}$ gives

$$\bar{s}_{1,2} = \left(-1 \pm \sqrt{1 + j2\Omega}\right)/2 \tag{6}$$

where $\bar{s}_1$ is an unstable root with positive real part and $\bar{s}_2$ is a stable root with negative real part. The way in which the real parts of these roots (which indicate the stability of the system) change with frequency can be seen in **Figure 3**. The overall behaviour of the internal moments over a speed range can be determined by substituting the unstable and stable roots from equation (6) into equation (5) and plotting the moments in the Argand plane for increasing speed as shown in **Figure 4**. The real and imaginary parts of the moments due to the unstable root are plotted in **Figure 4a**, and the corresponding moments for the stable root in **Figure 4b**.

It can be seen that for the unstable root, the damping and inertial moments have real and imaginary parts, while the cross-coupled stiffness moment is purely imaginary and negative. The imaginary parts of the inertia and damping moments combine to counter the cross-coupled stiffness moment, and the real parts of the damping and inertia moments counter each other. However, the real part of the damping moment is positive and it is this that makes the system unstable.

Conversely, moments from the stable root give damping with a negative real part and inertia with a positive real part as shown in **Figure 4b**. The cross-coupled stiffness moment is again purely imaginary and negative. Because the real part of the damping moment is negative, the system is stable.

To gain further physical insight into this behaviour it is interesting to examine the system in terms of the internal moments, at low and high speeds. For low speeds, when $\Omega \ll 1$ i.e. $\omega \ll \omega_k$, $\bar{s}_1 \approx -1 - (j\Omega/2)$ and $\bar{s}_2 \approx j\Omega/2$, where $\bar{s}_1$ is a stable root with negative real part and $\bar{s}_2$ is on the verge of instability, i.e. the real part is zero. For high speeds, when $\Omega \gg 1$ i.e. $\omega \gg \omega_k$, $\bar{s}_1 \approx \left(-1 - \sqrt{\Omega} - j\sqrt{\Omega}\right)/2$ and $\bar{s}_2 \approx \left(\sqrt{\Omega} + j\sqrt{\Omega}\right)/2$, where $\bar{s}_1$ is a stable root with negative real part and $\bar{s}_2$ is an unstable root with positive real part. The unstable root for low speed, $\bar{s}_2 \approx j\Omega/2$, is substituted into equation (5) to give

700

$$\underbrace{-\frac{\Omega^2}{4}}_{\text{inertial moment}} + \underbrace{\frac{j\Omega}{2}}_{\text{damping moment}} \underbrace{-\frac{j\Omega}{2}}_{\substack{\text{cross-coupled} \\ \text{stiffness moment}}} \approx 0 \qquad (7)$$

The inertial moment, $-\Omega^2/4$, is negligible compared to the other moments, since $\Omega \ll 1$. Hence, in this speed range, the system vibrates at a natural frequency, when the damping moment and the cross-coupled stiffness moments balance each other. Similarly for the high speed range, substituting the unstable root $\overline{s}_2 \approx \left(\sqrt{\Omega} + j\sqrt{\Omega}\right)/2$ gives

$$\underbrace{\frac{j\Omega}{2}}_{\text{inertia moment}} + \underbrace{\frac{\sqrt{\Omega}(1+j)}{2}}_{\text{damping moment}} \underbrace{-\frac{j\Omega}{2}}_{\substack{\text{cross-coupled} \\ \text{stiffness moment}}} \approx 0 \qquad (8)$$

The damping moment $\left(\sqrt{\Omega}(1+j)\right)/2$ is small compared to the other moments because $\Omega \gg 1$. Hence, in this speed range, the system vibrates at a natural frequency, when the inertia moment and the cross-coupled stiffness moments balance each other.

The characteristic equation of the system with gyroscopic effects is given by equation (5) with the term $-j\beta\Omega\overline{s}$ added to give

$$\overline{s}^2 + \overline{s} - j\beta\Omega\overline{s} - j\Omega/2 = 0 \qquad (9)$$

The solution to equation (9) is given by

$$\overline{s}_{1,2} = \frac{-1 + j\beta\Omega \pm \sqrt{\left(1 - j\beta\Omega\right)^2 + 2j\Omega}}{2} \qquad (10)$$

The interesting question is how the gyroscopic moment affects the internal moments of the system and hence the stability. The Argand diagram showing the internal moments is plotted in **Figures 5a** and **5b** for $\beta = 0.1$ and 0.25 respectively. By comparing **Figure 4a** and **Figures 5a,b** it can be seen that the gyroscopic moment has both real and imaginary parts, and has a tendency to reduce the positive real part of the damping moment and to increase the imaginary part; it also has a greater influence on the damping moment as β increases. It is notable that except for the cross-coupled stiffness moment, *all* the other moments are affected by the change of β.

Figures 6a and **b** show the Argand diagrams for the internal moments when $\beta = 0.5$ and 0.75 respectively. It can be seen that when $\beta = 0.5$ the real part of the damping moment is zero; it is imaginary and positive, and balances the cross-coupled stiffness moment. Similarly, the gyroscopic moment balances the inertia moment. Thus, when $\beta = 0.5$ the conical mode of vibration is stabilised. It can be seen that when $\beta > 0.5$ as in **Figure 6b**, then the real part of the damping moment is negative and hence the system is stable.

To obtain more physical insight, it is helpful to examine the asymptotic behaviour as before. For low speeds, when $\Omega \ll 1$ i.e. $\omega \ll \omega_k$, $\bar{s}_1 \approx -1 + j\beta\Omega - (j\Omega/2)$ is a stable root and $\bar{s}_2 \approx j\Omega/2$ is on the verge of instability (purely imaginary). For high speeds, when $\Omega \gg 1$ i.e. $\omega \gg \omega_k$, $\bar{s}_1 \approx -1/2$ is a purely real root and $\bar{s}_2 \approx \left(-1 + j\beta\Omega + \sqrt{-\beta^2\Omega^2}\right)\big/2 \approx j\beta\Omega$ is on the verge of instability. Substituting the root $\bar{s}_2 \approx j\Omega/2$ into equation (9) for low speeds, gives

$$\underbrace{-\frac{\Omega^2}{4}}_{\substack{\text{inertial} \\ \text{moment}}} + \underbrace{\frac{j\Omega}{2}}_{\substack{\text{damping} \\ \text{moment}}} + \underbrace{\frac{\beta\Omega^2}{2}}_{\substack{\text{gyroscopic} \\ \text{moment}}} \underbrace{-\frac{j\Omega}{2}}_{\substack{\text{cross-coupled} \\ \text{stiffness moment}}} \approx 0 \qquad (11)$$

Equation (11) shows that the damping moment and the cross-coupled moment balance each other, while the gyroscopic moment could balance with the inertial moment if $\beta = 1/2$. Now, substituting the root $\bar{s}_2 \approx j\beta\Omega$ into equation (9) for high speeds gives

$$\underbrace{-\beta^2\Omega^2}_{\substack{\text{inertial} \\ \text{moment}}} + \underbrace{j\beta\Omega}_{\substack{\text{damping} \\ \text{moment}}} + \underbrace{\beta^2\Omega^2}_{\substack{\text{gyroscopic} \\ \text{moment}}} \underbrace{-\frac{j\Omega}{2}}_{\substack{\text{cross-coupled} \\ \text{stiffness moment}}} \approx 0 \qquad (12)$$

where the inertial moment balances gyroscopic moment, while the damping moment counteracts the cross-coupled stiffness moment. Similar to the low speed range, here the damping moment perfectly balances the cross-coupled stiffness moment when $\beta = 1/2$. Now considering $\bar{s}_{1,2}$ from equation (10), it is clear that for $\beta = 1/2$, the roots factorise easily into

$$\bar{s} = \frac{-1 + j\Omega/2 \pm \sqrt{(1 + j\Omega/2)^2}}{2} \qquad (13)$$

giving $\bar{s}_1 = j\Omega/2$, $\bar{s}_2 = -1$ (purely real and purely imaginary root). Thus it can be seen that $\beta = 1/2$ is a threshold value where the unstable root with positive real part becomes purely imaginary. Physically this means that the turbocharger whirls in its perturbed position without any further growth of amplitude.

Finally, **Figure 7** shows the way in which the real part of the unstable root of the characteristic equation changes with speed. It is clear from this graph, that the conical mode is stabilised when $\beta = 1/2$ and that for values of $\beta \geq 1/2$ the system is stable.

4 CONCLUSIONS

This paper has described the way in which it is possible for the gyroscopic moment to stabilise the conical mode of an idealised model of a high speed turbocharger. In the model it is assumed that the system is symmetric, gravitational effects are ignored and there is a full oil film. The key parameter is the ratio of polar to transverse inertia of the rotor ($\beta = J/I$). It has been shown that the threshold value of β for which there is stability of the conical mode is $\beta = 1/2$. The effect of increasing β is to change the magnitudes and relative phases of the internal moments acting on the rotor system due to the oil filled bearings. The reason why it stabilises the system is that it changes the phase of the damping moment such that it is in anti-phase with the cross-coupled stiffness moment and thus the real part of the root to the characteristic equation is zero.

REFERENCES

1 Holmes, R. Turbocharger vibration – a case study. *Institution of Mechanical engineers Conference on Turbochargers and Turbocharging*, 2002, 91-100.
2 Holmes, R., Brennan, M.J. and Gottrand, B. Vibration of an automotive turbocharger – a case study. *Institution of Mechanical engineers Conference on Vibrations in Rotating Machinery*, 2004, 445 – 455.
3 Holmes, R. Oil whirl characteristics of a rigid rotor in 360° journal bearings. *Proceedings of the Institution of Mechanical Engineers*, 1963, 177(No.11), 291-299.

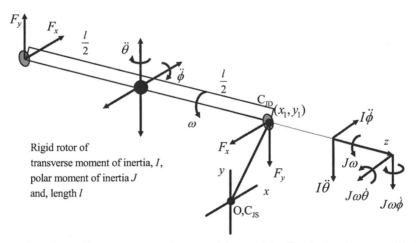

Figure 1. Co-ordinate system of symmetric model of a turbocharger with two identical journal bearings with rigid housings. O is the centre of the stationary housing, C_{JS} is the static journal centre, C_{JD} is the dynamic journal centre. θ and ϕ are the angular co-ordinates about y and x respectively. $J\omega$ is the angular momentum of the rotor about its spin axis z, where J is the polar moment of inertia of the rotor and ω is the rotor spin speed. l is the distance between the bearings.

703

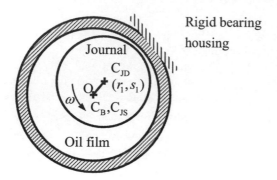

Figure 2 Schematic diagram of a turbocharger with journal bearing in rigid bearing housing C_B is the bearing centre, C_{JS} is the static journal centre, C_{JD} is the dynamic journal centre and O is the centre of the stationary casing.

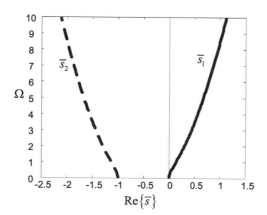

Figure 3 Plot of the relationship between non-dimensional speed and the real part of the roots of the characteristic equation of tilt motion of a turbocharger with journal bearings in rigid housings.

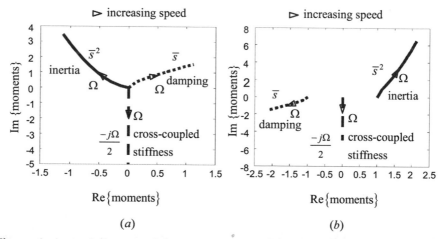

Figure 4 Argand diagram of the moments from (a) the unstable root and (b) the stable root of the characteristic equation of tilt motion of a turbocharger with journal bearings in rigid housings without gyroscopic effects. ($0 \leq \Omega \leq 10$)

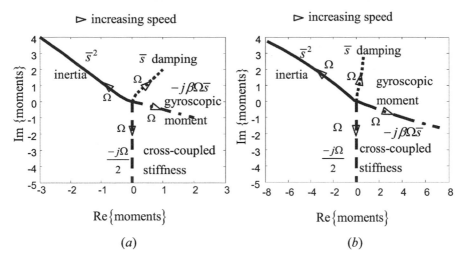

Figure 5 Real vs. imaginary parts of the moments from the unstable root of the equation of tilt motion of a turbocharger with journal bearings in rigid housings with gyroscopic effects, (*a*) $\beta = 0.1$ (*b*) $\beta = 0.25$ ($0 \le \Omega \le 10$)

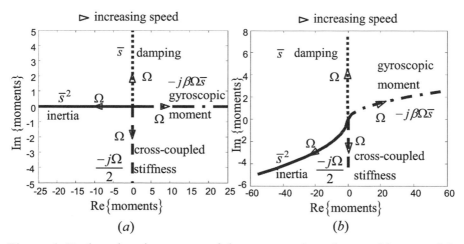

Figure 6 Real vs. imaginary parts of the moments from the unstable root of the equation of tilt motion of a turbocharger with journal bearings in rigid housings with gyroscopic effects, (*a*) $\beta = 0.5$ (*b*) $\beta = 0.75$ ($0 \le \Omega \le 10$)

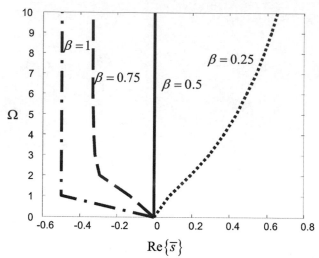

Figure 7 Real part of the sub-synchronous root of the equation of tilt motion of a turbocharger with journal bearings in rigid housings plotted against rotor spin speed for different values of β

DIAGNOSTICS

Bearing condition monitoring using feature generated by genetic programming based on Fisher criterion

H Guo, Q Zhang, A K Nandi
Signal Processing and Communications Group
Department of Electrical Engineering and Electronics
The University of Liverpool, UK

ABSTRACT

The issue of feature extraction and dimensionality reduction is addressed using Genetic Programming (GP) for bearing condition monitoring. A series of experiments are conducted using real-world bearing vibration data for a two-class problem and a six-class problem. A comparison is carried out between the GP-based approach and other popular feature extraction methods, including Kernel Generalized Discriminant Analysis (KGDA), Kernel Principal Component analysis (KPCA) and Genetic Algorithm (GA). In these experiments, the proposed GP-based method for the feature extraction typically achieves an improvement of 10% or more in average performance over the other methods reported here.

Keywords: **Bearing Condition Monitoring, Feature Extraction, Genetic Programming, Fisher Criterion**

1 INTRODUCTION

Feature extraction and dimensionality reduction is one of the most important tasks in pattern recognition problems. Improvements in classification accuracy can be achieved after carefully selecting the features. The reduction in the number of features can possibly avoid the implementation of sophisticated classifier, thus reducing the computational cost. The two major phases of feature extraction are: first, a large number of relevant features are extracted from raw data to a parameter vector X with m dimensions; second, feature vector Y with n dimensions ($n < m$) is extracted from the parameter vector X. The reduced feature vector Y should retain the discriminating information during the feature extraction process.

Principal Component Analysis (PCA) and Fisher Linear Discriminant Analysis (FLDA) are the most common techniques for linearly separable problems. These have low computation demand. Brunzell and Eviksson [1] have developed an extension of FLDA by minimizing a criterion function, namely an upper bound on the Bayes classification error. Aladjem [2] presents a method for the linear two-class problems by maximizing the Patrick-Fisher distance between the projected class-conditional densities. Chen and Yang [3] have proposed an alternative FLDA (AFLDA) with a new scatter measure for binary-class problems. Loog *et al.* [4] have introduced a weighted variant of the Fisher criterion for linear multiclass problems.

Kernel versions [5] of PCA and GDA have the ability to cope with nonlinearities. Cooke [6] presented two variations on FLDA for extracting nonlinear features for two-class problems. A generalized kernel discriminant analysis was developed by Baudat and Anouar [7] for multi-class problems. Mika *et al.* [8] proposed nonlinear generalization of Fisher's discriminant and oriented PCA using support kernel functions. Yang *et al.* [9] reformulated the multiclass problems into a two-phase process: kernel principal component analysis (KPCA) and LDA.

In recent years, machine learning and evolutionary computation techniques have been gaining increasing attention in the problems of feature extraction and dimensionality reduction. Methods include neural networks [10, 11], fuzzy systems [12], genetic algorithm (GA) [13-15] and genetic programming (GP) [16-19]. GA-based feature selection was carried out in [14] and [15] and proved to improve the classification accuracy considerably for real-world problems. GP was first introduced by Koza [20] and applied to multi-class pattern classification problems in [16]. In six medical diagnosis problems, it exhibited promising results over neural networks [19]. Zhang et al. applied GP for two-class normal/fault recognition in [21] and developed a method of Bundled-GPs for multi-class classification problems in machine condition monitoring [22]. Chien *et al.* have used GP to generate the discriminant functions using arithmetic operations with fuzzy attributes in [17] and applied Z-value measure as the objective function of GP in [18]. An effective discerning mechanism is required in [18]. However, the system needs c runs of GP for solving a c-class ($c > 2$) problem [16-19]. A method in [23] uses a step-wise learning process to design a multi-tree classifier consisting of c trees for c classes.

GP is explicitly used for the feature extraction in [24], but the system is unable to sample adequately the search space for high-dimensional problems and demands relatively high computation. Kotani et al. [25] used a hybrid of GP and a KNN classifier, as an effective tool for feature extraction. A GP-based feature generation method for machine condition monitoring was proposed in [26], where multiple features were generated and used as the inputs to ANN for the classification of bearing conditions. This current work is an extension of our previous research [26], with a reduction of required feature number to one. The motivation comes from industry applications, where a threshold-based classifier is so much more convenient.

2 GP-BASED FEATURE EXTRACTION MODEL

2.1. Fitness function

The fitness function plays a key role in the system performance. It should have the ability to rate the performance of each population member effectively and accurately. The fitness function used in this work is a tradeoff between accuracy and efficiency. There have been some successes [24, 25] in using the misclassification error as the fitness measure for multi-class problems. As these belong to a wrapper type approach, computational costs tend to be high.

In this work, a Fisher-criterion based fitness measure is designed for the task. It tests the rate of between-class scatter over the within-class scatter for two adjacent classes in the histogram, with the smallest rate indicating the separability of classes with the feature space.

For any two classes (i and j), the Fisher criterion is defined as, $t_{i,j} = \dfrac{|\mu_i - \mu_j|}{\sqrt{v_i + v_j}}$ (1)

where μ is the sample mean and v is the sample variance for class i or j. The value of $t_{i,j}$ provides an indication of the separability of class i and j in the feature space as larger value of $t_{i,j}$ is due to distant and less-overlapped classes and vice versa.

For a two-class problem, this measure can be used directly to the fitness function by assigning a threshold T for termination, $f^{(2)} = t_{1,2} / T$ (2)

When the fitness reaches one, the distribution of two classes are deemed as sufficiently separate thus the corresponding feature yields to be the evolution result.

The treatment of multi-class ($c > 2$) problems incorporates an individual-saturation mechanism. A significant increase of fitness between two classes will contribute to the overall fitness even if this improvement is not necessary or not beneficial as the separation between those two classes has been good enough ($t_{i,j} > T$). Therefore, a saturation condition is introduced in Equation (3),

$$f_i^{(c)} = \begin{cases} 1, & t_{i,j} > T \\ t_{i,j} / T & t_{i,j} \leq T \end{cases} \tag{3}$$

where i and j are sorted class label indices in terms of the sample mean of each class. The overall fitness value is the mean over all classes.

$$f^{(c)} = \frac{1}{c-1} \sum_i f_i^{(c)} \tag{4}$$

Apparently the computation cost of this design is less compared to standard approaches, which usually demands the order of N^2 or N^3 primitive mathematical operations (N is the sample number). The proposed fitness measure merely requires $2N$ sums and N multiplications.

2.2. Terminators and operators

Terminators act as the interface between GP and the experimental data by collecting relevant information from the raw data. Here, the full terminator set includes the original feature set and a series of numerical values randomly generated at the construction cycle of new individuals. These numerical values can be either integer or floating point numbers, both ranging from 1 to 100. The operator pool stores all the functions available for use in the nodal tree. Once assigned to the node tree, it performs a mathematical operation on one or more inputs from the sub-branch and reports the result to the parents. Table 1 lists the function names in the pool.

Table 1 - Function sets of GP

Symbol	No. of inputs	Description	Symbol	No. of inputs	Description
+, -	2	addition, subtraction	asin, acos	1	trigonometric functions
x, ÷	2	multiplication, division	tan, tanh	1	trigonometric functions
square, sqrt	1	square, square root	reciprocal, log	1	reciprocal, logarithm
sin, cos	1	trigonometric functions	abs, negator	1	absolute, change sign

2.3. Primitive operations

There are three primitive operations in GP to reproduce new individuals. 1) Crossover - GP carries out a crossover operation to create new individuals with a probability P_c. Two new individuals are generated by selecting compatible nodes randomly from each parent and swapping them, as illustrated in Fig. 1(a). 2) Mutation – This is performed by the creation of a subtree at a randomly selected node with the probability P_m. First, for a given parent, there is an index assigned to each node for identification. A random index number is generated to indicate the place where mutation will happen. The tree downstream from this node is deleted and a new subtree is generated from this node (see Fig. 1(b)), exactly in the same way as growing initial population. 3) Reproduction - This is performed by copying individuals to the next generation without any change with the probability P_r. These operations occur with equal probabilities through out the evolution,

$$P_{crossover} = P_{mutation} = P_{reproduction} = \frac{1}{3} \tag{5}$$

      (a) An example of crossover operation   (b) An example of mutation operation

Fig. 1 Examples of Primitive Operation

2.4. Tree model

As a tree model has been adopted in the GP paradigm, the mathematical operation carried out by each population member can be formulated as a polynomial expression consisting of the functions listed in the operator pool. For instance, a formula *TRoot = tanh(feature1) + feature2* corresponds to the tree shown in Fig. 2.

2.5. Termination criterion

The evolution is terminated by reaching either the maximum possible value of the fitness or the maximum number of generations.

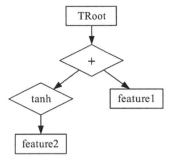

Fig. 2 Tree Model

3. FEATURE EXTRACTION AND CLASSIFICATION RESULTS

To provide a fair comparison to the conventional feature extraction methods, a series of experiments are conducted using real-world bearing vibration data.

3.1. Data description

A bearing vibration dataset [14, 21, 26] has been used for the evaluation. Six conditions of bearings can be grouped into normal and faulty, which constitutes a two-class problem. If the normal condition is sub-divided into brand-new (NO) and slightly worn (NW), and the faulty condition is sub-divided into Inner Race (IR) fault, Outer Race (OR) fault, Rolling Element (RE) fault and Cage (CA) fault, a six-class problem is formulated. The experimental data are obtained by running a roller bearing machine over a series of sixteen different speeds and taking ten examples of data at each speed. This gives a total of 160 examples of each condition, and a total of 960 raw data examples over six conditions to work with. The experimental data are further processed by taking a simple 32 point FFT of the raw vibration data for each of the two channels sampled. A 66 by 960 matrix forms the terminator set to the GP. For each given matrix in the experimental data set, a corresponding target matrix is to reflect the actual condition of the machine. All data are normalised prior to use.

3.2. Feature extraction results

The proposed GP-based feature extractor produces a single feature. Other feature extraction methods included in this study are KPCA and KGDA [27]. Figure 3 provides an illustrative demonstration of the different distributions of data by three feature extraction methods using the bearing data for six conditions monitoring. The solid line is the histogram of the feature generated by GP with population size 100, maximum tree depth 10 and threshold $T = 4$ for termination. The six clusters, which are well separated in the one dimensional feature space, belong to the six bearing conditions. It is remarkable that, by setting thresholds at these local minima, it does not require a sophisticated classifier to tell the class labels.

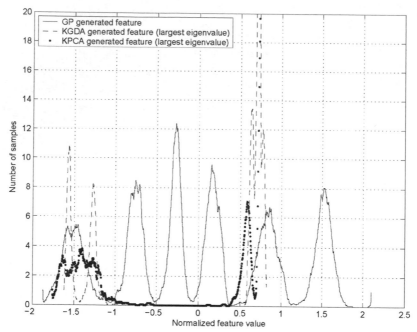

Fig. 3 Plots of feature values extracted by the KPCA, KGDA and GP-based approaches

The dashed line (KGDA histogram) has only three apparent clusters while the data from the KPCA feature is evenly distributed and no apparent cluster can be identified. However what we want to point out here is the promising attribute presented in the GP-extracted features.

3.3. Classification results

The feature extraction methods included in our comparison are, KPCA, KGDA and GA. Kernel function plays an important role in reducing the nonlinearity in the original feature space. In this work, a Gaussian kernel is used for KPCA approach, with the kernel parameter varying from 1 to σ_{max} ($\sigma_{max} = 20$). The converted features corresponding to the M ($M = 1, 2, ..., 6$) largest eigenvalues are chosen as the input to the classifier. Consequently, for a specific classifier, a total of $20 \times 6 = 120$ experiments are conducted and only the highest classification performance among the experiments is deemed as the representative of the KPCA feature extraction capability. The KGDA adopts the same scheme except that the $\sigma_{max} = 8$ and the maximum number of features in use is limited by the number of classes. Among the three classifiers, MLP requires some special attention as the number of neurons used in the hidden layer also varies within the range of 1 to N_{max} ($N_{max} = 14$). Hence the total number of experiments for each MLP application is $\sigma_{max} \times M_{max} \times Nmax$.

To further provide a fair comparison, three different classifiers are utilised for the classification task. The implementation of MDC and K-NN, in our case 1-NN, is rather straightforward [28]. The MLP is a double-layer neuron network with the

714

number of neurons for the hidden layer varying from 1 to 14 in order to provide different levels of nonlinearity handling capability. In all experiments, the data set is equally divided into three groups for classification training, validation and testing respectively. The test result is used to indicate the performance.

Table 2
The best classification performance for the two-class problem.

Solution	NOE	σmax	M_{max}	N_{max}	R	Best Perf.	Ave. Perf	NOF	Kernel	NON
Orig. Feat. + MDC	1	N/A	1	N/A	1	66.6%	N/A	66	N/A	N/A
Orig. Feat. +1-NN	1	N/A	1	N/A	1	99.2%	N/A	66	N/A	N/A
Orig. Feat. +MLP	700	N/A	1	14	50	97%	96.7%	66	N/A	9
KPCA+MDC	120	20	6	N/A	1	96.3%	79.1%	5	1	N/A
KPCA+1-NN	120	20	6	N/A	1	96.3%	87.8%	6	15	N/A
KPCA+MLP	84000	20	6	14	50	98.0%	88.0%	6	1	9
KGDA+MDC	8	8	1	N/A	1	99.2%	86.5%	1	1	N/A
KGDA+1-NN	8	8	1	N/A	1	99.5%	88.7%	1	1	N/A
KGDA+MLP	1120	8	1	14	50	99.8%	89.8%	1	1	14
GA+MLP[a]	N/A	N/A	N/A	N/A	N/A	100%	N/A	9	N/A	N/A
GP+MDC	50	N/A	1	N/A	50	100%	99.8%	1	N/A	N/A
GP+1-NN	50	N/A	1	N/A	50	100%	99.5%	1	N/A	N/A
GP+MLP	700	N/A	1	14	50	100%	99.9%	1	N/A	7

NOE=Total Number of Experiments = $\sigma_{max} \times M_{max} \times N_{max} \times R$, R=number of Run,
NOF = Number of Features, NON=Number of Neurons
[a] The result of GA/MLP is taken from [14]

3.3.1. Two-class problem
The experimental results are presented in Table 2 for the two-class problem. The results from the classifiers using the original feature set are given as a reference. Using the original 66 features, the 1-NN and MLP classifiers performs rather well, with more than 30% success than that of the MDC. Apparently, the simple structure of MDC cannot handle the complicated distribution in the 66 dimensions. The MLP requires N neurons in the hidden layer to achieve the 97% success, meaning that there exists considerable nonlinearity within the data, which is handled particularly well by the 1-NN classifier.

The performance of GA is rather good with a 100% success rate using 9 features selected during the evolutionary process. This data is quoted from [14], in which only MLP is tested for the classification task. The number of hidden neurons in use is 12. The GP-based approach achieves 100% success rate in all three classifiers, using a single feature. Overall, the GP-based approach outperforms other methods by providing high success rate independent of classifiers. This approach also reduces the computation demand in the classification block, an important attribute in industry applications providing simple and effective classification.

715

3.3.2. Six-class problem

Experimental results for the six-class problem are recorded in Table 3. The GP-extracted feature outperforms other features in both classification success rate and robustness. The improvement is significant as compared to the KPCA features. The 1-NN and MLP classifiers perform very well using original 66 features and the MDC produces poor results. The results using KPCA and KGDA are similar, with around 96% and 97% success rate respectively. The number of features required for this problem is also similar. Overall, GP produces the most satisfactory results, with all three classifiers offering high classification success.

Table 3
The best classification performance for the six-class problem.

Solution	NOE	σmax	M_{max}	N_{max}	R	Best Perf.	Ave. Perf	NOF	Kernel	NON
Orig. Feat. + MDC	1	N/A	1	N/A	1	57.1%	N/A	66	N/A	N/A
Orig. Feat. +1-NN	1	N/A	1	N/A	1	98.4%	N/A	66	N/A	N/A
Orig. Feat. +MLP	700	N/A	1	14	50	97%	87.2%	66	10	9
KPCA+MDC	120	20	6	N/A	1	93.0%	79.7%	2	10	N/A
KPCA+1-NN	120	20	6	N/A	1	96.6%	83.8%	6	14	N/A
KPCA+MLP	84000	20	6	14	50	96.3%	85.6%	8	N/A	7
KGDA+MDC	40	8	5	N/A	1	97.9%	80.6%	5	3	N/A
KGDA+1-NN	40	8	5	N/A	1	98.2%	86.8%	5	3	N/A
KGDA+MLP	28000	8	5	14	50	98.9%	90.8%	5	5	11
GA+MLP[a]	N/A	N/A	N/A	N/A	N/A	99.7%	98.6%	8	N/A	11
GP+MDC	50	N/A	1	N/A	50	99.7%	98.2%	1	N/A	N/A
GP+1-NN	50	N/A	1	N/A	50	99.9%	98.8%	1	N/A	N/A
GP+MLP	700	N/A	1	14	50	99.9%	98.7%	1	N/A	3

NOE=Total Number of Experiments = $\sigma_{max} \times M_{max} \times N_{max} \times R$, R=number of Run,
NOF = Number of Features, NON=Number of Neurons
[a] The result of GA/MLP is taken from [14]

4 CONCLUSIONS

A GP paradigm is proposed for the multi-class nonlinear feature extraction problem based on Fisher criterion. Under this framework, an effective feature extraction can be achieved without the explicit knowledge of probabilistic distribution of data. A number of experiments are conducted using a real-world bearing vibration dataset for a two-class problem and a six-class problem respectively. In all experiments, the GP-based solution exhibits superior optimization capability and produces extremely promising results. It performs either the best or equally the best among all methods. Typically it achieves an improvement of 10% or more in average performance; also robustness is another remarkable achievement as all three classifiers using the GP-generated features produce high classification success. We believe this attribute will have substantial benefit to industrial applications.

5 ACKNOWLEDGEMENTS

The authors would like to thank Weir Pumps Ltd., of Glasgow, Scotland, for the loan of the machine set used in this paper. H. Guo would like to acknowledge the financial support of the Overseas Research Studentship Committee, UK, the University of Liverpool and the University of Liverpool Graduates Association (HK).

REFERENCES

[1] H. Brunzell, J. Eriksson, Feature extraction for classification of multidimensional data, Pattern Recognition 33 (2000) 1741-1748.

[2] M. E. Aladjem, Nonparametric discriminant analysis via recursive optimization of Patrick-Fisher distance, IEEE Trans. Syst., Man and Cybern. Part B 28 (2) (1998) 292-299.

[3] S. Chen, X. Yang, Alternative linear discriminant classifier, Pattern Recognition 37 (2004) 1545-1547.

[4] M. Loog, R. P. W. Duin, R. Haeb-Umbach, Multiclass linear dimension reduction by weighted pairwise Fisher criteria, IEEE Trans. on Pattern Analysis and Machine Intelligence 23 (2001) 762-766.

[5] B. Scholkopf, A. Smola, K. R. Muller, Nonlinear component analysis as a kernel eigenvalus problem, Neural Computation (1998) 1299-1309.

[6] T. Cooke, Two variations on Fisher's linear discriminant for pattern recognition, IEEE Trans. on Pattern Analysis and Machine Intelligence 24 (2002) 268-273.

[7] G. Baudat, F. Anouar, Generalized discriminant analysis using a kernel approach, Neural Computation 10 (2000) 2385-2404.

[8] S. Mika, G. Ratsch, J. Weston, B. Scholkopf, A. Smola, K. Muller, Constructing descriptive and discriminative nonlinear features: Rayleigh coefficients in kernel feature spaces, IEEE Trans. on Pattern Analysis and Machine Intelligence 25 (5) (2003) 623-628.

[9] J. Yang, Z. Jin, J. Y. Yang, D. Zhang, A. F. Frangi, Essence of kernel Fisher discriminant: KPCA plus LDA, Pattern Recognition 37 (2004) 2097-2100.

[10] P. D. Heerman, N. Khazenie, Classification of multispectral remote sensing data using a back propogation neural network, IEEE Trans. Geosci. Remote Sensing 30 (1) (1992) 81-88.

[11] M. J. Chang, A. K. Jain, Artificial neural networks for feature extraction and multivariate data projection, IEEE Trans. on Neural Networks 5 (2) (1995) 296-317.

[12] K. P. Philip, E. L. Dove, D. D. McPherson, N. L. Gotteiner, W. Stanford, K. B. Chandran, The fuzzy Hough transform feature extraction in medical images, IEEE Trans. on Medical Imaging 13 (2) (1994) 235-240.

[13] J. Yang, V. Honvar, Feature subset selection using a genetic algorithm, IEEE Intelligent Systems 13 (1998) 44-49.

[14] L. B. Jack, A. K. Nandi, Genetic algorithms for feature selection in machine condition monitoring with vibration signals, IEE Proc. Vision, Image and signal Processing 147 (3) (2000) 205-212.

[15] M. L. Raymer, W. F. Punch, E. D. Goodman, L. A. Kuhn, A. K. Jain, Dimensionality reduction using genetic algorithms, IEEE Trans. on Evolutionary Computation 4 (2000) 164-171.

[16] J. K. Kishore, L. M. Patnaik, V. Mani, V. K. Arawal, Application of genetic programming for multicategory pattern classification, IEEE Trans. on Evolutionary Computation 4 (3) (2000) 242-258.

[17] B. C. Chien, J. Y. Lin, T. P. Hong, Learning discriminant functions with fuzzy attributes for classification using genetic programming, Expert Syst. with Applications 23 (2002) 31-37.

[18] B. C. Chien, J. Y. Lin, W. P. Yang, Learning effective classifiers with Z-value measure based on genetic programming, Pattern Recognition 37 (2004) 1957-1972.

[19] M. Brameier, W. Banzhaf, A comparison of linear genetic programming and neural networks in medical data mining, IEEE Trans. on Evolutionary Computation 5 (1) (2001) 17-26.

[20] J. R. Koza, Genetic Programming: On the Programming of Computers by Means of Natural Selection, MIT Press, Cambridge, 1992.

[21] L. Zhang, L. B. Jack, A. K. Nandi, Fault detection using genetic programming, Mechanical Systems Signal Processing 19 (2005) 271-289.

[22] L. Zhang, A. K. Nandi, Fault classification using genetic programming, Mechanical Systems Signal Processing 21 (2007) 1273-1284.

[23] D. P. Muni, N. R. Pal, J. Das, A novel approach to design classifiers using genetic programming, IEEE Trans. on Evolutionary Computation 8 (2004) 183-196.

[24] J. R. Sherrah, R. E. Bogner, A. Bouzerdoum, The evolutionary pre-processor: Automatic feature extraction for supervised classification using genetic programming, in: Proc. 2nd Int. Conf. Genetic Programming (GP-97), 1997, pp. 304-312.

[25] M. Kotani, S. Ozawa, M. Nasak, K.Akazawa, Emergence of feature extraction function using genetic programming, in: Knowledge-Based Intelligent Information Engineering Systems, Third International Conference, 1997, pp. 149-152.

[26] H. Guo, L. B. Jack, A. K. Nandi, Feature generation using genetic programming with application to fault classification, IEEE Trans. Syst., Man and Cybern. Part B 35 (1) (2005) 89-99.

[27] K. R. Muller, S. Mika, G. Ratsch, K. Tsuda, B. Scholkopf, An introduction to kernel-based learning algorithms, IEEE Trans. on Neural Networks 12 (2) (2001) 181-201.

[28] B. D. Ripley, Pattern Recognition and Neural networks, Cambridge University Press, Boston, 2004.

Fan diagnosis in the field

A El-Shafei
RITEC, Egypt

ABSTRACT

Fans are probably the simplest types of rotating machinery. However, the diagnosis of Fan problems in the field may require the use of elaborate techniques for the proper diagnosis of fan malfunctions. These techniques range from the use of spectral analysis to the use of operating deflection shape analysis. This paper presents a diagnosis procedure and several case studies for the field diagnosis of fans from the cement, fertilizer, and petrochemical industries. These cases include resonance, rotor unbalance, gearbox wear, flexible supports, and bearing faults. The cases illustrate the use of particular diagnosis technologies to identify fan malfunctions.

1 INTRODUCTION

Fans are used extensively in many industries, and in some cases a fan stoppage would cause a complete plant shutdown. This is the case with the kiln ID fan in the cement industry, or the boiler FD fan in the power industry. Therefore an accurate diagnosis of fan faults becomes imperative in these cases.

Fans are probably the simplest types of rotating machinery. However, even with this simplicity, the diagnosis of fan faults can be an elaborate exercise that may require the use of more advanced diagnosis tools. This paper introduces the diagnosis of fans as a step-by-step procedure describing the diagnosis process to guide the direction in performing the diagnosis and reaching an accurate and reliable diagnosis in a reasonable time frame to minimize production disruption. This step-by-step approach in a flow-chart format has been adopted by the author for a long time, and was published in conjunction with the diagnosis of installation faults [1]. More recently, this approach has been adopted by ISO TC108/SC2/WG10 in developing subsequent parts of the international standards ISO 13373 [2]. This step-by-step diagnosis procedure is illustrated by a number of case studies from different industries, where the diagnosis was performed according to the procedure outlined. These cases illustrate the effectiveness of the diagnosis procedure.

2 FAN DIAGNOSIS PROCEDURE

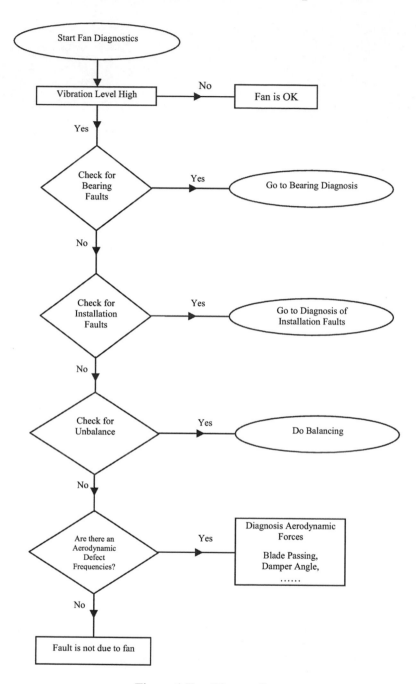

Figure 1 Fan Diagnostics

Figure 1 illustrates the fan diagnosis procedure. The most common faults in fans are bearing faults, installation faults, unbalance and aerodynamic excitation. The diagnosis procedure first asks if the vibration level in any band is high. If not, the fan is acceptable, otherwise check for bearing faults. The flow chart for bearing faults is not shown for brevity, but it includes check for the rolling element bearing frequencies: BPFO, BPFI, BSF, and FTF [3], as well as checking for bearing looseness or excessive bearing clearance. Also checks for faults in fluid film bearings are included. If the bearings are normal, then the user needs to check for installation faults. The flow chart for installation faults is illustrated in Figure 2.

Basically, it is suggested that before any testing of installed machinery be performed, a visual inspection of the machine and the site be completed. In many cases, the presence of skid looseness would be evident to the naked eye. Actually, it is suggested that all skid and anchor bolts be tightened before testing an installed machine.

Spectral analysis is the core of the diagnosis of rotating machinery. Spectral data are usually taken as velocity data, which should be measured on all bearings on the fan, in all three directions, horizontal, vertical and axial. The purpose of the spectral analysis is to identify the frequencies causing the machine to vibrate. If all vibration amplitude levels are within acceptable limits [4], then the machine would be accepted as normal. However, if any of the spectral components has a high amplitude, then spectral analysis is used to correlate the frequency of the high amplitude vibration to a machine frequency.

The result of the spectral analysis of the high amplitude vibration is one of three cases: a) at a known frequency, b) at an unknown frequency, or c) at the running speed. By a known (unknown) frequency it is meant that the reason of the presence of this frequency in the spectrum is known (unknown). The first case is the easiest to analyze. If the high amplitude vibration is at a known frequency, then the problem is correlated to that known frequency. For example, if 2x vibration is identified then this is usually correlated to misalignment. If decreasing harmonics of the running speed are present in the spectrum, then this spectrum shape is usually correlated with looseness in the bearings or the skid. If however, there were unknown frequencies in the spectrum, then additional testing would be required to determine the source of the unknown frequencies. Amongst the additional testing that may be required are: resonance testing (including impact (bump) test and critical speed test), modal testing, and flow characteristics testing. The purpose of the resonance testing is to correlate the unknown frequency to natural frequencies (stationary components) or critical speeds (rotating components) of the machine. Modal testing is a more advanced form of resonance testing, where all the modal characteristics of the machine are determined, including natural frequencies, damping ratios, and mode shapes. Modal testing is rarely used in the field, as it is an elaborate testing method, and is usually time consuming and costly. However, when justified, it can be a very powerful tool to obtain the machine characteristics and both identify clearly the unknown frequency in the spectrum, and suggest a solution to the problem.

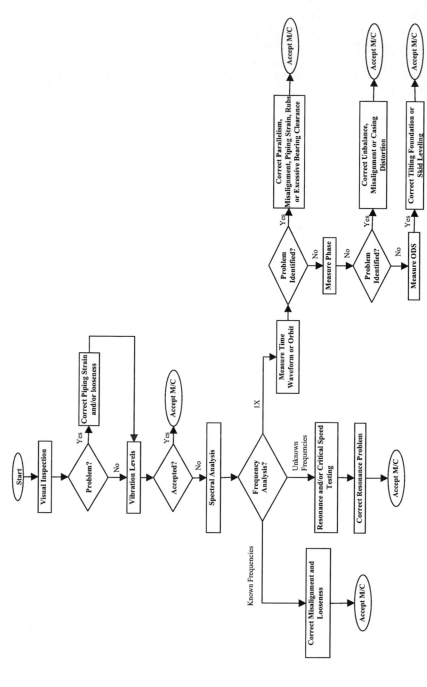

Figure 2 Flow Chart for Diagnosing Installation Faults

The most difficult case occurs when the spectral analysis reveals high 1x vibration. There are many faults, related to installation problems, that lead to high 1x vibration. Amongst these faults are unbalance, misalignment, casing distortion, tilted foundation, skid leveling, piping strain and excessive bearing clearance. In this case, special vibration measurements have to be conducted on the machine to describe the nature of this 1x vibration, and to distinguish between the different 1x faults. These measurements include: time waveform measurement, phase measurement, and measurement of the Operational Deflection Shape (ODS).

The time-waveform measurement can be used to distinguish between misalignment, piping strain and excessive bearing clearance. For piping strain, it will be quite clear that the forcing on the machine is directional, usually in the horizontal direction, and this directional force would be acting on the whole machine. Inappropriate bearing clearance also results in directional forces, however this would be localized at the bearing with the inappropriate clearance. This is particularly true for special geometry bearings, such as lemon bore or multi-lobe bearings.

The phase analysis is quite important to diagnose unbalance, misalignment, bent shaft and casing distortion. In many cases, misalignment (the main installation anomaly) manifests itself as vibration at 1x only. One of the best ways to distinguish between 1x vibration due to unbalance and 1x vibration due to misalignment is to measure phase across the coupling. If there is a 180° phase shift across the coupling, then the problem is misalignment [5]. If no phase shift occurs across the coupling, then the problem is unbalance. A bent shaft would produce a 180° phase shift in the axial direction across the machine (end-to-end), corrected for transducer orientation. Casing distortion can be easily identified by 180° phase shift across the machine (side –to-side or end-to-end) in the horizontal, vertical and/or axial directions. A cocked bearing can be identified by measuring phase around the bearing housing and noticing the phase shift due to the wobbly action of the cocked bearing [6]. In many cases a coupled time-waveform-phase analysis is quite useful in visualizing the vibration pattern and identifying the problem.

If the 1x vibration problem is still not solved after the timewaveform and phase analysis, then an ODS should be measured. The ODS is useful in identifying problems of tilted foundation, skid leveling, skid looseness, and shaft parallelity. In the ODS measurement, phase-referenced 1x vibration is measured at grid points on the machine structure or skid. This reveals the actual deflection shape of the machine under the operating load, and at the operating speed. Note that the ODS is not a mode shape of the machine or structure, unless the machine is in resonance, but it can be considered as a summation of the contribution of all of the modes of vibration. ODS analysis can be quite useful in identifying installation problems, as it provides a visualization of the actual vibration pattern of the machine and /or skid. In particular, if a machine skid exhibits a node in its ODS, then this is a clear indication of a tilted foundation or a leveling problem in the skid. Accurate measurement of skid and/or foundation levels would then be required to confirm the results of the ODS analysis.

The above steps should guide the user to diagnosing installation faults. If no installation fault is found then the next step according to the flow chart of Figure 1 is to determine if the fan is unbalanced. If the fan exhibits high 1x vibration with in-phase vibration at the coupling, then the fan needs cleaning and/or balancing. The final step in the step-by-step procedure shown in the flow chart of Figure 1 is to check for aerodynamic defect frequencies, usually the blade passing frequency. In this case the user needs either to correct for the controlling damper angle or check for appropriate axial clearances in the fan.

If the user follows the procedure in Figure 1, and no defect has been identified, then most probably the problem is not with fan, perhaps with the driving motor, and the user should check the motor diagnosis procedure.

The remaining sections of this paper describe cases where the diagnostic procedure of Figure 1 was used. These cases illustrate the application of the diagnostic procedure and clearly show the effectiveness of the step-by-step approach.

3 4000 RPM AIR BLOWER, FERTILIZER PLANT: SPECTRAL ANALYSIS

The main air blower in a sulfur plant was experiencing frequent motor bearing failure for the last two years. The 600 KW AC motor operating at 1500 rpm, was mounted on two rolling element bearings. The axial load was taken on the non-drive end angular contact bearing, while the drive-end bearing was the one experiencing the frequent failures. Most failures were in the form of a shearing cage. The motor was torn apart several times, without finding any cause for this problem.

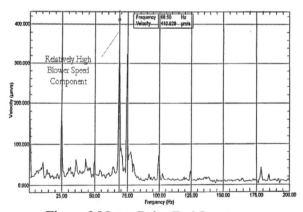

Figure 3 Motor Drive End Spectrum

On investigating this machine, the bearing failure rate was at a rate of once every two to three weeks. Spectral measurement on the machine confirmed that the motor had no malfunction, but the axial vibration is relatively high at the speed of the blower. Figure 3 shows the spectrum of the motor drive end bearing with high vibration at the speed of the blower. The motor was driving through a flexible coupling, a speed increaser gearbox mounted on journal bearings. The gearbox was driving the blower at 4000 rpm through a flexible coupling. The blower was overhung and mounted on two journal bearings. Survey of the blower and the gearbox showed a high axial component at both the gearbox and blower bearings at the speed of the blower. Figure 4 shows a typical gearbox spectrum. The gearbox showed a high gear mesh frequency of 1966 Hz and a high twice gear mesh frequency. Also, impacting was quite clear at a natural frequency of 2365 Hz with side bands. The highest vibration levels were at the vertical and axial directions of the blower bearings and were reading 5.2 mm/s.

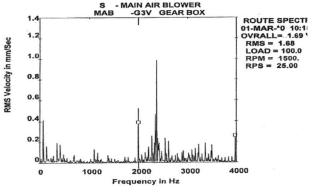

Figure 4 Gearbox Spectrum

The diagnosis of this sensitive machine was as follows: The blower is experiencing a couple unbalance producing an axial force that is providing an impacting on the gearbox and deteriorating the gear meshing. This axial force is then transmitted to the electric motor causing the bearing failure. The recommendation was to inspect the gearbox, correct any problems, and to balance the blower.

The gearbox was dismantled; the drive end journal bearing on the gearbox was completely destroyed. The axial thrust collar on the gearbox and on the blower bearings showed clear signs of impact. The blower bearings showed excessive clearance and rubbing on the rotor and bearings.

The gearbox was replaced, the blower bearings and shaft were replaced, and the motor bearings were replaced. The machine showed reduced overall vibration, but balancing was still recommended. The vibration vectors showed the need for a couple shot. A trial weight of 10 gm was used with this sensitive machine that was trim balanced with a 3 gm shot. The highest vibration level was reduced to 1.2 mm/s. No motor bearing failures have been reported since.

This case history illustrates the need to clearly identify the cause of vibration from spectral analysis. These problems should be corrected before attempting to balance the machine. The diagnostic procedure in Figure 1 was followed. No bearing frequencies were identified, however spectral analysis indicated deterioration of the gearbox, and a clear 1x indicated unbalance in the blower. 1x axial vibration due to unbalance occurs only in overhung machines. This axial vibration was transmitted to the motor, and its presence at the blower speed on the motor drive side was key to diagnosis of this machine train.

4 10.5 TON OVERHUNG CEMENT KILN ID FAN: BALANCING AND ODS

A Cement factory south of Cairo had a kiln fan that was experiencing high vibration. The fan has been installed some 12 years earlier, and had since experienced one catastrophic failure, and a major design modification to overcome its problems. In recent years, the fan has been experiencing high vibration resulting in the need to reduce its speed to reduce the vibration. This adversely affected production output.

On inspection of the machine it was found to be a huge 10.5 ton overhung fan on rolling element bearings with an impeller of diameter over 5 meters, and driven by a 1600 KW DC motor. The maximum speed is 490 rpm. Production required the machine to be operating at 95 % of its max. speed, while because of the high vibration the machine was operating at 82 % of its max. speed. The 4 meter high foundation showed visible cracks. Vibration measurements on the non-drive bearing showed a whopping 290 mm/s level, mainly at 1x.

Applying the diagnostic procedure of Figure 1, no bearing faults were found. Clearly the machine foundation required attention, however due to production requirements it was determined to balance the machine. This is unusual, and the author recommends that other problems should be resolved before balancing, but in this case the plant had requested a temporary solution, pending the resolution of the problem of the skid and foundation.

Before balancing it was requested to clean the fan blades, the vibration levels were reduced to 50 mm/s. A run-up test was performed to try to determine if any resonant conditions occurred and to mitigate any undesirable effects from the skid and foundation. Table 1 illustrates the results of the run-up test.

Table 1 Run-up Test on Kiln ID Fan

Speed rpm	Drive-end bearing		Non-Drive-end-bearing	
	Magnitude (mm/s)	Phase (deg.)	Magnitude (mm/s)	Phase (deg.)
284	6.74	-160	7.27	-159
309	3.36	-149	5.67	-144
355	3.52	173	2.72	172
380	3.74	162	5.2	155
415	3.73	140	10.9	135
425	5.8	126	16.5	125
440	10.6	112	31.8	111

From this Table it is clear that the machine is approaching resonance as its speed increases. A computer model was developed on RIMAP, a rotordynamics package [7], and it was determined that resonance occurs at the maximum speed of the machine. The mode shape is shown in Figure 5. Yet, it was decided to balance the machine. Inspection of the results of the run-up test show that the machine exhibits vibration that is in phase at both bearings at nearly all speeds, even though it exhibits a conical mode at resonance.

726

It was decided that the machine only needs a Static balance because of the in-phase measurement. A trial weight of 2 kg barely made the machine respond, but was used to obtain the balance sensitivity. The machine was finally balanced by a 6.1 kg weight, as shown in Figure 6, after removal of the trial weight. The vibration was reduced to 2.7 mm/s, and has been performing satisfactorily for over a year now. Production can now attain whatever speed they wish to load the kiln. Figure 7 shows the vibration spectra before and after balancing.

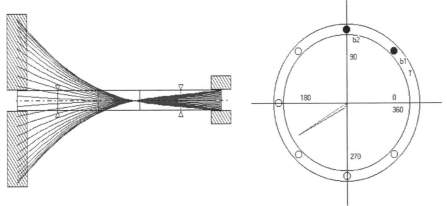

Figure 5 Mode Shape of the Fan.   **Figure 6 Static Balancing of the Fan**

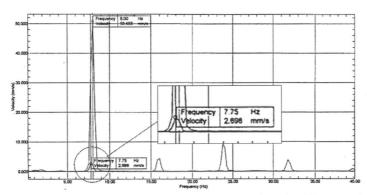

Figure 7 Spectra before and after balancing showing significant reduction

Upon successful balancing, it was required to investigate the skid resonance problem. An ODS of the skid was measured. Figure 8 shows a clear rocking motion for the skid in the vertical direction, while Fig. 9 shows a clear swaying in the skid in the horizontal direction. These measurements increase in amplitude as the machine speed increases and approaches resonance. Apparently the long period at which the fan was operating with severe unbalance resulted in the I-beams, embedded 0.8 m in the foundation and carrying the skid, separating from the concrete. A finite element model of the skid and foundation was developed and a suitable supporting structure was designed and implemented to prevent the rocking and swaying and to increase the natural frequency well beyond the operating speed of the fan.

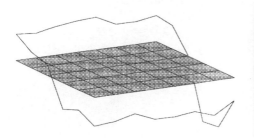
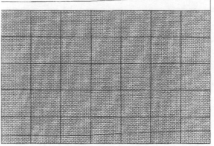

Figure 8 Vertical ODS Showing Rocking of Skid

Figure 9 Horizontal ODS Showing Swaying of the Skid

5 VIBRATION OF FAN SUPPORTING STRUCTURE: ODS ANALYSIS

This case study is from the petrochemical industry, and is for the supporting frame of a vertical air cooling fan supported on a horizontal steel frame with a horizontal motor and a bevel gearbox (see Figure 10). The unit is one of 12 units supported on an elevated steel frame, used to cool air for 3 Gas Turbines 9 MW each. The units were experiencing frequent shear failures of the shaft connecting the motor and the gearbox at the input shaft of the gearbox. The manufacturer suggested the use of a larger diameter shaft. However, this resulted in accelerated wear in the gearbox.

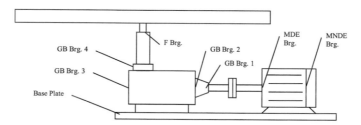

Figure 10 Schematic Layout of the Machine

Upon investigation of the problem, and applying the diagnostic procedure shown in Figure 1, a spectral analysis indicated that no bearing faults were present, and no unbalance, but the main exciting frequency is at 6x (blade passing frequency). However, investigating the operating conditions, it was clear that the blade angles were correct and the blade clearance was also set correctly. Therefore aerodynamic excitation was eliminated as a cause of the problem. A bump test was conducted, but was not conclusive because of transmitted vibration and it was not possible to shut down all coolers at the same time. An ODS was conducted at 6x, the exciting frequency, and the results are shown in Figures 11 and 12. It is seen quite clearly that the main deflection occurs in the middle of the frame, right where the repeated failure occurs, both in the horizontal and vertical directions. The reason of the failures is quite clear: the flexibility of the supporting structure at the gearbox input shaft position results in shaft failure. Increasing the shaft diameter only transmitted the problem to the gearbox, resulting in the accelerated wear. The root cause of the problem is the flexibility of the supporting structure. This supporting structure should be stiffened.

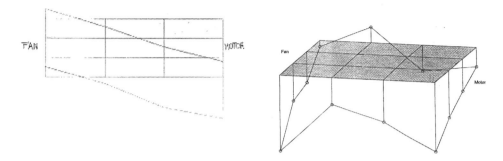

Figure 11 Base Plate ODS-Horizontal Direction   **Figure 12 Base Plate ODS-Vertical Direction**

6 BEARINGS PROBLEM

This is an unusual case. During acceptance testing of a cement plant, the vibration levels were relatively high for a new fan operating on rolling element bearings. The spectra are shown in Figures 13 and 14.

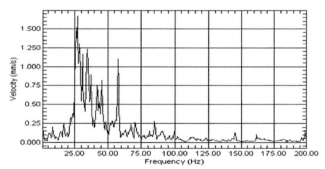

Figure 13 Drive End Bearing Spectrum

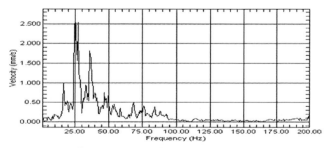

Figure 14 Non-Drive End Bearing

Applying the diagnosis procedure shown in Figure 1, no bearing frequencies were found, no installation faults were found, and the fan was well balanced. Also no vibration at the blade-pass frequency was seen. However, the fan bearings were noisy. The vibration pattern shown in Figures 13 and 14 is not usually associated with bearing faults. Although the spectrum is noisy with significant energy and has many frequencies, yet none of the frequencies matches with the bearing frequencies, namely BPFI, BPFO, BSF and FTF. However, discussion with plant personnel revealed that welding was performed near the fan. This immediately made us reconsider our diagnosis. The pattern shown in Figures 13 and 14 is an indication of electric discharge on the bearings, and occurs when welding is performed without proper electrical isolation. The fan was stopped and the bearings were removed, and clearly the discharge marks were evident. The fan was running smoothly, after the bearings change, allowing its acceptance by the customer.

7 CONCLUSIONS

A step-by-step diagnostic procedure for the diagnosis of fans was presented in this paper. The flow chart of Figure 1 is the basic tool for the diagnosis procedure. A user can follow the application of this procedure to diagnose fan problems. The case studies presented cover the full application of the diagnosis procedure, including installation faults, balancing and bearing faults. Various technologies are used including spectral analysis, phase analysis and ODS analysis.

REFERENCES

[1] El-Shafei, A., 2002, "Diagnosis of Installation Problems of Turbomachinery", presented at ASME Turbo Expo, June 2002, Amsterdam, ASME paper No. GT-2002-30432.
[2] ISO standard 13373, Condition Monitoring and Diagnostics of Machines - Vibration Condition Monitoring of Machines, International Organization for Standardization, Geneve, Switzerland.
[3] Eshleman, R. L., 1999, *Basic Machinery Vibrations*, VIP Press, Clarendon Hills, IL.
[4] ISO standard 10816-1, 1995, Mechanical Vibration-Evaluation of Machine Vibration by Measurements on Non-Rotating Parts, International Organization for Standardization, Geneve, Switzerland.
[5] Mitchell, J. S., 1993, *Introduction to Machinery Analysis and Monitoring*, 2nd edition, Penn Well Books, Tulsa, OK.
[6] Eshleman, R. L., 1998, *Machinery Vibration Analysis II Notes*, The Vibration Institute, Willowbrook, IL.
[7] RIMAP, vesion 1.3, RITEC, Cairo, Egypt.

DAMPING

Active bearing unit for damping of bending vibrations in rotating machinery

B Petermeier, H Springer
Dept. of Mechanical Engineering, Vienna University of Technology, Austria

ABSTRACT

In this paper a new actuation concept for active suppression of bending vibrations in rotating machinery is proposed. One or both passive rolling bearings of a rotor are complemented by particular active bearing units that contain four piezo actuators each. Numerical investigations are performed using a Finite Element model of a test cylinder system composed of beam- and lumped mass/stiffness elements. Simulation is carried out to calculate time series of the transient cylinder deflections. For control design the $\mathcal{H}_\infty$ technique is employed using μ-synthesis. Furthermore, for experimental prove a simple test rig system is used to practically demonstrate active damping. It is shown, that numerical simulations and experimental results coincide well and show considerable effectiveness in suppressing transient vibrations in rotating or non-rotating cylinders.

1 INTRODUCTION

In various mechanical applications structural vibrations are highly undesired and frequently faced by considerable effort. It is well known, that an increase of the damping capacity leads to a faster attenuation of the vibration amplitudes and a faster dissipation of the vibration energy. Beside the practically limited enhancement of material damping, there are several more or less established techniques available to enlarge structural damping. On the one hand, there are traditional passive damping methods, in principle based on the employment of viscoelastic dampers. On the other hand, active damping techniques are becoming more and more relevant. One of the most advanced system in rotating machinery is active magnetic bearing technique. Moreover, semi-active vibration damping technique has become an interesting alternative to purely passive or active damping methods.

Active vibration damping of rotating machine elements has already been extensively studied in the past. Many publications in literature propose active control devices for

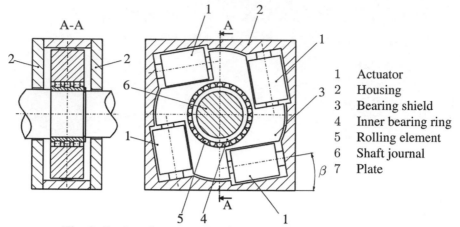

Fig. 1: Design elements of a piezoelectric active bearing unit.

1   Actuator
2   Housing
3   Bearing shield
4   Inner bearing ring
5   Rolling element
6   Shaft journal
7   Plate

hydrodynamic bearings using actuators of hydraulic, electric, pneumatic or different type, see [6], [8], [10]. The use of piezo actuators in active hydrodynamic bearings was proposed by Carmignani et al. in 2001 [3]. Furthermore, Palazzolo et al. and Barret et al. proposed an active vibration control using piezo actuators [9], [2]. Especially, several investigations have been carried out by using rolling bearings that are supported by piezo actuators. Recent investigations have been carried out by Ehmann et al. [5] and Alizadeh [1]. They obtained various results in testing different (robust) control techniques for both, collocated and non-collocated systems.

In 2005 Li [7] proposed a bearing design for active damping, containing three radial acting piezo actuators mounted in a flexible housing. He used robust control theory as well as adaptive control to suppress vibrations in any kind of rotational system.

The basic purpose of this work was to show the feasibility of active damping in a simple test rig system and to test various active bearing design possibilities. Numerous simulations and experiments have been carried out to develop an active bearing unit that could be an effective alternative to active magnetic bearings.

2   ACTIVE BEARING AND TEST RIG DESIGN

An active bearing unit that is functionally adequate, strongly depends on its design and the obedience of demanded specifications. Hence, the bearing design needs to combine advantages like low actuator load, low heat production and high lifetime with requirements like small size and high static stiffness. Fig. 1 shows one invented solution of an active bearing structure containing the cylinder journal 6 of the damped rotor. The intention of this design is to arrange the actuators 1 in a tangential order, to provide an evenly distributed actuator force to the rolling elements 5. As a consequence to the demanded space and stiffness specifications, the outer bearing ring is replaced by a bearing shield 3 including the contact surface on the inner diameter. The inner bearing ring and the rolling elements may still be the common parts of a conven-

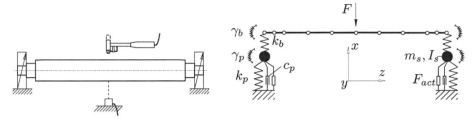

Fig. 2: Basic outline and the corresponding mechanical 26 dof-model of the test rig.

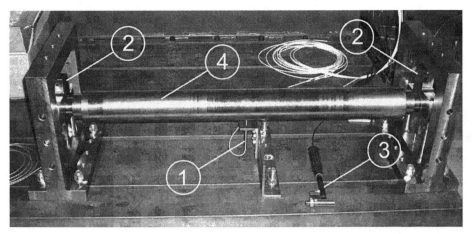

Fig. 3: Picture of the test rig. 1 measurement, 2 active/passive bearings, 3 impact hammer, 4 cylinder.

tional rolling bearing. The actuators are internally prestressed and are supported by the housing body 2. Each parallel pair of actuators are of the same type and generates the necessary damping force calculated by the controller in its specific orthogonal direction. That means that one actuator of a pair generates a compression force and the second one generates a tension force of the same amount. In principle, the resulting damping force is directed in opposition to the current velocity vector of the journal. The design angle β can be customized according to a distinctive vibration orientation. For a more precise description of the active bearing unit, further design possibilities and optional actuator protection modifications, see the corresponding SKF documentation.

The main requirements for the test rig design were the simplicity and flexibility with respect to several active bearing design possibilities. Hence, the experimental setup for verification and evaluation of individual active bearing designs was intentionally kept as simple as possible. Fig. 2 shows a schematic drawing of the test rig and its corresponding mechanical model. Discretization of the cylinder is carried out by 10 finite elements of Hermitian type. The model consists of 13 nodes in the $x - z$ plane,

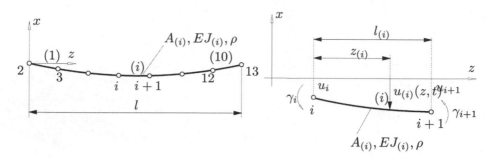

Fig. 4: Planar beam model of the cylinder consisting of 10 finite elements.

possessing two degrees of freedom each: one translational along the x-axis and one rotational about the y-axis. The actuators are represented by a piezo stiffness k_p, a passive damping factor c_p and an active actuator force F_{act} that are connected by a tilting stiffness γ_p to the bearing shield. The bearing shield is modelled by a mass m_s with mass moment of inertia I_s and is connected to the (tilting) stiffness of the rolling bearing k_b and γ_b, respectively. The impact hammer applies a disturbance force F to the cylinder and consequently induces bending vibrations in the system.

3 ANALYTICAL STUDY AND NUMERICAL SIMULATION

Fig. 4 shows the cylinder that is represented by a beam which performs planar vibrations in the x-z plane. The beam is assembled by 10 finite bending elements with 11 nodal points $i = 2, 3, \ldots, 13$. The adjacent picture shows a finite beam element (i) of length $l_{(i)}$, constant cross section $A_{(i)}$ and constant bending stiffness $EJ_{(i)}$. The mass density of the beam material is denoted by ρ. The lateral displacement field $u_{(i)}(z, t)$ of a beam element (i) is approximated by Rayleigh-Ritz shape functions in terms of the nodal point displacements u_i, u_{i+1} and the nodal shape angles γ_i, γ_{i+1} in the form

$$u_{(i)}(\xi, t) = [N_1(\xi), N_2(\xi), N_3(\xi), N_4(\xi)] \begin{Bmatrix} u_i \\ \gamma_i \\ u_{i+1} \\ \gamma_{i+1} \end{Bmatrix}_{(i)}$$

$$= \mathbf{N}(\xi)\mathbf{q}_{(i)}, \tag{1}$$

where $\xi = z/l_{(i)}$ and $\mathbf{N}(\xi)$ represents Hermitian polynomials of third order. The axial displacement $w_{(i)}(z, t)$ of the beam element is neglected as well as torsional effects. The equations of motion can be derived from d'Alembert's principle

$$\delta W_I + \delta W_C + \delta W_D + \delta W_A = 0, \tag{2}$$

where δW_I, δW_C, δW_N, and δW_A, respectively, represent the virtual works of inertia forces, conservative and non-conservative forces and other applied forces in the system. Employing the virtual works into Eq. (2) the equations of motion finally lead to

$$\mathbf{M}\ddot{\mathbf{q}} + \tau_0 \mathbf{K}_S \dot{\mathbf{q}} + \mathbf{K}_S \mathbf{q} = \mathbf{f}, \tag{3}$$

where $\mathbf{M}$ and $\mathbf{K}_S$ are the symmetric global mass and stiffness matrices of dimension (26×26), $\mathbf{f}$ is the global excitation vector of dimension (26×1) and

$$\mathbf{q} = \{u_1, \gamma_1, \ldots, u_i, \gamma_i, \ldots, u_{13}, \gamma_{13}\}^T \tag{4}$$

is the global beam displacement vector containing lateral displacements u_i and slope angles γ_i at the finite element nodal points. From structural dynamics it is well-known that material damping in the beam may be approximated by internal stiffness-proportional viscous forces. Therefore, the element damping matrix is estimated by $\tau_0 \mathbf{K}_S$ with τ_0 being an empirical time constant. Nonlinear terms in the mass matrix caused by axial inertia effects are considered to be small and consequently are neglected. Furthermore, the rotary inertia of the beam cross section and gyroscopic effects as well are neglected for the same reason. Internal damping of the rotating cylinder may generate non-conservative displacement forces and destabilize the bending vibrations of the open-loop system at high angular speed. These effects are not considered at low speed values.

4 CONTROL SYNTHESIS

Active damping is commonly carried out by feedback control techniques as well as by feedforward control concepts. Nevertheless, the latter technique is only effective in case of predictable or measurable disturbances. However, only feedback control concepts are suitable for unknown disturbances and to provide stability to an inherently unstable system. Most controllers typically use proportional-integral-derivative (PID) feedback. Nevertheless, using a PID controller may not be the best choice when facing parameter uncertainties of the plant. One technique that has proved a valuable tool for many kinds of application is $\mathcal{H}_\infty$ control design. In most applications an ideal model of the controlled system is not available, but a mathematical model that is inaccurate compared to the real system. Those inaccuracies are unavoidable for many reasons like unknown dynamics, neglected nonlinearities, parameter variations, etc. The $\mathcal{H}_\infty$ and the μ- synthesis are powerful design techniques that can expand the model by some defined parameter uncertainties. $\mathcal{H}_\infty$ control provides robust stability as well as robust performance even in case of not exactly known system parameters. Transforming the differential equations of motion (3) into state space description and defining the measurement outputs lead to

$$\dot{\mathbf{x}}(t) = \mathbf{A}\mathbf{x}(t) + \mathbf{B}\mathbf{u}(t) \tag{5}$$

$$\mathbf{y}(t) = \mathbf{C}\mathbf{x}(t) + \mathbf{D}\mathbf{u}(t), \tag{6}$$

where $\mathbf{A}$, $\mathbf{B}$, $\mathbf{C}$, $\mathbf{D}$ denote the system matrix, input matrix, output matrix and the feedthrough matrix. Furthermore, $\mathbf{x}(t)$, $\mathbf{u}(t)$ and $\mathbf{y}(t)$ denote the state vector, the input vector and the output vector. The specifications for the considered S/KS mixed sensitivity problem (see [11]) are defined by an upper bound for the sensitivity functions $\mathbf{S} = (\mathbf{I} + \mathbf{G}\mathbf{K})^{-1}$ and for the transfer functions $\mathbf{KS}$, where $K(s)$ denotes the

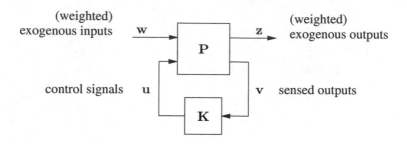

Fig. 5: General control configuration.

$\mathcal{H}_\infty$-controller transfer functions and $\mathbf{G}(s)$ denotes the transfer function description of the test rig. The frequency characteristic of the weighting functions is chosen in a way, that the solution of the $\mathcal{H}_\infty$ design method fulfills the specifications of the control loop. In Fig. 5 the general formulated control problem is depicted which contains a generalized plant $\mathbf{P}$ with

$$\begin{bmatrix} \mathbf{z} \\ \mathbf{v} \end{bmatrix} = \mathbf{P}(s) \begin{bmatrix} \mathbf{w} \\ \mathbf{u} \end{bmatrix}. \tag{7}$$

In order to obtain the closed loop transfer functions from $\mathbf{w}$ to $\mathbf{z}$, a lower linear fractional transformation

$$\mathbf{z} = \mathcal{F}_l(\mathbf{P}, \mathbf{K})\mathbf{w} \tag{8}$$

has to be performed [11]. The transfer matrix $\mathbf{T}_{zw}$ of the general formulated control problem shown in Fig. 5 contains the weighting functions $\mathbf{W}_e, \mathbf{W}_u$ of the functions $\mathbf{S}$ and $\mathbf{KS}$ and is given by

$$\mathbf{T}_{zw}(s) = \mathcal{F}_l(\mathbf{P}, \mathbf{K}) = \begin{bmatrix} \mathbf{W}_e \mathbf{S} \\ \mathbf{W}_u \mathbf{KS} \end{bmatrix}. \tag{9}$$

The basis of the $\mathcal{H}_\infty$ design procedure is to obtain a solution, for which the $\mathcal{H}_\infty$-norm of the transfer matrix $\mathbf{T}_{zw}$ drops below a certain value γ

$$\|\mathbf{T}_{zw}(s)\|_\infty < \gamma, \tag{10}$$

where the $\mathcal{H}_\infty$-norm is defined under consideration of the highest singular value $\bar{\sigma}$ by

$$\|\mathbf{T}_{zw}(s)\|_\infty = \sup_\omega \bar{\sigma}(\mathbf{T}_{zw}(j\omega)). \tag{11}$$

In order to finally determine this $\mathcal{H}_\infty$ problem, a pair of algebraic Riccati equations have to be solved, see [4]. Depending on the employed robust control technique, the obtained controller $\mathbf{K}$ may be of high order, what generally leads to a considerable numerical effort. Nevertheless, several modes do only have minor effect on the control performance and consequently can be reduced step-by-step as long as there is no significant effect on the sensitivity, the complementary sensitivity, the Nyquist curve or the step-response. As a consequence, the control task can be carried out consuming less calculation resources, which is especially important for experimental investigations and real-world applications.

Model	Mode	Measurement
$f_1 = 134$ Hz		$f_1 = 134$ Hz
$f_2 = 289$ Hz		$f_2 = 285$ Hz
$f_3 = 502$ Hz		$f_3 = 342$ Hz
$f_4 = 1028$ Hz		$f_4 = 926$ Hz

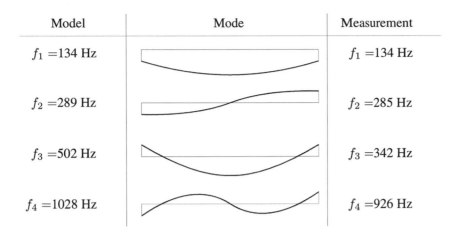

Fig. 6: Lowest bending modes and corresponding natural frequencies in the $x - z$ plane.

5 RESULTS

Fig. 6 shows a comparison between calculated and measured natural frequencies of the test rig and their corresponding modes. The nominal values of the uncertain model parameters such as the actuator stiffness and tilting stiffness of the active bearing are adjusted to the measured frequency $f_1 = 134$ Hz of the lowest bending mode. All higher natural frequencies calculated from the finite element model are significantly larger than the corresponding values obtained by the experimental modal analysis. This is because of the number and the type of the finite elements, that consequently lead to a more or less stiff model description.

Fig. 7 shows the simulation of an impact applied to the cylinder in the vertical bending plane. The radial displacement of the mid span of the test rig cylinder is observed. The vibration displacement of the passive structure is shown in the left-hand diagram of Fig. 7. The cylinder shows an initial deflection of about 33μm. The vibration is attenuating monotonous within a period of about 550ms. The damping capacity of the uncontrolled system is only governed by the material of the test rig components. The simulated, actively damped vibration is shown on the right-hand diagram of Fig. 7. It can be seen, that the vibration is already completely damped after less than 30ms. Already the first half wave of the vibration is damped and does not reach the maximum deflection value of the uncontrolled system. In comparison to the simulated results, Fig. 8 shows the vibration of the real test rig as a result of the same excitation of the cylinder. The system shows a similar reaction compared to the simulation results. The initial deflection and the attenuation times of both passive systems coincide well. Nevertheless, no significant reduction of the initial displacement in the actively damped case can be obtained. However, the effect of active damping is significant and can be seen clearly in both, simulation and experiment.

Fig. 9 shows the test rig vibration in the mid span of the cylinder, employing an active

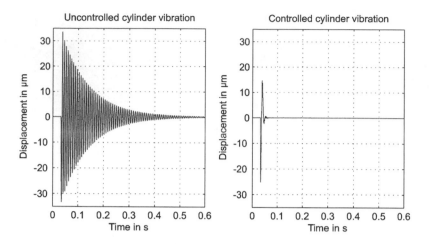

Fig. 7: Simulated passive and active system reaction to an impact. Measurement and excitation in the mid-span of the cylinder.

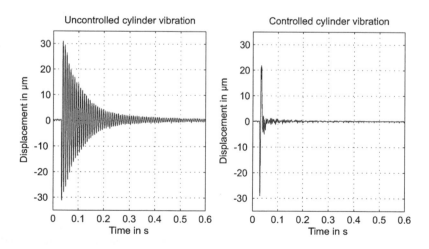

Fig. 8: Measured passive and active system reaction to an impact. Measurement and excitation in the mid-span of the cylinder.

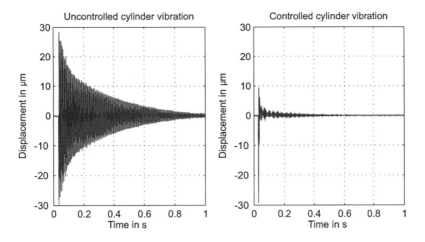

Fig. 9: Measured passive and active system reaction to an impact. Measurement and excitation in the mid-span of the cylinder.

bearing unit with significantly increased stiffness in bending direction. The left-hand diagram again shows the vibration of the cylinder mid span without active damping. The increase in stiffness causes a raise of the natural frequencies, leading the lowest bending mode to a frequency of $f_1 = 159$ Hz. The system approximately needs 1 second to entirely decay, what is about twice the time of the system as shown in Fig. 8. The controlled system is shown in the right-hand diagram of Fig. 9. Compared to the uncontrolled cylinder, the actively damped system obviously shows an effective reaction. The vibrations are quickly suppressed to a negligible and small amplitude. Nevertheless, the low rate of decay within the time range between 0.1 and 0.4 seconds seems to result from suboptimal control parameters. Fig. 10 shows a comparison between an active/active and active/passive bearing combination. It can be seen, that the system response is nearly equal in both cases. Consequently, the application of two active bearings seems to have no substantial benefit in the performance of this active damping task.

6 CONCLUSION

This study has shown a new concept for active damping of bending vibrations in slowly rotating or idle cylinders. The cylinder is mounted by one active and one passive bearing. Bending vibrations are excited by an impact hammer. A response action carried out by the active bearing prevents further vibrations even in case of a persistent disturbance. Hence, the stability and the vibration behavior of a rotor can significantly be improved. Numerical simulations were performed using a finite element model of a test cylinder system composed of beam- and lumped mass/stiffness elements. For control design a $\mathcal{H}_\infty$ synthesis has been carried out. The numerical simulation and experimental investigations indicate that the proposed methods are very effective in suppressing transient vibrations.

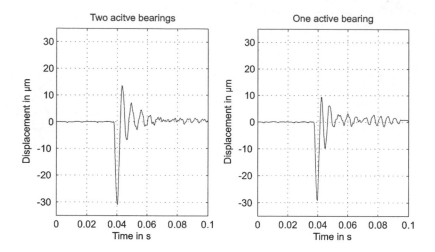

Fig. 10: Measured active vibration damping for an active/active and an active/passive bearing combination. Measurement and excitation in the mid-span of the cylinder.

Furthermore, it is shown that there is no significant improvement in the performance by applying two active bearings to the test rig cylinder.

ACKNOWLEDGEMENT

The authors gratefully acknowledge the financial and technical support of SKF Österreich AG in Steyr, Austria.

REFERENCES

[1] Alizadeh A.: *Robuste Regelung zur aktiven Schwingungsdämpfung elastischer Rotoren mit Piezo-Stapelaktoren.* Dissertation, TU-Darmstadt, 2005.

[2] Barret T. S., Palazzolo A. B., Kascak A. F.: *Active vibration control of rotating machinery using piezoelectric actuators incorporating flexible casing effects.* International Gas Turbine and Aeroengine Congress and Exposition, Transactions of the ASME, Cincinnati, Ohio.

[3] Carmignani C., Forte P., Rustighi E.: *Active Control of Rotor Vibrations by means of Piezoelectric Actuators.* ASME Design Engineering Technical Conference, Pittsburgh, PA, Sept. 2001.

[4] Doyle J., Glover K., Khargonekar P. Francis B.: *State-Space Solutions to Standard H_2 and H_∞ Control Problems.* IEEE Transactions on Automatic Control, Vol. 34, Aug. 1989.

[5] Ehmann C., Alizadeh A., Nordmann R.: *Schwingungsdämpfung aktiv gelagerter Rotoren mit robuster Regelung.* Department of Mechanical Engineering, Darmstadt University of Technology, 2004.

[6] Krodkiewski J. M., Sun L.: *Stability control of rotor-bearing system by an active journal bearing.* Proc. Vibration and Noise, Venezia, pp. 217-225, 1995.

[7] Li W.: *Aktive Dämpfung und Kompensation von Rotorschwingungen über aktive Piezo-Stapel-Aktuator-Lager.* Shaker Verlag, Aachen, 2005.

[8] Muszynska A., Franklin W.D., Bently D.E.: *Rotor Active "Anti-Swirl" Control.* ASME Journal of Vibration, Acoustics, Stress, and Reliability in Design, 110, pp. 143-150, 1988.

[9] Palazzolo A. B., Jagannathan S., Kascak A. F., Montague G. T., Kiraly L. J.: *Hybrid active vibration control of rotorbearing system using piezoelectric actuators.* Journal of Vibration and Acoustics, Trans. of ASME, 115, pp. 111-119, 1993.

[10] Santos I.F.: *Design and evaluation of two types of active tilting pad journal bearings.* Proceedings of IUTAM Symposium: The Active Control of Vibration, University of Bath (UK), pp. 45-52, 1994.

[11] Skogestad S., Postlethwaite I.: *Multivariable Feedback Control.* John Wiley & Sons, Chichester 1996.

Damping properties of syntactic foams with nanoparticulate additives

J A Rongong, G R Tomlinson
The University of Sheffield, Department of Mechanical Engineering, UK
I Stepanova, A Ivanenko, S Panin
Tomsk Polytechnic University, Russia

ABSTRACT

Polymer-based damping materials can be filled with microballoons to create closed cell *syntactic* foams. As well as having an ability to dissipate vibration energy, syntactic foams are light and relatively strong. For this reason they make excellent damping materials when used as fillers in hollow cavities. It is shown that to damp vibration in hollow cavities, the foam needs to have a modulus that is significantly higher than that seen in common damping materials. While it is desirable to combine low density with high modulus and damping over a broad temperature range, it is rarely achievable with traditional syntactic foams. This paper reports work carried out to modify the properties of syntactic foams through the addition of nanoparticulates to the mix. It has been claimed that by reducing the average particle size from micro-scale to nano-scale, dramatic increases in stiffness and damping capability can be achieved. Modifications in the properties of syntactic foams that can be achieved using nanoparticulates are investigated experimentally Results are reported for both silica nanoparticulates and carbon nanofibres. The effect of these changes on the vibration properties of two different example structures are demonstrated numerically.

1 INTRODUCTION

Currently, important driving factors in the design of high performance machinery include weight minimisation for fuel efficiency and reduction in parts count for reduced assembly and maintenance costs. Over the last two decades, there has been a noticeable trend in which single-piece, lightweight components have replaced heavier multi-part assemblies. The reduction in mass and the removal of damping associated with the joints, have increased the susceptibility to resonant vibration and the associated problems of increased noise and reduced fatigue life. Advances in predictive capability relating to fluid and structural dynamics coupled with improved manufacturing techniques have, to a certain extent, allowed designers to alleviate dynamic forcing and reduce the interaction of excitation frequencies and structural resonances. At the same time, there has been a significant rise in the attention given to the use of damping technologies for dealing with resonant vibrations. The addition of damping often occurs as a retro-fit to an existing machine with a vibration problem. However, it is usually more successful if incorporated at the initial design stage and can be used to simplify the design process.

A very large variety of both passive and active damping technologies are currently available and many more are being investigated. Passive viscoelastic damping is the most common approach [1-4] although specific applications depend on economic, environmental and performance requirements. Vibrations in bladed rotors are particularly challenging for damping technologies. Reliable performance must be provided over wide temperature ranges and high rotation-induced accelerations whilst strict limitations exist on mass, shape, size and durability. For example, a fan blade in an aircraft turbofan engine might experience temperatures in the range -50 to 150 °C and centripetal accelerations exceeding 400,000 m/s^2. Survival under bird impact and sand ingestion is essential and performance is not expected to degrade over many years under normal operation. As engine performance is linked to the blade shape, modification of the external geometry is not desirable. Finally, increase in mass over existing metallic or carbon-composite blades would be a major drawback. Damping technologies studied for fan blade applications have included co-cured viscoelastic constrained layers (for composites) [5,6], ceramic thermally sprayed coatings [7] and for hollow blades, granular materials [8] and syntactic foams [9].

Syntactic foams, as discussed in this paper, involve a polymeric matrix filled with a large number of microballoons to produce a closed cell compound. Microballoons are typically 10 to 100 µm in diameter and can be made of glass, polymer or metal. The work reported here focuses on foams made using an epoxy matrix and glass microballoons but results are expected to be relevant to other foam types. Syntactic foams can provide significant damping when operating in the viscoelastic transition region. They can be used as a lightweight damping material and are particularly effective as fillers in hollow components [10]. The constitutive components of a particular foam can be adjusted to suit specific applications. However, as the properties of the foam also have significant influence on the strength and stiffness of the hollow component limitations exist to the modifications that can be made to improve damping.

Recent developments in the field of nanoparticulate enhanced materials have suggested that there are large benefits to be gained through reduction in scale from micro to nano in terms of particulate additives for polymers. For example, the effects of silica particulate size on the compound's mechanical properties are presented for epoxies [11] and thermoplastics [12]. For the large part though, current literature focuses on the structural properties of materials in their rigid, glassy state. This paper presents initial numerical and experimental work carried out to investigating the transition zone and the effect that additives have on the damping capabilities. The approach taken involves the following stages:

a) Identification of desirable properties for a syntactic foam used in damping are first identified with reference to typical applications (Section 2),
b) The damping properties micro particulate syntactic foams are investigated using a validated theoretical model (Section 3),

c) The expected and actual effects of adding silica nanoparticulates to a syntactic foam are reported (Section 4),
d) An attempt to assess the effectiveness of carbon nanofibre additives is reported (Section 5)
e) The damping performance of modified syntactic foams is examined in two numerical examples (Section 6).

2 CAVITY-FILL DAMPING MATERIALS

Damping technologies for hollow cavities fall into two categories. One category involves the use of wave or fluid-like motion in the damping medium and includes granular materials and viscous fluids. As the filler is not able to carry significant tensile or shear loading, its effect on the strength and stiffness of the structure is negligible. The second category involves filling of the void with materials (and structures) that carry significant load. The properties of the damping material, in these cases, can affect significantly, the properties of the entire structure. This section considers important features of load-bearing cavity fillers and attempts to identify desirable properties.

2.1 Modelling approach
Where energy dissipation arises from the dynamic straining of a viscoelastic material (VEM), the system loss factor η (ratio of energy dissipated per radian to peak stored energy) is,

$$\eta = \eta_v R \tag{1}$$

where η_v is the loss factor of the material used and R is the ratio of dynamic strain energy in the damping material to that in the entire structure. The strain energy ratio R is controlled by the configuration, dimensions, material properties and mode shapes of the total structure/damper system in question. A designer aiming to maximise system damping clearly needs to ensure that both the material loss factor and the strain energy ratio are high over the operating conditions of interest.

The true value of R depends on a large number of parameters including η_v. However, if one uses the Modal Strain Energy approach, it is possible to estimate R for each system eigenmode using only the elastic properties of the system. It has been shown that this approach give results that are almost identical to an exact solution when material loss factors are low ($\eta_v<0.3$) but some errors can arise if loss factors are very high ($\eta_v>1$) [13].

A second factor for consideration in the model is that properties of viscoelastic polymers change with frequency and temperature. Increases in frequency are equivalent to decreases in temperature where the exact correlation is defined by the temperature-frequency *shift curve* that is specific to each material. The use of the Modal Strain Energy approach to get accurate estimates with viscoelastic behaviour is described in detail elsewhere [9, 13]. For the design studies carried out here, it was considered sufficient to study the damping achievable over the range of modulus values that the viscoelastic foam might take. This is sufficient to show whether a modified syntactic foam might give improved performance.

For components in rotating systems, such as propeller and fan blades, an additional modelling issue relates to the constant strain applied by the body forces. To achieve accurate predictions a non-linear analysis step is usually required prior to the eigenvalue extraction routine to represent the rotation-induced stress field. When performing the MSE analysis, it is important to note that the dissipated energy depends on the vibrating strain energy only while the stored energy includes the rotation-induced effects. The result is usually a dramatic reduction in damping at high rotation speeds.

In this work, the commercial finite element (FE) software Abaqus was used. However, as the Modal Strain Energy approach is general, identical results could have been obtained using any other FE package. In the FE models, the structure and syntactic foam were represented as elastic, homogenous solids using quadratic brick elements with 20 nodes. The mesh density was selected to give a good compromise between accuracy and calculation speed, however for brevity mesh studies are not reported here.

2.2 Optimising damping in cavity fillers

The simplest dampers that employ viscoelastic materials are free-layer treatments – coatings of polymeric material on thin-walled structures. It is well-known that the highest strain energy ratio R (and hence potentially the best damping) for a given thickness of such a damper is achieved by maximising the Young's modulus of the coating E_v. For constrained layer dampers however, there exists an optimum shear modulus where the exact value depends on a number of material and geometric parameters [14, 15]. For more complicated systems, the strain energy optimum performance varies depending on mode shape. The equivalent optimisation curves for filled hollow cavities are not well understood and will vary depending on the exact application. However, by considering the optimisation curves for using two typical applications, some understanding of the nature of these curves can be gained.

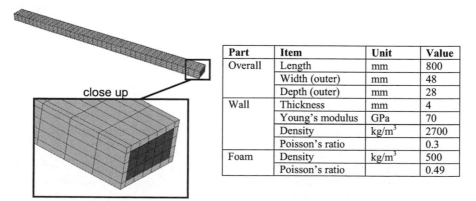

Part	Item	Unit	Value
Overall	Length	mm	800
	Width (outer)	mm	48
	Depth (outer)	mm	28
Wall	Thickness	mm	4
	Young's modulus	GPa	70
	Density	kg/m^3	2700
	Poisson's ratio		0.3
Foam	Density	kg/m^3	500
	Poisson's ratio		0.49

close up

Figure 1:   Sketch and model of hollow beam filled with viscoelastic syntactic foam

The first application considers the flexural vibrations of a hollow bar filled with damping foam. This represents structures seen in cars (e.g. door pillars and bumpers) and various sporting goods. A sketch of the finite element mesh and properties used are summarised in Figure 1.

Results showing modal properties for a range of VEM Young's modulus values are presented in Figure 2. It can be seen that for this type of structure, it is beneficial to have a very stiff filler – significant advantages exist in being able to increase Young's modulus above 1 GPa.

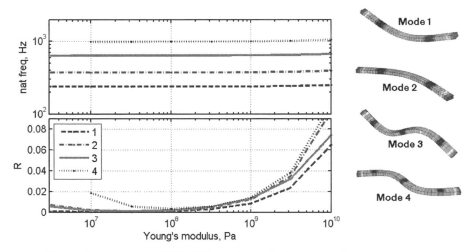

Figure 2:   Effect of VEM Young's modulus on strain energy ratio for the representative hollow bar

The second application considers a hollow fan blade from an aircraft engine. In this concept, the metallic stiffening ribs, sometimes known as line core [16], are replaced by syntactic foam [9]. The concept, the finite element model used and nominal dimensions are presented in Figure 3. Note that the dimensions do not relate to any specific fan blade – they have been chosen to have representative properties for a variety of actual blades.

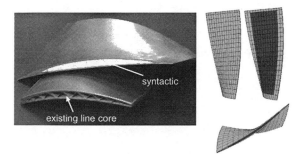

Item	Unit	Value
Length	mm	1000
Width	mm	400
Thickness (root)	mm	28
Thickness (tip)	mm	6
Twist	degrees	70
Camber	mm	35
Skin Young's modulus	GPa	115
Skin density	kg/m^3	4429
Skin Poisson's ratio		0.33
Core density	kg/m^3	500
Core Poisson's ratio		0.49

Figure 3:   VEM filled fan blade concept, model and novel polymer syntactic core

Results showing modal properties for a range of VEM Young's modulus values are presented in Figure 5. Note that for this study, the fan rotation speed was selected to give a tip acceleration of 200,000 m/s^2.

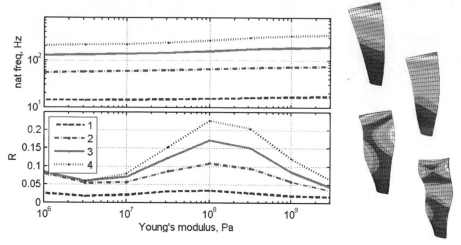

Figure 4:   Effect of VEM Young's modulus on strain energy ratio for the hollow fan blade

From Figure 4, it can be seen that the best Young's modulus for the fan blade is in the range 30 MPa to 1 GPa. Re-running this with different fan geometries shows that this optimum range is fairly consistent. The shape of the curves (showing a peak value) suggests that the deformation of the core is more similar to a constrained layer damper (in shear) than for the filled hollow bar.

Together, these results indicate that it is desirable for a syntactic foam used for damping as a cavity filler should to have a relatively high Young's modulus – at least 30 MPa for the fan and much higher for the hollow bar. As peak damping for most unfilled matrix polymers occurs around E_v=10 MPa, significant stiffening is required.

3   PERFORMANCE OF SYNTACTIC FOAMS

The syntactic foams considered in this paper involve a polymeric matrix filled with a high concentration of microballoons. Scanning electron microscope (SEM) images of the broken surface of a typical syntactic foam (50% volume fill of glass microballoons, overall density 700 kg/m^3) are presented in Figure 5. The typical wall thickness of a glass microballoon can be seen in the second image.

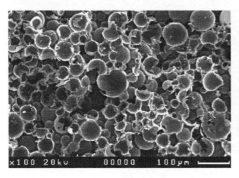 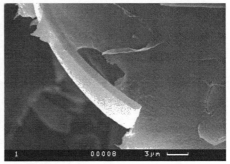

Figure 5: SEM images of syntactic foam

3.1 Syntactic foam model

Theoretical models have been developed over the years to estimate the effective properties of such materials. Here, an adaptation of the micromechanical model presented by Bardella and Genna [17] is used to illustrate some of the design considerations associated with the development of damping foams. In this approach, the morphology of the microballoons in the matrix are represented by a single, multi-layered, spherical inclusion located within a homogeneous medium as shown in Figure 6.

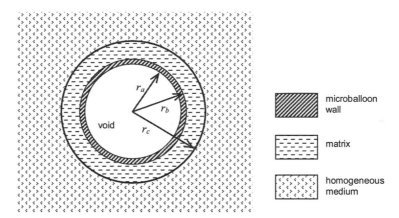

Figure 6: Four phase model

The distances r_a and r_b are used to represent the inner and outer radii of a typical microballoon and r_c the radius of the composite spherical inclusion. The magnitude of r_c is selected such that the volume fraction of microballoon in the inclusion matches that for microballoons in the syntactic foam. Thus,

$$f = \frac{r_b^3}{r_c^3} \qquad (2)$$

Expressions for homogenised shear and bulk moduli are obtained individually by considering the stresses generated by strains applied at infinity. In the Self Consistent model, the properties of the surrounding homogenous material are set to be equal to the effective properties of the inclusion. Expressions for the homogenised shear modulus ($G^{(s)}$) and bulk modulus ($K^{(s)}$) are given as,

$$G^{(s)} = \frac{1}{\varepsilon_\gamma^{(cs)}} \left(G^{(w)} f \left(1 - \frac{r_a^3}{r_b^3} \right) \varepsilon_\gamma^{(w)} + G^{(m)} \left(1 - f \right) \varepsilon_\gamma^{(m)} \right) \tag{3}$$

and

$$K^{(s)} = \frac{1}{\varepsilon_\theta^{(cs)}} \left(K^{(w)} f \left(1 - \frac{r_a^3}{r_b^3} \right) \varepsilon_\theta^{(w)} + K^{(m)} \left(1 - f \right) \varepsilon_\theta^{(m)} \right) \tag{4}$$

where ε_γ and ε_ϕ are the volume-averaged, shear and bulk strains respectively. The bracketed superscript terms (e.g. $G^{(w)}$) refer the property to the object considered: w for microballoon wall, m for matrix, s for surrounding homogenised medium and cs for the whole composite sphere. The equations used to compute the volume-averaged strains in an elastic syntactic foam can be found in reference [17]. Results for a viscoelastic syntactic foam can be obtained by replacing the relevant elastic moduli by complex values. As the strains are applied at infinity, strains applied to and within the inclusion depend on the modulus of the surrounding medium. Because of this, Equations 3 and 4 are implicit and must be solved iteratively.

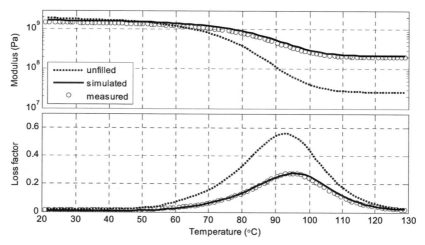

Figure 7: Measured and simulated complex moduli for B-Flex filled ScotchLite A20 microballoons (70% by volume)

The predicted complex modulus of a syntactic foam produced using the model was compared with measured values. For this study, the matrix epoxy used is a structural adhesive called B-Flex (manufactured by Soundcoat). The syntactic foam was made by adding ScotchLite A20 glass microballoons to B-Flex. The microballoons used have a nominal density of 200 kg/m³ while the properties of the glass walls were assumed to have a Young's modulus of 65 GPa, Poisson's ratio of 0.23 and density of 2500 kg/m³. Comparisons of predicted and measured properties can be seen in Figure 7. Note that a very high volume fill (70%) was used as it was assumed that the model would be least accurate under these conditions. The close correlation achieved between theory and experiment show that the model can be used in these studies.

3.2   Damping performance of syntactic foams

This section contains a brief numerical study showing the sensitivity of syntactic foam properties to various parameters. For this (and subsequent) work, the matrix material was chosen to provide give high pre-cure fluidity, good postcure stability and a minimum of particulate additives. The matrix, referred to here simply as "epoxy", was based on the epoxy system Epophen EL5 (Borden Chemicals) which was reacted with Jeffamine D-400 and Jeffamine D-2000 using with accelerator 399 (all from Huntsman). The microballoon properties were based on those of Eccosphere SID 311Z, a glass microballoon with nominal density of 320 kg/m^3.

The effect of increasing volume fraction is shown in Figure 8. The most obvious change is that the modulus is increased dramatically – up to a factor of 3 in the rubbery zone. The peak loss factor drops significantly and the temperature at which this occurs increases slightly.

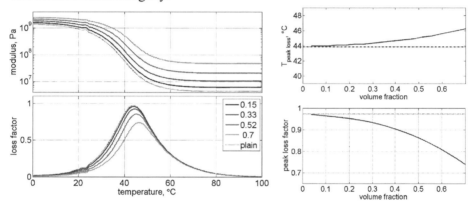

Figure 8:   Effect of volume fill on properties of syntactic foam

The effect of the wall thickness on the complex modulus is shown in Figure 9. Note this is effectively a measure of the stiffness of the inclusion.

The notation "solid" and "air" relate to walls that are very thick or vanishingly thin respectively while the term "microballoon" reflects the existing Eccosphere microsphere. In this figure it can be seen that very flexible or very stiff microballoons do not change the loss factor but do affect the modulus. A condition occurs when the presence of microballoons decreases the loss factor the most – this is the condition where a significant portion of the dynamic strain energy is in the microballoon and is therefore not dissipated.

The results from this study show that microballoons can increase modulus significantly. However, syntactic foams still cannot achieve the very high values required for optimum damping using cavity fillers. Additionally, it is undesirable to increase microballoon volume fraction much above 50% as it can result in inadequate wetting and hence poor quality microballoon/matrix contact leading to reduced strength and durability.

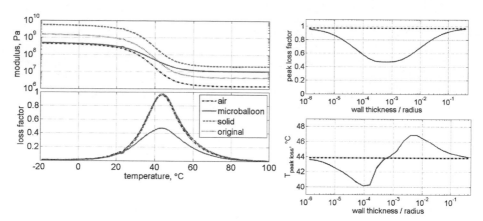

Figure 9:   Effect of microballoon wall thickness on properties of syntactic foam (50% volume fill)

4   FOAM MODIFICATION USING SILICA NANOPARTICULATES

A common explanation for changes that nanoparticulates cause, and one of the main driving factors for their use is that the particulates are of the same order of magnitude as the polymer chains and can therefore directly affect their mobility. Also, as the effective surface area of a nanoparticulate is greater than that for an equivalent volume fraction of microparticulate, it is believed that this can offer better interface connection. In this study, the effects different concentrations of 10 nm diameter silica on the properties of filled epoxy are considered. Silica is readily available in nanoparticulate form and has been used by other researchers. Additionally, as silica in bulk is not particularly reactive, it is easier to attribute changes seen when adding nanoparticulates to the matrix material to the size of the particle rather than its chemical composition.

A range of different volume fractions of the nanoparticulates were added to the epoxy matrix. Dispersion of particles we encouraged through the use of a homogeniser during mixing. Measurements of the density of resulting specimens showed that entrapment of air was relatively low. The complex modulus of different volume fractions is presented in Figure 10.

As seen with the microballoons, addition of the high modulus filler increases modulus and reduces the peak loss factor. An interesting departure from the micro-scale understanding is the change in the temperature at which peak loss occurs. With a microparticulate, one would expect a small increase in temperature while the test results with the nanoparticulate clearly show a significant decrease.

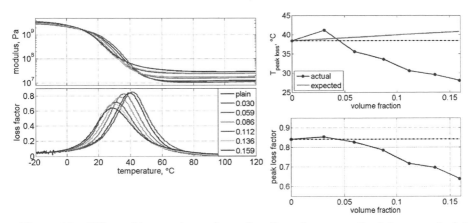

Figure 10: Effect of increasing volume fraction of nanoparticulate on material properties

A second notable difference occurs in the magnitude of the measured Young's modulus. The relative value (scaled by the unfilled matrix) for different concentrations of nanoparticulates is presented in Figure 11. In the glassy (cold) zone, this is relatively close to that expected for microparticulates. In the rubbery zone however, the nanoparticulates increase the modulus by more than three times that which might be expected for an equivalent volume fill of microparticulate. This appears to agree with the claims in the literature that the larger contact surface is beneficial. It is interesting to note from the TEM image in Figure 11 that the nanoparticles are not fully dispersed – they can be seen to form clumps that are 50 to 100 nm in diameter. Improved dispersion would increase stiffness further.

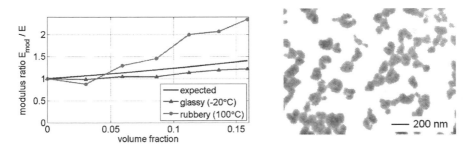

Figure 11: Nanoparticulate additives to epoxy

The effect of the addition of nanoparticles to the syntactic foam was next studied. A number of specimens were manufactured that contained equal quantities (by mass) of microballoons and nano silica. The nanomaterial was mixed into the matrix using the homogenizer before the addition of microballoons – this was an important step carried out to avoid fracturing the microballoons. Results showing the effect of 10% (by volume) nanoparticulate additives to a 50% 9by volume) syntactic foam are presented in Figure 12.

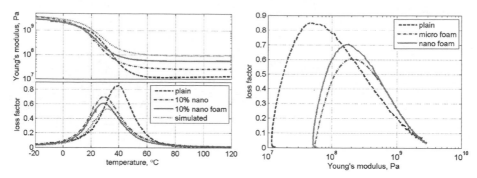

Figure 12: Nanoparticulate modified syntactic foams (final density 700 kg/m³)

Results of a simulation showing the expected effect of adding microballoons to a 10% nanoparticulate enhanced epoxy are also plotted on the same figure. Here, the prediction does not exactly match the measured behaviour – measure properties for the nano foam are lower than might be expected from a simple superposition rule.

Figure 12 also shows a comparison between the plain material, the standard syntactic foam and the nano-enhanced foam. To remove the effect of the temperature change, this is plotted as loss factor against Young's modulus. On this curve, it can be seen that the nanoparticulates have a similar effect to that which might be seen with extra microparticulates. However, as the physical scale between the two components is very different, the pre-cure mixture has better handling properties.

5   MODIFICATION OF EPOXY USING CARBON NANOFIBRE

One concern with work presented in the previous section is that it is unclear how well the nano particles bond with the polymer. It was considered valuable to carry out a comparative study by considering the properties of the same epoxy, this time modified using carbon nanofibres. The nanofibres used in this study were approximately 50 nm in thickness and several micrometers in length. The resulting complex modulus information is presented in Figure 13.

Again, examination of Figure 13 reveals that the presence of the nanofibres has reduced the temperature at which peak loss occurs but has not significantly increased the Young's modulus. It is particularly noticeable that there is no noticeable advantage in terms of modulus for higher concentrations. This can be partially explained by considering the mass density of the resulting specimens – addition of the nanofibre does not show a complementary increase in density. This shows that the process has entrapped air.

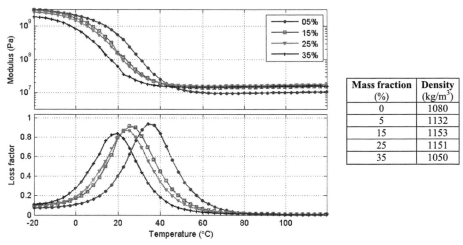

The following table appears to the right of the figure:

Mass fraction (%)	Density (kg/m³)
0	1080
5	1132
15	1153
25	1151
35	1050

Figure 13: Effect of nanofibre on material properties (quoted as mass fraction)

6 PERFORMANCE OF NANOFOAM ENHANCED STRUCTURES

This section presents simulations of damping that could be achieved in the structures considered in Section 2 through the use of nanoparticulate enhanced foams. Note that the aim is not to present high levels of damping in the system – for this, a better combination of matrix and particulates would have been chosen – but to show the difference between a standard syntactic foam and one enhanced using nanoparticulates.

Results for the third mode of the hollow bar structure are presented in Figure 14 and for the second mode of the fan blade in Figure 15.

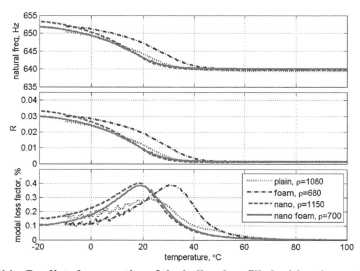

Figure 14: Predicted properties of the hollow bar filled with polymer (Mode 3)

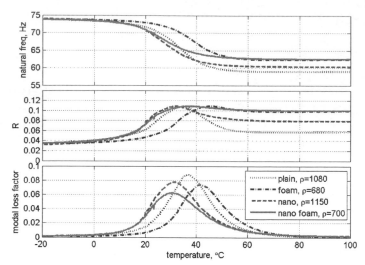

Figure 15: Predicted properties for the hollow fan blade (Mode 2)

Both sets of results show that the principal effect of the nanoparticulate is to alter the temperature at which peak damping occurs.

7   CONCLUSIONS

The work presented in this paper considered the use of nanoparticulates to modify the properties of syntactic foams used to damp vibration in hollow structures. Experimental work was carried out to characterise standard syntactic and nano particulate modified foams. The nanoparticulate additives included silica and carbon nanofibre. Numerical predictive work has indicated that for the constituents considered, the main contribution of the nanoparticulate additive was to modify the temperature at which peak damping occurs. The work also noted that the current fabrication procedures were not able to disperse the nanoparticulates adequately – they clumped together and in some cases entrained air. Improved results may therefore be achievable with different material manufacturing methods.

8   ACKNOWLEDGEMENTS

The work presented here was financially supported by the British Council through the BRIDGE grant "Design of damping coatings across the length scales". The authors would also like to thank Mr Les Morton for his assistance with specimen preparation and operation of the viscoanalysers.

9   REFERENCES

[1]   Nashif A D, Jones D I G and Henderson J P, 1985, *Vibration damping*, John Wiley & Sons.

[2]   Sun C T and Lu Y P, 1995, *Vibration damping of structural elements*, Prentice Hall, USA.

[3]   Mead D J, 1999, *Passive Vibration Control*, Wiley, New York.

[4]   Rao M D, 2003, "Recent applications of viscoelastic damping for noise control in automobiles and commercial airplanes", *Journal of Sound and Vibration*, 262, 457–474.

[5]   Kosmatka J B, Lapid A J and Mehmed O, 1996, "Passive vibration control of advanced composite turbo-fan blades using integral damping materials", *AIAA/ASME/ASCE/AHS/ASC Structures, Structural Dynamics and Materials Conference and Exhibit*, Salt Lake City, Apr. 15-17, Paper AIAA-1996-1598.

[6]   Garay G, Schorr D, Turner A, Weaver M, McCormick M, "Design and subcomponent testing of a viscoelastic constraint layer damping system for fan airfoils, *9th National Turbine Engine HCF Conference*, March 18, 2004.

[7]   Patsias S, Tomlinson G R and Jones A M, 2001, "Initial studies into hard coatings for fan blade damping", *6th National Turbine Engine High Cycle Fatigue Conference*, Jacksonville, Florida, CD-ROM, 5th – 8th March 2001.

[8]   Fowler B L, Flint E M and Olsen S E, 2000, "Effectiveness and predictability of particle damping", Proceeding of *SPIE Smart Structures and Materials Conference*, Vol. 4331, pp. 186-197.

[9]   Rongong J A and Williams R J, 2004, "Lightweight, polymeric cavity filler for damping hollow fan blade vibrations", *9th Annual High Cycle Fatigue Conference,* Pinehurst, North Carolina, 16-19 March.

[10] Rongong J A, 2003, "Passive damping materials for the future", *2nd Int. Conference on Damping – Materials and devices for the next decade*, 24-26 March 2003, Stellenbosch, South Africa.

[11] Nakamura Y, Yamaguchi M, Okubo M, Matsumoto T, 1992, "Effects of particle size on mechanical and impact properties of epoxy resin filled with spherical silica", *Journal of Applied Polymer Science*, 45, 1281-1289.

[12] Y Liu, M Kontopoulou, 2006, "The structure and physical properties of polypropylene and thermoplastic olefin nanocomposited containing nanosilica", *Polymer*, 47, 7731-7739.

[13] Torvik P J and Runyon B, 2007, "Modifications to the method of modal strain energy for improved estimates of loss factors for damped structures", Shock and Vibration, 14(5), 339-353.

[14] Ungar E E and Kerwin E M, 1964, "Loss factors of viscoelastic systems in terms of energy concepts", *Journal of the Acoustical Society of America*, 34, 954-957.

[15] Torvik P J and Runyon B D, 2006, "On estimating the loss factors of plates with constrained layer treatments", 47th AIAA/ASME/AHS/ASC Structures, Structural Dynamics and Materials Conference, 1-4 May 2006, Rhode Island, Newport, USA.

[16] Lloyd A D and Fitzpatrick G A, 1998, Diffusion-bonded / superplastically formed wide chord fan blade, *IMechE Conference Civil Aerospace Technologies FITEC '98*, C545/023/98.

[17] Bardella L and Genna F, 2001, "On the elastic behaviour of syntactic foams", *International Journal of Solids and Structures*, 38, 7235-7260.

Design of electromagnetic damper for aero-engine applications

A Tonoli, M Silvagni, N Amati
Mechatronics Laboratory, Politecnico di Torino, Italy
B Staples, E Karpenko
Rolls-Royce plc, UK

ABSTRACT

The vibration control of rotors for gas or steam turbines is usually performed using passive dampers mostly based on the use of viscous fluid. Their nature and the variability of their performance with temperature and frequency represent the main disadvantages. Dampers with magneto-rheological and electro-rheological fluid and even active magnetic bearings allow solving only a part of the above mentioned drawbacks. Electromagnetic dampers seem to be a valid alternative due to the absence of all fatigue and tribology issues motivated by the absence of contact; the small sensitivity to the working environment; the wide possibility of tuning even during operation and the predictability of the behaviour.

The aim of the present paper is to describe the design procedure adopted to develop electromagnetic dampers to be installed in an aero-engine. This procedure has been validated using a reduced scale laboratory test rig and then adopted for real civil aircraft engines. Results in terms of achievable vibration reductions, mass and overall dimensions are hence presented.

1   INTRODUCTION

The control of rotor vibration is essential in aero engine design. In addition to the safety requirement of avoiding rotor lateral critical speeds within the engine running range, the response in the many other modes of the engine structure system must be controlled to ensure acceptable levels of vibration. Control of bearing loads, structural fatigue loads, rotor/casing tip clearances, casing and engine external responses and the transmission of vibration to the airframe are all required. In most engines, squeeze film dampers are used to help control vibration. Squeeze films are very effective in reducing response for the moderate range of residual unbalance expected after the rotor balancing process. In addition they are self-sufficient and robust in the event of high unbalance levels e.g. due to rotor blade damage. However, being of passive nature, they are designed as a compromise for a range of conditions. Their performance can be affected by engine oil temperature, supply pressure and thermal growth of the bearing housing; these parameters vary throughout a flight. Active or semi-active electro-hydraulic systems have been proposed to allow some forms of online tuning or adaptive behavior. More recently, electro-rheological ([1]) and magneto-rheological ([2]) semi-active damping systems have shown attractive potentialities for the adaptation of the damping force to the operating conditions. However, these devices cannot avoid drawbacks related to the ageing of the fluid and to the tuning required for the compensation of the temperature and frequency effects.

Active Magnetic Bearings (AMBs) have received much attention. They offer the greatest possibilities to isolate the rotors from the engine structure and to optimize the system dynamics. The conventional rolling element bearings and the whole engine oil supply system might be dispensed with. The main problem is that AMBs must provide stiffness and damping forces. The requirements to support the rotor weight define the size and mass of the coils. If damping only is required a much lighter electro-magnetic device (actuator/damper) may be devised. Electromechanical dampers seem to be a valid alternative to visco-elastic and hydraulic ones due to, among the others: a) the absence of all fatigue and tribology issues motivated by the absence of contact, b) the small sensitivity to the operating conditions, c) the wide possibility of tuning even during operation, and d) the predictability of the behavior. Following the experience accumulated in magnetic suspensions, Kasarda et al. ([3]) proposed active magnetic bearings in conjunction with conventional supports to damp the lateral vibration of reduction gears. Vance et ali ([4]) underlined how the modulation of the damping with the operating conditions is a very powerful means to improve the dynamic behavior of aero engines. However, the midterm applicability of active configurations suffers the lack of a consolidated design practice and the availability of low cost and reliable power and signal electronics.

Passive or semi-active eddy current dampers seem to be an interesting alternative in the near future. Compared to active closed loop devices, eddy current dampers have a simpler architecture thanks to the absence of power electronics and position sensors and are intrinsically not affected by instability problems due to the absence of a fast feedback loop. The simplified architecture guarantees more reliability and lower cost, but allows less flexibility and adaptability to the operating conditions. The working principle of eddy current dampers is based on the magnetic interaction generated by a magnetic flux linkage's variation in a conductor ([5], [6]). Such a variation may be generated using two different strategies: moving a conductor in a stationary magnetic field that is variable along the direction of the motion; changing the reluctance of a magnetic circuit whose flux is linked to the conductor. In the first case, the eddy currents in the conductor interact with the magnetic field and generate Lorenz forces proportional to the relative velocity of the conductor itself. Graves, Toncich and Iovenitti ([7]) define this kind of damper as "motional" or "Lorentz" type. In the second case, the variation of the reluctance of the magnetic circuit produces a time variation of the magnetic flux. The flux variation induces a current in the voltage driven coil and, therefore, a dissipation of energy. This kind of damper is defined in [8] as "transformer", or "reluctance" type.

The literature on eddy current dampers is mainly focused on the analysis of "motional" devices. Karnopp and Margolis in [9] and [10] describe how "Lorentz" type eddy current dampers could be adopted as semi-active shock absorbers in automotive suspensions. The application of the same type of eddy current damper in the field of rotordynamics is described in [11] and [12]. Additionally, a dedicated research activity ([7], [10], [14]) has been devoted to model the effects of the electric circuit dynamics on the damping performances and therefore on the design of eddy current dampers.

Being usually less efficient than "Lorentz" type, "transformer" eddy current dampers are less common in industrial applications. However they may be preferred in some areas for their flexibility and construction simplicity. If driven with a constant voltage they operate in passive mode while if current driven they become force actuators suitable for active configurations. A promising application of the "transformer" eddy current dampers seems to be their use in aero-engines as a non rotating damping device in series to a conventional rolling element bearing that is connected to the main frame with a mechanical flexible support. Similarly to a squeeze film damper, the device acts on the non rotating part of the bearing. As it is not rotating, there are no eddy currents in it due to its rotation but just to its whirling. The coupling effects between the whirling motion and the torsional behavior of the rotor can be considered negligible in balanced rotors ([15]).

In a previous paper [19], the authors presented the mathematical models governing the dynamic behaviour of such devices. The models where validated experimentally on laboratory test rigs.

The aim of the present paper is to illustrate a design procedure to apply electromagnetic dampers of active and transformer type to rotating machines in order to reduce vibrations. The proposed dampers will be applied and validated on a scale test rig representing the dynamic behaviour of a real aircraft engine. The same procedure will be also applied in a feasibility study on a real aero-engine to evaluate achievable vibration reductions, added mass and overall damper dimensions.

2   ELECTROMAGNETIC DAMPERS

A trade-off analysis on electromagnetic dampers between AMBs, Active Magnetic Dampers (AMDs) and Passive or Semi-Active "transformer" eddy current dampers (SAMDs) has been performed in [18] to address the advantages and drawbacks of such technology. Beside this quantitative evaluation a preliminary qualitative investigation has been performed comparing the electromagnetic damping solution with other technologies such as electro-active material (piezoelectric, electro-active polymer), SMA (shape memory alloy) and magnetic strained material. These solutions will probably lead to similar results (in term of achievable damping level and added mass) but seem to be reliable in terms of high-stress/temperature operation and fatigue; furthermore these materials and technologies are suited to an appropriate technology level for engine application. Build technologies for AMD or SAMD are more common being equivalent to typical rotating electrical machines.

AMDs and SAMDs seem to be promising and the study focused on these.

2.1 Active Magnetic Dampers (AMDs)

Active magnetic dampers (AMDs, Figure 1) work in the same way as active magnetic bearings, the only difference being that the force generated by the actuator is not aimed at supporting the rotor, but just at supplying damping. In the case that an AMD is integrated into the support of a rolling element bearing, it is proposed to connect the bearing to the housing with mechanical springs providing the required stiffness, while the damping is due to the electromagnetic actuator. Both the spring and the damper act on the non rotating part of the support. It is worth noting that the stiffness of the spring must take into account also the open loop negative stiffness of a typical Maxwell actuator, this allows to decrease the requirements on the proportional action of the closed loop feedback. Like all active systems, stability must be assessed by properly designing the control law which works in a proportional-derivative mode. Sensors and a controller are then required.

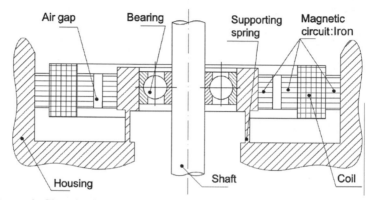

Figure 1: Sketch of an Active Magnetic Damper in conjunction with a mechanical spring. They both act on the non rotating part of the bearing

2.2 Semi-Active Magnetic Dampers (SAMDs)

The Semi Active Magnetic damper ([19]), of the "transformer" type, may use the same configuration as an AMD with the difference that now the coils are supplied with a constant voltage.

The displacement with speed $\dot{q}$ of the rotor (Figure 2a) changes the reluctance of the magnetic circuit and produces a variation of the flux linkage so generating a back electromotive force and ultimately eddy currents in the coils. The currents in the coils have two contributions: a fixed one due to the voltage supply and a variable one induced by the back electromotive force. The first contribution produces a negative stiffness, while the damping force is generated by the second contribution. It is then a semiactive device, since it requires some power to generate the magnetic field, but it doesn't need to be controlled in closed loop and hence may have no sensors. Energy dissipations take place in the stator providing nonrotating damping. In the case of an already supported structure, the mechanical stiffness could be that of the structure, alternatively, in case of a structure supported by the dampers themselves, the springs have to be installed in parallel to them.

As shown in Figure 2b, it is possible to identify three different frequency ranges:
- Equivalent stiffness range (low frequencies): the system behaves as a spring.
- Damping range (mid frequencies): the system behaves as a viscous damper.
- Mechanical stiffness range (high frequencies): the system behaves as a spring.

It is interesting to note (Figure 2c) that the resulting model is the same as the Maxwell's model of viscoelastic materials. At low frequency the system is dominated by spring K_{eq} while the lower arm of the parallel does not contribute. At high frequency the viscous damper "locks" and the stiffnesses of the two springs add. The viscous damping dominates at intermediate frequencies.

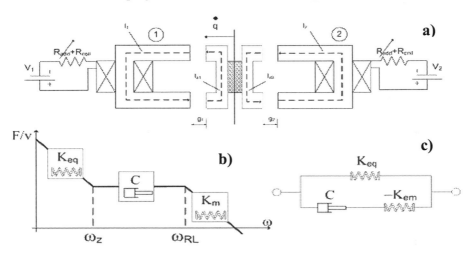

Figure 2: a) Sketch of Semi Active Magnetic Damper (the elastic support is omitted); b) mechanical impedance of a transformer eddy current damper in parallel to a spring of stiffness; c) Mechanical equivalent.

3   CASE I: SCALED PROTOTYPE TEST RIG

The models presented in the previous section have been experimentally validated by means of specific test rigs representing the dampers themselves ([19]). A deeper investigation of concepts is needed to verify the assumptions when applied to a rotating system. In the following the application of electromagnetic dampers to a general purpose rotordynamic test rig will be presented. The same procedure has been used to investigate the feasibility of applying the electromagnetic dampers in a real aircraft engine.

3.1   Test Rig Description and Specifications
Figure 3a shows the *Engine-like Rotordynamics Test Rig* (RTR). The rig is intended to work at speeds up to 6000 r/min. The main requirement for initial studies was for

the stator dynamic behaviour not to interact with that of the rotor up to the 2X component of the vibration. Such a requirement was fulfilled by designing a very stiff structure based on four columns and thick end plates. The rotor must be representative of an engine low pressure shaft taking into account the rotational speed limitation (Ω_{max}=6000 r/min) and the dimensional constraints (max axial length=700 mm, max diameter of the disks=200 mm). The first two critical speeds (ω_{cr1}, ω_{cr2}) should occur at about 2000-2500 r/min and 4000-5000 r/min while the third critical speed (ω_{cr3}) should be out of the speed range (about twice the max rotational speed) to avoid the contribution of higher order harmonics in the working range. Moreover, the first and the second critical speeds should be characterized by the bouncing of the LPC and LPT supports respectively, that means a high value of the strain energy related to the supports. Additionally, the vibration amplitude should not be higher than 1 mm in the whole speed range to cope with standard range position sensors. These specifications lead to an architecture based on: the low pressure turbine (LPT) and the low pressure compressor (LPC) disks (1 and 6 in Figure 3b) connected by a small diameter hollow shaft (7) supported by a roller (2) and a ball bearing (5) on the turbine and compressor side respectively. Both supports are connected to the stator by two sets of four bars (3 and 4) subject to bending when the rotor moves laterally.

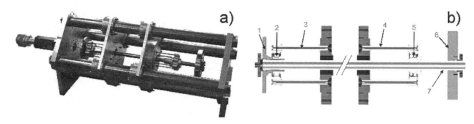

Figure 3: a) Photograph of Rotordynamics test rig b) Section view of the rotor of the machine and of the elastic supports. (1) Low pressure turbine (LPT) disk, (2) LPT roller bearing, (3) LPT beam support, (4) Low pressure compressor (LPC) beam support, (5) LPC ball bearing, (6) LPC disk, (7) hollow shaft.

3.2 Mechanical and electromechanical design of dampers

According to the previous specification the design of the test rig and the design of dampers starts with the identification of mechanical parameters, then, once the mechanical structure is defined, the electromechanical part (actuators) is devised.

The rotor and its supports have been designed through the following steps:
- Design the free-free rotor to satisfy rotordynamics requirements (gyroscopic effects, rotor bending modes…)
- Consider the rotor supported by means of springs and determine the first two (LPC and LPT) critical speeds
- Check the strain energies related to the supports (over all the rotation speed range) to assure adequate values associated to them.

Figure 4 shows the Campbell diagram of the rotor on supports (springs only): the first two modes are the support modes (characterized by a low gyroscopical behaviour) while the third mode is the rotor bending mode. The first two critical speeds are in the operating range while the third is far above. In this diagram, in addition to the natural frequencies as a function of the spin speed, the support and rotor strain energies (as percentage of the total amount of strain energies) are also reported. According to the design requirements a high value of strain energy has been obtained in the supports.

Following the above specification, the design leads to a rotor with a total mass of 4.15 kg and a polar/transverse moment of inertia ratio of 1/27. The required stiffness of 300 kN/m at both the supports is provided by a set of four bars having a length of 75 mm and a diameter of 4 mm (Figure 3b).

The rotor lateral dynamic behaviour was assessed through use of DYNROT, an FE code specific for rotordynamics.

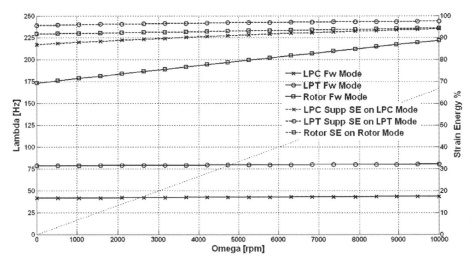

Figure 4: Campbell Diagram of the RTR (supported rotor).
With reference to the right scale the strain energies are reported.

3.2.1 AMD Design

As the strain energy related to the first two modes is mostly in the supports, during the crossing of the first two critical speeds, the elastic and inertia forces balance each other while the damping forces must counteract the unbalance forces.

The design criterion for AMD is based on the assumption that the actuator forces should be able to counteract the unbalance ones.

To address the dimensioning of the actuator a sensitivity analysis has been performed. A set of unbalance responses of the rotor supported by springs and viscous dampers have been computed by varying the damping level (both on LPC and LPT supports) for a static unbalance of G6.3 on both disks. For each damping value, the maximum of the forces exerted by the viscous dampers has been computed. According to these results a suitable design point could be the lowest value of damping that achieves the acceptable displacements and forces. In this case, 10Ns/m for LPT and 20Ns/m has been chosen; the relative total force is about 4 N. Figures 6 show the unbalance response (displacements and velocities) and the relative forces exerted by the springs and dampers. The dampers design force is about 4N with an airgap that should be larger than the displacements at the supports; a value of 0.5mm is suitable.

Once the maximum force and the relative airgap has been identified the design of the actuator can be perfomed as a standard Maxwell force actuator, in order to identify mechanical dimensions and winding parameters.

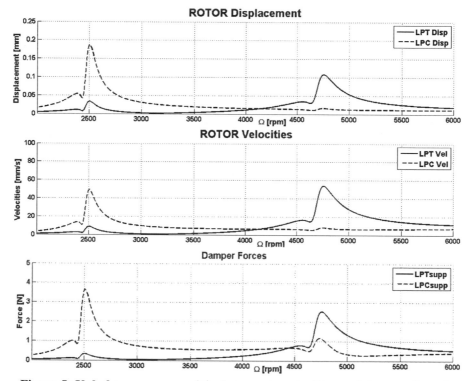

Figure 5: Unbalance response (displacements, velocities and damper forces) of the RTR with the identified amount of damping (viscous).

3.2.2 SAMD Design

The design method of the SAMD is quite different from the active damper even if it starts from similar assumptions. Once the unbalance response has been computed and the relative values of damping have been chosen (as performed for the AMD design), instead of using the amount of force exerted by the dampers the damping coefficients themselves have to be considered. The viscous damping coefficient should be translated into the equivalent modal damping ζ computed at the design frequencies. In addition to the damping factor, SAMDs are characterized by the frequency range where the device performs as a damper. The frequency range can be selected considering Figure 2b: the flat part of the characteristic should coincide with the working frequency. As far as the critical speeds are crossed the damper should enter in the spring range (high frequencies) so as to decrease the transmissibility. The selection of the supply voltage deals only with electrical parameters such as available voltage source and the type of winding. These values will limit vibration amplitude within the required range while crossing the critical speed. (For the detailed design procedure refer to [19]).

3.3   Experimental Validation

The application of the design procedure has been validated by testing the dampers on the RTR. In order to build a reliable actuator with a size compatible with the test bench the dampers (both AMDs and SAMDs) have been oversized relative to those resulting from the design process.

Only one support has been completely instrumented and monitored during the experimental validation. During the tests a standard P-D control has been selected for the AMD, while a supply power of approximately 50W has been used for the SAMD. The following results report the behaviour measured at the LPC support.

The first tests were performed at standstill; the rotor was struck with an instrumented hammer at the LPC node while the displacement at the damper level is measured. The undamped system peak at about 38 Hz is highly reduced when the dampers are switched on. The typical Unbalance response of the rotor while crossing the first critical speed (related to the LPC mode) is reported in Figure 6. The peak amplitude is reduced about 10 times, from 0.25mm to less than 0.03mm.

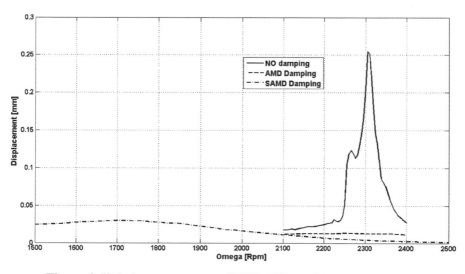

Figure 6: Unbalance response of RTR: Effect of AMD and SAMD applied on LPC support compared with the undamped system

769

4   CASE II: REAL AERO-ENGINE

The design methodology adopted in the previous test case has been used to investigate the feasibility of applying the electromagnetic dampers on a real aircraft engine. At this stage only the AMD solution has been investigated.

4.1   Engine Description and Specifications

The gas turbine engine used in this study is shown in Figure 7. An FE model was used to represent the dynamics of the LP rotor and stator in a test bed configuration with test bed intake and nacelle components. In the aircraft case, transmission of engine vibration into the airframe is most possible for the lowest forcing frequency, so the study focused on vibration of the LP rotor. Also, in many engine designs, there tends to be more space available around the LP rotor bearings where AMDs might be installed.

The front, mid and rear bearing housings have been considered for the AMDs application; they are highlighted in Figure 7. Parameters such as the loads exerted at typical flight conditions, temperatures, cooling and space availability have been assessed. The most suitable locations were chosen to be the Front Bearing Housing (FBH) and the Tail Bearing Housing (TBH). Considering the maximum operating temperature of AMDs being 250°C, cooling would be essential at both rotor supports.

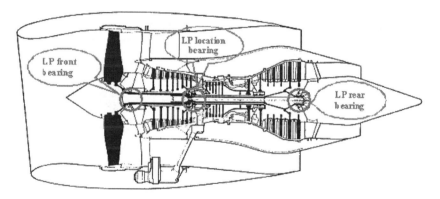

Figure 7: Gas turbine engine

4.2   Mechanical Design

The design procedure requires both rotor and stator models: a lumped parameter finite element model of the LP rotor was constructed using Dynrot by means of beam elements and lumped mass/inertia. A NASTRAN model of the rest of the engine was substructured to release the space for AMD inclusion and then reduced using Component Mode Synthesis. The AMD connections were modelled separately as a parallel arrangement with the mechanical supporting stiffness. The LP rotor and the engine model were assembled together via the AMD connections; in this case the AMD springs work in series to the support (bearing housing) stiffness.

770

As for the RTR case, a preliminary run of the rotor model was performed with a rigid connection between rotor and stator. According to these results the strain energies of the supports modes were low, this would lead to an oversizing of the AMD in order to cope with the reduced efficiency. To increase the strain energy the support stiffness was decreased: this led to higher value of energies but the increased compliance had to be limited to avoid rotor touch down during high manoeuver loads on a too compliant a support.

4.3 Electromagnetic Design

Once the AMD stiffness has been identified the whole engine model has been assembled with viscous dampers in parallel to the spring connecting the stator to the rotor. A sensitivity analysis has been performed on the viscous damping coefficient of those viscous dampers to identify the values that allow comply with the maximum displacement limit when crossing the critical speeds. The AMD stiffness allows maximum displacements at support level of about 0.3mm, when the aircraft manoeuver loads are applied. Hence the actuator airgap has been chosen to be 0.5mm.

Starting from the above mentioned data the actuators have been designed. The electromechanical requirements should dictate the design, rather than the mechanical, especially as to the required space inside the engine. A preliminary design showed that the mass of both actuators should be about 10-15 kg; however in this computation the effect of the thermal constraint is not taken into account: usually working at higher temperature decreases the force density of the actuator, so the mass should be increased. On the other hand this design considers an application duty cycle of 100% that is clearly conservative. Besides the actuator mass, the spring mass (about 3-4 kg) and the power electronic mass (not yet investigated but probably a little lower than the actuator mass) should be added. The whole damping system could have a total mass of about 25kg.

4.4 Numerical Results

The lowered stiffness of the supports, including the added springs, allows the increment of strain energy of a factor of 2 from the original value. Critical speeds are subsequently lowered but the effect of damping allows the displacements to be reduced inside the allowed range.

Figure 8a and Figure 8b reports a set of unbalance response (velocities and force) of various reference points along the structure of the engine; all displacements and velocities are well inside the maximum allowed quantities. The response levels are comparable to those expected from the conventional engine design; furthermore the active nature of the damper allows high tunability of the response. The damper forces are in the range of 1000N as specified during the design.

Figure 8c shows a preliminary sketch of the AMD installed in the available locations inside the engine.

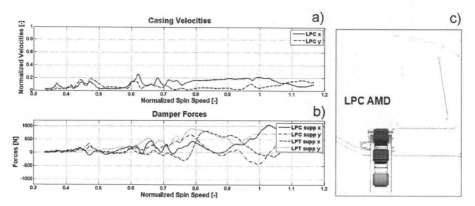

Figure 8: a) Unbalance response (NOT to scale): velocities of reference node on stator ; b) Unbalance response: forces exerted by the dampers; c) Sketch of the designed AMD inserted in the real engine for low pressure compressor support.

5   CONCLUSIONS

This paper presents an overview of feasibility studies within the ADVACT EC project on electro-magnetic damping devices for aerospace application. Although it is mainly focused on active magnetic dampers and semi-active magnetic dampers, some other passive, active and semi-active electro-magnetic damping devices were considered in this work. The paper shows the basic principles of how the dampers work, and describes the methodology adopted for the design of the damping system in order to comply with the specification about rotor response. The electro-magnetic damping devices were assessed on the scaled prototype test rig before being proposed for further computational studies using a real aero engine model. It represents some possibility for installation on the LP rotor within the front and rear structure of the engine and preliminary computational results look very promising. Furthermore the proposed solutions offer some advantages and capabilities, such as more consistent performances, tunability, reliability, online monitoring and diagnostics, compared with the traditional devices. Nevertheless, further design and experimental studies with a full aero engine rotor would be required to validate the technology.

6   ACKNOWLEDGMENTS

The present research has been carried out within the ADVACT project supported by the sixth research framework program of the European Union. The authors thanks all the researchers involved in the project.

7 BIBLIOGRAPHY

[1] Y. K. Ahn, B-S. Yang, S. Morishita, 2002, "Directional Controllable Squeeze Film Damper Using Electro-Rheological Fluid", ASME Journal of Vibration and Acoustics, vol. 124, pp. 105-109.

[2] J. M. Vance, D. Ying, 2000, "Experimental Measurements of Actively Controlled Bearing Damping with an Electrorheological fluid", ASME Journal of Engineering for Gas Turbines and Power, vol. 122, pp 337 - 344.

[3] M. Kasarda, M. Johnson, J. Imlach, H. Mendoza, G. Kirk, H. Bash, 2000, "Applications of a Magnetic Bearing Acting as an Actuator In Conjunction with Conventional Support Bearings", Proceedings of the 9th Int. Symposium on Magnetic Bearings, Lexington, USA, pp. 1 – 6.

[4] J. M. Vance, D. Ying, , J. L. Nikolajsen, 2000, "Actively Controlled Bearing Dampers for Aircraft Engine Applications", ASME Journal of Engineering for Gas Turbines and Power, vol. 122, pp. 466 - 472.

[5] J. Meisel, 1984, Principles of Electromechanical Energy Conversion, Robert Krieger, Malabar, Florida.

[6] S. H. Crandall, D. Karnopp, E. F Kurtz, E. C. Pridmore-Brown, 1968, Dynamics of Mechanical and Electromechanical Systems, New York: McGraw-Hill.

[7] K. E., Graves, D. Toncich, P. G. Iovenitti, 2000, "Theoretical Comparison of the Motional and Transformer EMF Device Damping Efficiency" Journal of Sound and Vibration, Vol. 233, No. 3, pp 441-453.

[8] A. Tonoli, N. Amati, M. Padovani, 2003, "Progetto di Smorzatori e Giunti Elettromagnetici per Alberi a Gomiti", Proceedings of the XXIII AIAS Conference, Salerno, 3-6.

[9] D. Karnopp, 1989, "Permanent Magnets Linear Motors used as Variable Mechanical Damper for Vehicle Suspension", Vehicle System Dynamics, vol. 18, pp. 187-200.

[10] D. Karnopp, D. L. Margolis, R. C. Rosenberg, 1990, System Dynamics: a Unified Approach, J. Wiley & Sons.

[11] Y. Kligerman, O. Gottlieb, 1998, "Dynamics of a Rotating System with a Nonlinear Eddy-Current Damper", ASME Journal of Vibration and Acoustics, Vol. 120, pp. 848-853.

[12] Y. Kligerman, A. Grushkevich, M. S. Darlow, 1998, "Analytical and Experimental Evaluation of Instability in Rotordynamics System with Electromagnetic Eddy-Current Damper", ASME Journal of Vibration and Acoustics, Vol. 120, pp. 272-278.

[13] A.Tonoli, S. Carabelli, N. Amati, 2003, "Design of Semiactive Vibration Dampers for Automotive Applications", Proceedings of the International Workshop ROBOTICS AND MEMS FOR VEHICLE SYSTEMS, Roma, pp. 99-116.

[14] A.Tonoli, 2007, "Dynamic characteristics of eddy current dampers and couplers", Journal of Sound and Vibration, vol. 301, pp. 576-591.

[15] G. Genta, 2004, Dynamics of Rotating Systems, Springer Verlag.

[16] S. Carabelli, P. Macchi, M. Silvagn, A. Tonoli, M. Visconti, 2005, Rotordynamics signature for embedded systems, ISCORMA 3, Cleveland, Ohio.

[17] Holmes R., Sykes J.E.H., 1996, The Vibration of an aero-engine rotor incorporating two squeeze-film dampers, Proceedings of the Institution of Mechanical Engineers, Part G: Journal of Aerospace Engineering, vol. 210, pp. 39-51.

[18] N. Amati, S. Carabelli, G. Genta, P. Macchi, M. Silvagni, A. Tonoli, More electric aero engines tradeoff between different electromagnetic dampers and supports, Proceedings of the 10th ISMB Conference, Martigny, Switzerland, 21-23 August 2006

[19] N. Amati, M. Silvagni, A. Tonoli, Transformer eddy current dampers in rotating machines vibration control, accepted for publication on Journal of Dynamic Systems, Measurement, and Control

Damping of a flexible rotor by time-periodic stiffness and damping variation

F Dohnal, B R Mace
Institute of Sound and Vibration Research, University of Southampton, UK

ABSTRACT

Rotor systems can become unstable due to self-excitation for various reasons. Self-excitation in a rotor system may occur due to internal damping in a flexible rotor, from rotor/stator interaction or bearing forces. It has been shown recently that a self-excited rigid rotor can be stabilised and vibrations can be suppressed by an open-loop control of the stiffness parameter in the bearing mounts, a stabilisation by parametric stiffness excitation. In this paper this approach is investigated further for more realistic bearing characteristics and for an unstable flexible rotor.

Parametric excitation of bearing stiffness and damping is considered. To understand the potential and limitations of damping by parametric excitation, the stability boundary curves in the system parameter space are investigated. Both direct numerical simulations and analytical predictions are performed to calculate ranges for control and system parameters where damping by parametric excitation is effective. Results from both methods are compared and agree very well. Finally, analytical conditions are given under which the method of damping by parametric stiffness and damping excitation is effective and the damping effect is guaranteed.

1. INTRODUCTION

Self-excited vibrations in rotor systems may arise for a number of reasons, see e.g. [1, 2, 3, 4]. For instance, they may be caused by internal damping of a flexible rotor shaft or by non-conservative forces generated by interacting forces between rotor and stator in seals and gaps. Self-excited vibrations in general lead to large vibration amplitudes which may damage or destroy rotating machinery. The source of self-excitation is not important for the concept under investigation and in this study only one source is considered.

This paper examines an open-loop strategy to suppress self-excited vibrations, which is based on a parametric *anti*-resonance phenomenon that can occur in parametrically excited systems, see [5]. In previous studies [6, 7] a Jeffcott-rotor was investigated with parametric excitation induced by time-periodic change in the stiffness of the bearing mounts. Therein, a rotor with a flexible as well as a rigid shaft was analysed. It was shown that a periodic change in the stiffness may be employed to increase the critical speed of the rotor.

In the present work this approach is investigated further for more realistic bearing characteristics. A realisation of periodic, open-loop control of the stiffness coefficients of a bearing almost certainly also induces a synchronous change of its damping coefficients with the same phase as the stiffness. The interaction of parametric stiffness and damping variation is of importance. In previous studies (e.g. [8]), it was shown that the effect of a periodic stiffness variation acts in opposition to a periodic damping variation: for example, the stabilising effect of stiffness excitation may be reduced by a destabilising effect of the synchronous damping excitation, which may lead to an overall loss of stability and, henceforth, a decrease of the critical rotational speed of the rotor.

In the first section, the model equations for a flexible rotor are derived. The source of self-excitation is not important for the method of damping by parametric excitation and in this study only one source is considered. Then a general stability analysis is performed employing a singular perturbation technique. Finally, an example rotor is discussed numerically and analytically.

2. MODEL EQUATIONS

The mechanical model of the rotor is shown in Fig. 1 and is based on the classical Jeffcott rotor where both the shaft and the bearings are considered as deformable bodies. Although the Jeffcott rotor model is an oversimplification of real-world rotors, it retains some basic characteristics and allows a qualitative insight into important phenomena typical to rotordynamics to be gained. Since we want to investigate the stability of the rotor system in its undeflected position a linear(ised) model is sufficient for this study.

A massless, flexible shaft whose axis is vertically rotating at a constant rotational speed ω is considered, $\dot{\omega} = 0$. The shaft is assumed to be torsionally rigid and isotropic. A lumped mass m_d, representing a rigid disk, is attached to the rotating shaft at a distance of l_1 from bearing $b1$. The disc is assumed to be balanced. A basic numerical investigation of the interaction of an unbalance excitation and a Jeffcott rotor supported by equal bearings with time-periodic stiffness can be found in [9].

The rotor is supported at two bearing supports $b1$ and $b2$. Lumped mass parameters m_{bj} are assumed to represent mass concentrations of the rotating counterparts of the bearings. Each bearing possesses a radial stiffness $k_{bj}(t)$ and damping $c_{bj}(t)$, respectively. The displacements of the disc and the bearings are described by two translational degrees of freedom x_k and y_k, where index k denotes the disc d or a bearing $b1$, $b2$.

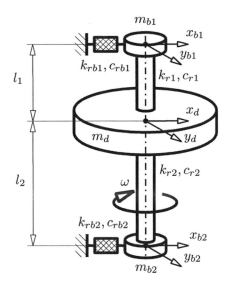

Figure 1: Symmetric rotor model.

In this study, these properties are time-dependent and are changed periodically according to

$$k_{bj}(t) = k_{bj}(1 + \varepsilon_{kj} \cos \nu t), \tag{1a}$$

$$c_{bj}(t) = c_{bj}(1 + \varepsilon_{cj} \cos \nu t), \tag{1b}$$

with the parametric excitation frequency ν and amplitude ratios ε_{kj}, ε_{cj}. Note that the stiffness and damping properties of both bearings are varied with the same frequency and with no phase lag, which is referred to as a synchronous parametric stiffness and damping excitation [8].

The complex notation

$$z_k = x_k + j y_k \tag{2}$$

is introduced where $j = \sqrt{-1}$. The linear elastic shaft generates a restoring force proportional to the relative deflection of the shaft

$$F_d = -k_{r1}(z_d - z_{b1}) - k_{r2}(z_d - z_{b2}). \tag{3}$$

In rotating flexible rotors, damping effects are divided into two categories: internal and external damping [3]. External refers to stationary elements, e.g. energy dissipation due to micro-stick-slip friction in rotor supports or labyrinth seals. These forces depend on the absolute velocities. Internal refers to the rotating elements, e.g. structural damping of the shaft (c_{ri}). These forces depends on the relative velocities. Forces resulting from internal damping of the shaft are generated by a relative motion of the disc with respect to the bearing supports. For our model in Fig. 1, internal damping leads to a force described by [2]

$$F_i = -c_{r1}(\dot{z}_d - \dot{z}_{b1}) - c_{r2}(\dot{z}_d - \dot{z}_{b2}) + j\omega \left[c_{r1}(z_d - z_{b1}) + c_{r2}(z_d - z_{b2}) \right]. \tag{4}$$

777

The linearised forces acting at the bearing supports are

$$F_{bj} = -k_{bj} z_{bj} - c_{bj} \dot{z}_{bj} \tag{5}$$

Applying Newton's law, three linear differential equations of second order in complex notation are obtained. Written in matrix notation, these equations become

$$\mathbf{M}\ddot{\mathbf{z}} + (\mathbf{C} + \mathbf{C}_\nu(t))\dot{\mathbf{z}} + (\mathbf{K} + \mathbf{K}_\nu(t) - j\omega\mathbf{C})\mathbf{z} = \mathbf{0}, \tag{6}$$

with the complex position vector $\mathbf{z} = [z_d, z_{b1}, z_{b2}]^T$, the mass matrix $\mathbf{M} = \mathrm{diag}(m_d, m_{b1}, m_{b2})$ and the remaining coefficient matrices

$$\mathbf{K} = \mathbf{H}(k), \quad \mathbf{C} = \mathbf{H}(c), \quad \mathbf{K}_\nu(t) = \mathbf{H}_\nu(t; k), \quad \mathbf{C}_\nu(t) = \mathbf{H}_\nu(t; c), \tag{7a}$$

$$\mathbf{H}(h) = \begin{bmatrix} h_{r1} + h_{r2} & -h_{r1} & -h_{r2} \\ -h_{r1} & h_{r1} & 0 \\ -h_{r2} & 0 & h_{r2} \end{bmatrix}, \quad \mathbf{H}_\nu(t; h) = \mathrm{diag}(0, h_{b1}(t), h_{b2}(t)) \tag{7b}$$

Note that eq. (6) is homogeneous since the disc in Fig. 1 is assumed to be balanced and no additional external forces are acting on the rotor.

For simulation, these equations are rewritten in real notation resulting in six linear differential equations of second order. According to eq. (2), the equations of motion in eq. (6) yield

$$\mathbf{M}_r\ddot{\mathbf{x}} + \mathbf{C}_r(t)\dot{\mathbf{x}} + \mathbf{K}_r(t)\mathbf{x} = \mathbf{0} \tag{8}$$

with the real position vector $\mathbf{x} = [x_d, x_{b1}, x_{b2}, y_d, y_{b1}, y_{b2}]^T$ and the coefficient matrices

$$\mathbf{M}_r = \begin{bmatrix} \mathbf{M} & \mathbf{0} \\ \mathbf{0} & \mathbf{M} \end{bmatrix}, \quad \mathbf{K}_r(t) = \begin{bmatrix} \mathbf{K}(t) & \omega\mathbf{C} \\ -\omega\mathbf{C} & \mathbf{K}(t) \end{bmatrix}, \quad \mathbf{C}_r(t) = \begin{bmatrix} \mathbf{C} + \mathbf{C}_\nu(t) & \mathbf{0} \\ \mathbf{0} & \mathbf{C} + \mathbf{C}_\nu(t) \end{bmatrix}. \tag{9}$$

The coefficient matrices in eqs. (7) are symmetric with respect to the complex position vector. However, the real counterpart of the symmetric imaginary matrix $i\omega\mathbf{C}$ is skew-symmetric, which corresponds to the matrix of non-conservative forces leading to self-excitation.

3.   ANALYTICAL STABILITY ANALYSIS

In general, analytical solutions to linear differential equations with periodic, time-dependent coefficients do not exist. The exact solution can be approximated by performing a singular perturbation as used in the literature in various fields of physics. Well-known examples of these techniques are harmonic balance (or two-timing), Poincaré-Lindstedt method, averaging method, method of multiple scales or successive approximation. General review of these methods can be found in [10], [11] and [12]. The method chosen here is the averaging method, see [13], and represents an extension of the results presented in [8]. A local approximation to the solution will be derived for a first order perturbation.

The system matrices in eq. (9) are split into a constant and a time-periodic part, i.e.

$$\mathbf{K}_r(t) = \mathbf{K}_0 + \varepsilon \mathbf{S}_0 + \varepsilon \mathbf{K}_t \cos(\nu t), \quad \mathbf{C}_r(t) = \varepsilon \mathbf{C}_0 + \varepsilon \mathbf{C}_t \cos(\nu t), \qquad (10)$$

where ε is a parameter and the indices 0, t denote the constant and time-dependent part of the corresponding matrix, respectively. The matrix $\mathbf{S}_0$ represents a skew-symmetric matrix of non-conservative forces. Rescaling by a parameter ε enables an approximation to the slow dynamics of the system to be found by a first order perturbation that is valid on the time scale $1/\varepsilon$. The coordinate transformation $\mathbf{x} = \mathbf{T}\mathbf{z}$ is introduced, where $\mathbf{T}$ is the modal matrix of the undamped, time-independent system $\mathbf{M}_r \ddot{\mathbf{x}} + \mathbf{K}_0 \mathbf{x} = \mathbf{0}$. The equations of motion are transformed to the normal form

$$\mathbf{z}''(t) + \boldsymbol{\Omega}^2 \mathbf{z}(t) = -\varepsilon \left\{ \boldsymbol{\Theta} + \mathbf{R} \cos(\nu t) \right\} \mathbf{z}'(t) - \varepsilon \left\{ \mathbf{N} + \mathbf{Q} \cos(\nu t) \right\} \mathbf{z}(t), \quad (11)$$

where

$$\boldsymbol{\Omega}^2 = \mathbf{T}^{-1} \mathbf{M}_r^{-1} \mathbf{K}_0 \mathbf{T}, \quad \mathbf{N} = \mathbf{T}^{-1} \mathbf{M}_r^{-1} \mathbf{S}_0 \mathbf{T}, \qquad (12a)$$

$$\boldsymbol{\Theta} = \mathbf{T}^{-1} \mathbf{M}_r^{-1} \mathbf{C}_t \mathbf{T}, \quad \mathbf{Q} = \mathbf{T}^{-1} \mathbf{M}_r^{-1} \mathbf{K}_t \mathbf{T}, \quad \mathbf{R} = \mathbf{T}^{-1} \mathbf{M}_r^{-1} \mathbf{C}_t \mathbf{T}. \qquad (12b)$$

From now on the scaling factor ε is assumed to be small.

The time transformation $\nu t \mapsto \tau$, $\mathbf{z}(t) \mapsto \mathbf{z}(\tau)$ is applied to eq. (11) in order to normalise the frequency ν to become one on the chosen time scale. A small detuning of first order near ν_0 is introduced in the form of

$$\nu = \nu_0 + \varepsilon \sigma + \mathcal{O}\left(\varepsilon^2\right). \qquad (13)$$

For a first order approximation all terms of higher order than ε are neglected. The coordinate transformation

$$z_i = u_i \cos \varpi_i t + v_i \sin \varpi_i t, \qquad \dot{z}_i = -u_i \varpi_i \sin \varpi_i t + v_i \varpi_i \cos \varpi_i t, \qquad (14)$$

with $\varpi_i = \Omega_i / \nu_0$, is performed that satisfies the unperturbed equations ($\varepsilon = 0$). This results, see [8] for details, in a system of first order differential equations of the form

$$\dot{\mathbf{u}}(\tau) = \varepsilon \mathbf{f}(\mathbf{u}, \tau), \qquad (15)$$

where $\mathbf{u} = [u_1, v_1, \ldots, u_n, v_n]$ and $\mathbf{f}$ is a vector field. For this system the averaging method in the general case from [13] can be applied, where the difference between the solutions $\mathbf{u}$ of the original and $\hat{\mathbf{u}}$ of the averaged system is of order ε, $\hat{u}_i(t) - u_i(t) = \mathcal{O}(\varepsilon)$, on the time scale $1/\varepsilon$.

The following study will be performed for the case of a parametric combination frequency of first order defined by

$$\nu_0 = |\Omega_k \mp \Omega_l| \neq 0, \qquad (16)$$

where Ω_i is the ith natural frequency of the system defined by eq. (12a). Combination resonances of higher order may be applied to suppress self-excited vibrations, too, but in general their effectiveness is negligible.

The complex abbreviations for the solution $\hat{\mathbf{u}}$ averaged over τ is introduced:

$$\hat{w}_i = \hat{u}_i + j\hat{v}_i \quad \text{for} \quad (i \neq l), \quad \text{and} \quad \hat{w}_l = \hat{u}_l \pm j\hat{v}_l. \tag{17}$$

Averaging eq. (15) results in

$$\begin{bmatrix} \dot{w}_k \\ \dot{w}_l \end{bmatrix} = \frac{\varepsilon}{\nu_0^2} \begin{bmatrix} -\frac{\nu_0}{2}\Theta_{kk} + j\Omega_k\sigma - j\frac{\nu_0}{2\Omega_k}N_{kk} & \mp\frac{\nu_0}{4\Omega_k}\left(\Omega_l R_{kl} \pm jQ_{kl}\right) \\ \mp\frac{\nu_0}{4\Omega_l}\left(\Omega_k R_{lk} + jQ_{lk}\right) & -\frac{\nu_0}{2}\Theta_{ll} \pm j\Omega_l\sigma - j\frac{\nu_0}{2\Omega_l}N_{ll} \end{bmatrix} \begin{bmatrix} w_k \\ w_l \end{bmatrix}, \tag{18a}$$

$$\dot{w}_i = \frac{\varepsilon}{\nu_0^2}\left\{ -\frac{\nu_0}{2}\Theta_{ii} + j\Omega_i\sigma - j\frac{\nu_0}{2\Omega_i}N_{ii} \right\} w_i \quad \text{for} \quad (i \neq k, l) \tag{18b}$$

where the upper signs correspond to $\nu_0 = |\Omega_k - \Omega_l|$ and the lower signs to $\nu_0 = \Omega_k + \Omega_l$. Setting $N_{ij} = 0$, eqs. (18) coincide with these given in [8].

3.1 Stability conditions

First, the stability of the system without parametric excitation is analysed, i.e. $Q_{ij} = 0 = R_{ij}$. In this case, eq. (18a) simplifies to eq. (18b) for all i. This system of equations is stable if and only if

$$\Theta_{ii} > 0. \tag{19}$$

Now it will be analysed whether these condition can be relaxed with parametric excitation present in the system. A parametric excitation, $Q_{ij}, R_{ij} \neq 0$, is introduced with a frequency close to ν_0 as defined in eq. (16). The characteristic equation of the coefficient matrix in eq. (18a) is a complex polynomial of order two. The stability of this polynomial is determined by applying the extended Routh-Hurwitz criterion for complex polynomials in e.g. [14]. For the general case of $\sigma \neq 0$ in eq. (13), the following two conditions have to be satisfied for the system in eq. (18a) to be stable:

$$\Theta_{kk} + \Theta_{ll} > 0, \tag{20a}$$

$$a_0\sigma^2 + a_1(Q_{ij}, R_{ij})\sigma + a_2(Q_{ij}, R_{ij}) > 0, \tag{20b}$$

with scalar functions $a_i(Q_{ij}, R_{ij})$. The condition in eq. (20b) defines the critical values

$$\sigma_1 = \sigma_s - \hat{\sigma}_{kc} \quad \text{and} \quad \sigma_2 = \sigma_s + \hat{\sigma}_{kc} \tag{21}$$

of the detuning σ in eq. (13). The frequency shift σ_s and the frequency width $\hat{\sigma}_{kc}$ are

$$\sigma_s = \frac{(\Theta_{kk} - \Theta_{ll})}{2\Theta_{kk}\Theta_{ll}}\frac{\beta}{2} + \frac{N_{kk}}{\Omega_k} - \frac{N_{ll}}{\Omega_k}, \qquad \hat{\sigma}_{kc} = \frac{(\Theta_{kk} + \Theta_{ll})}{2\Theta_{kk}\Theta_{ll}}\sqrt{d_{kc}}, \tag{22}$$

with the parametric excitation term

$$d_{kc} = -\Theta_{kk}\Theta_{ll}\left(\Theta_{kk}\Theta_{ll} \pm \frac{Q_{kl}Q_{lk}}{4\Omega_1\Omega_2} - \frac{R_{kl}R_{lk}}{4}\right) + \left(\frac{\beta}{2}\right)^2, \tag{23}$$

780

and the interaction term

$$\beta = \mp \frac{Q_{kl}R_{lk}}{4\Omega_2} - \frac{R_{kl}Q_{lk}}{4\Omega_1}. \qquad (24)$$

Note that the interaction term β vanishes in case of a pure stiffness excitation, $R_{ij} = 0$, as well as in case of a pure damping excitation, $Q_{ij} = 0$. For a system without non-conservative forces, $N_{ij} = 0$, the conditions in eq. (22) simplify to the ones found in [8]. A necessary condition for the condition in eq. (20b) to be met is that the parametric excitation term in eq. (23) is positive leading to real-valued critical values $\sigma_{1/2}$ in eq. (21). These critical values determine the stability border in the system parameter space. If this condition is fulfilled then a parametric *anti*-resonance near η_0 with a width of $\widehat{\sigma}_{kc}$ in eq. (22) is obtained. Otherwise no damping by parametric excitation is possible and the vibration amplitudes grow without restriction.

These conditions for vibration suppression can be simplified for the case of a pure parametric stiffness excitation, $R_{ij} = 0$. Consequently, the interaction term β in eq. (24) vanishes, while a frequency shift σ_s in eq. (22) remains for $N_{ii} \neq 0$. The frequency width $\widehat{\sigma}_{kc}$ simplifies to

$$\widehat{\sigma}_{kc}|_{R_{ij}=0} = \widehat{\sigma}_k = \frac{\Theta_{kk} + \Theta_{ll}}{2} \sqrt{-\frac{\Theta_{kk}\Theta_{ll} \pm \dfrac{Q_{kl}Q_{lk}}{4\Omega_k\Omega_l}}{\Theta_{kk}\Theta_{ll}}}, \qquad (25)$$

an expression derived in [5], [15] and [16].

Summarising, for a system with a parametric excitation at $\nu_0 = |\Omega_k - \Omega_l|$, the stability conditions in eq. (19) can be relaxed if $d_{kc} > 0$ and the condition in eq. (20b) is satisfied, i.e. the parametric excitation frequency lies within the interval

$$\nu_0 + \varepsilon\sigma_s - \varepsilon\widehat{\sigma}_{kc} \leq \nu \leq \nu_0 + \varepsilon\sigma_s + \varepsilon\widehat{\sigma}_{kc}, \qquad (26)$$

according to the definitions in eqs. (13, 21). Then, the stability conditions in eq. (19) relaxes for $i = k, l$ to the weaker condition in eq. (20a). The conditions for $i \neq k, l$ in eq. (19) remain unaltered.

4.   EXAMPLE

To verify the results of the analytical analysis an exact method based on Floquet's theorem [13] is used in combination with numerical time integration. For fixed system parameters, the system equations in eqs. (8,9) are integrated numerically for one period of the parametric excitation for linear dependent initial conditions. The eigenvalues of the monodromy matrix determine the stability of the system (see [17] for details).

For parameter studies it is convenient to introduce non-dimensional parameters, with respect to the characteristic frequency of the disc-shaft subsystem $\omega_r =$

$\sqrt{(k_{r1} + k_{r2})/m_d}$, as defined in e.g. [6],

$$M_j = \frac{m_d}{m_{bj}}, \quad Q_j = \frac{1}{\omega_r}\sqrt{\frac{k_{bj}}{m_{bj}}}, \quad \varpi = \frac{\omega}{\omega_r}, \quad \eta = \frac{\nu}{\omega_r}, \tag{27a}$$

$$\kappa_{bj} = \frac{c_{bj}}{m_{bj}\omega_r}, \quad \kappa_r = \frac{c_r}{m_r\omega_r}, \quad \tilde{\Omega}_i = \frac{\Omega_i}{\omega_r}. \tag{27b}$$

Table 1: Non-dimensional model parameters.

M_1	M_2	Q_1	Q_2	κ_{r1}	κ_{r1}	κ_{b1}	κ_{b2}
4.0	4.0	2.0	2.0	0.05	0.05	0.5	0.5

The flexible rotor system in Fig. 1 is examined for a given set of the non-dimensional parameters defined in eq. (27). The chosen values of the non-dimensional system parameters are listed in Table 1, representing a symmetric arrangement with highly damped support stations. The corresponding Campbell diagram for the damped rotor system, with all system parameters fixed, is shown in Fig. 2. Due to the self-excitation induced by internal damping $\kappa_r = \kappa_{r1} = \kappa_{r2}$ of the flexible shaft, the rotor becomes unstable in the first mode at the non-dimensional critical speed $\varpi_{crit} \approx 1.49$, which is indicated by circles in Fig. 2. For activated parametric excitation, this critical speed can be changed as will be shown in the following.

Numerically generated stability domains are shown in Fig. 3 as functions of the non-dimensional rotor speed ϖ and the control parameters $\varepsilon_k = \varepsilon_{k1} = \varepsilon_{k2}$ and η, as introduced in eqs. (1,27). Figure 3a shows the result for a rotor system with a parametric stiffness excitation at the support stations, while Fig. 3b shows the result for a parametric stiffness excitation with an additional damping variation at a large constant variation of $\varepsilon_c = \varepsilon_{c1} = \varepsilon_{c2} = 0.8$. Combinations of ϖ, ε_k and η, which are enclosed by the shaded surface (indicating the

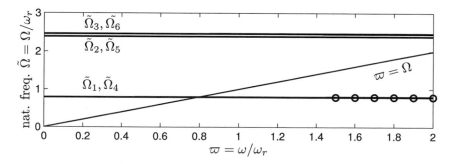

Figure 2: Campbell diagram of model 1 in Table 1 for system without parametric excitation.

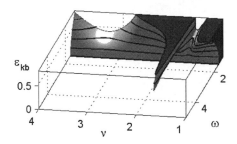 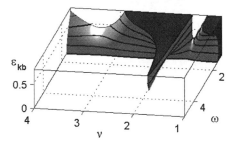

(a) Parametric stiffness excitation. (b) Parametric stiffness and damping excitation.

Figure 3: Numerical stability domains for different kinds of parametric excitation of model 1.

stability boundary) lead to a stable trivial solution. Highly complex behaviour is observed. From Fig. 3 it can be seen that the system without parametric excitation, $\varepsilon_{kb} = 0$, is stable if the rotor speed remains below its critical speed ϖ_{crit}, which corresponds to the instability described in Fig. 2. For higher rotor speeds, $\varpi > \varpi_{crit}$ the trivial solution of the rotor system becomes unstable.

Activating a parametric stiffness excitation, $\varepsilon_k > 0$ in Fig. 3, generates a major spike close to $\eta \approx 1.73$ that reaches far into the region of $\varpi > \varpi_{crit}$. This gain in stability enables the rotor to be driven safely far beyond its critical speed if the control parameters ε_k and η are adjusted accordingly. A significant decrease in critical speed is encountered in a wide range around $\eta \approx 3.3$ and in a narrow range near $\eta \approx 1.6$. Hence, setting the control parameters incorrectly may lead to a loss of stability. These stability losses and gains are now analysed in more detail.

Slices for constant values of the control parameter ε_k of this stability domain are shown in more detail in Fig. 4. These numerically obtained results are compared to the analytical prediction derived in the previous section. A key to the interpretation are the parametric resonance frequencies of first order defined in eq. (16). As predicted by the analytical analysis, the main area of stability occurs near a combination resonance frequency. Since the system matrices in eq. (9) are symmetric, the parametric anti-resonance occurs only in the vicinity of the parametric combination frequency of difference type, $\eta_0 = |\tilde{\Omega}_1 - \tilde{\Omega}_2|$, while the frequency of summation type, $\eta_0 = \tilde{\Omega}_1 + \tilde{\Omega}_2$, corresponds to a true parametric resonance. A primary parametric resonance occurs close to the frequencies $2\tilde{\Omega}_1$ and $2\tilde{\Omega}_2$. The natural frequencies are determined according to eq. (12aa) and are independent of the rotor speed η. For a specific system parameter p, the local stability boundary eq. (26) in the system parameter space is projected to the p-η-plane. The projection of the stability boundary in the vicinity of $\eta_0 = |\tilde{\Omega}_1 - \tilde{\Omega}_2|$ is plotted as a solid line in Fig. 4 and agrees well

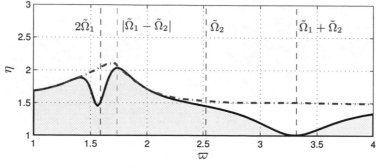

(a) Parametric stiffness excitation.

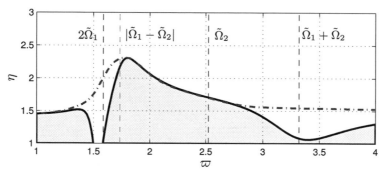

(b) Parametric stiffness and damping excitation.

Figure 4: Comparison of numerical simulation (shaded region) and analytical prediction (broken line) at $\varepsilon_k = 0.3$ and $\varepsilon_c = 0.8$.

with the exact numerical results in the vicinity of the investigated parametric anti-resonance. The projection of the stability boundary in the vicinity of $\eta_0 = \tilde{\Omega}_1 + \tilde{\Omega}_2$ is of similar accuracy but is omitted here.

Figures 3 and 4 summarise the interaction between a time-periodic stiffness variation and a synchronous damping variation of the support stations. According to Fig. 3, for large amplitudes, $\varepsilon_k \approx 0.7$ of the parametric stiffness excitation and far away from the trivial stability boundary $\varpi \approx 1.49$ the stable parameter ranges of the control parameters ε_k and η tend to be independent of an additional damping variation. According to Fig. 4, significant differences of the stable parameter ranges in the parameter domain are observed close to the trivial stability boundary. Despite the fact that a pure parametric damping variation always destabilises the system, the interaction between a stiffness and damping variation may lead to wider stable regions in the vicinity of the parametric anti-resonance, see Fig. 4. Consequently, the stabilising effect of a parametric stiffness variation may be even amplified by a synchronous damping variation.

According to [8], a non-dimensional excitation ratio r between the stiffness and damping excitation can be defined and expressed by the parameters introduced in eq. (12a) as

$$r_{kl} = \frac{\Omega_k Q_{kl} R_{lk} + \Omega_l R_{kl} Q_{lk}}{Q_{kl} Q_{lk} - \Omega_k \Omega_l R_{kl} R_{lk}}.$$ (28)

This ratio is an indicator of the strength of the damping variation R_{ij} with respect to the stiffness excitation Q_{ij}. The effects of the stiffness excitation exceed the effects of the damping excitation if $r \lessapprox 1$ and vice versa if $r \gtrapprox 1$. For $r \lessapprox 1$ the stability gained by pure stiffness variation is conserved while for $r \gtrapprox 1$ the destabilising damping variation outweighs the otherwise stable pure stiffness variation.

5. CONCLUSIONS

A self-excited flexible rotor is analysed having support stations with time-periodic open loop control of the stiffness and/or damping coefficients. Due to self-excitation by internal damping, the flexible rotor becomes unstable if the rotational speed exceeds the critical speed. A stability analysis of the time-periodic equations of motion is performed by applying the concept of singular perturbation. For the first time, analytical conditions for stable ranges of the control parameters are presented that include the skew-symmetric matrix of non-conservative forces. These analytical approximations are benchmarked against numerical results and show very good agreement in the vicinity of the investigated parametric anti-resonance. The difference between a pure stiffness variation and a synchronous stiffness and damping variation is discussed for an example system. Despite the fact that a pure parametric damping variation always destabilises the system, it was shown that the stabilising effect of a parametric stiffness variation may even be amplified by a synchronous damping variation.

ACKNOWLEDGEMENT

The financial support of the EPSCR Platform Grant in Structural Acoustics (EP/E006450/1) is gratefully acknowledged.

REFERENCES

[1] T. Yamamoto and Y. Ishida, Linear and nonlinear rotordynamics, John Wiley, 2001.

[2] R. Gasch, R. Nordmann and H. Pfützner, Rotordynamik, 2nd ed., Springer, 2002.

[3] A. Muszuńska, Rotordynamics, CRC Press, 2005.

[4] G. Genta, Dynamics of rotating systems, Mechanical Engineering Series, Springer, 2005.

[5] A. Tondl, To the problem of quenching self-excited vibrations, *Acta Technica CSAV*, 43 (1998), 109–116.

[6] A. Tondl, Self-excited vibration quenching in a rotor system by means of parametric excitation. Acta Techn. CSAV 45, Institute of Electrical Engineering, Acad. Sci. Czech Republic, 2000.

[7] H. Ecker and A. Tondl, Stabilization of a rigid rotor by a time-varying stiffness of the bearing mounts, 8th Int. Conf. on Vibration in Rotating Machinery, Suffolk, UK, 2004.

[8] F. Dohnal, Suppressing self-excited vibrations by synchronous and time-periodic stiffness and damping variation, *Journal of Sound and Vibration*, 306 (2007), 136–152.

[9] A. Tondl and H. Ecker, Self-excited vibration quenching using parametric excitation: The effect of an additoinal external excitation. In Proc. of 6th Int. Conf. on Vibration Problems (ICOVP2003), Liberec, Czech Republic, Sept. 2003.

[10] R. Mickens, *An introduction to nonlinear oscillations*, Cambridge Press, 1981.

[11] A.H. Nayfeh and D.T. Mook, *Nonlinear Oscillations*, John Wiley and Sons, New York, 3rd edition, 1995.

[12] J.J. Thomsen, *Vibrations and stability – Advanced theory, analysis and tools*, 2nd edition, Springer Verlag, Berlin Heidelberg New York, 2003.

[13] F. Verhulst, *Nonlinear differential equations and dynamical systems, Texts in applied mathematics 50* Springer-Verlag, Berlin Heidelberg New York, 2000.

[14] F. Gantmacher, *Matrizenrechnung Teil II, spezielle Fragen und Anwendungen* (in German), Hochschulbücher für Mathematik, VEB Deutscher Verlag der Wissenschaften, Berlin, 1966.

[15] Abadi, Nonlinear dynamics of self-excitation in autoparametric systems, PhD Thesis, Utrecht University, 2003.

[16] F. Dohnal, Damping by parametric stiffness excitation – resonance and anti-resonance, *Journal of Vibration and Control*, 14 (2008), 669–688.

[17] W.T. Wu, J.H. Griffin and J.A. Wickert, Perturbation method for the Floquet eigenvalues and stability boundary of periodic linear systems, *Journal of Sound and Vibration*, 182 (1995), 245-257.

Application of squeeze-film dampers with a centrifugal compressor

E A Memmott, K Ramesh
Dresser-Rand Company, USA

ABSTRACT

This paper discusses the application of squeeze-film dampers in series with the tilting pad bearings of a centrifugal compressor. The compressor was factory tested successfully both with damper bearings and with non-damper bearings and shipped with damper bearings. Analytical and test results will be presented for both rotor dynamic systems. Design considerations in the use of squeeze-film dampers with tilting pad journal bearings will be reviewed.

1. OBJECTIVE

One of the primary objectives in machinery selection is robustness. Critical unspared compressors are continuously operated for many years between maintenance activities, which is the case with the compressor described in this paper. As a result, the end user and equipment manufacturer required the machine design to be very conservative. It was recognized that the compressor discussed in this paper would have improved mechanical performance with damper bearings, but the end user required that the mechanical performance be acceptable even with the dampers inoperable or outside their design range. The requirement was that the compressor rotor be stable and have acceptable amplification factors in either case. This issue was pursued in the initial and final design reviews, as well as during the factory testing.

Damper bearings consist of placing a squeeze-film damper and support spring in series with the journal bearing, which in this case is a tilting pad type. The application of a squeeze-film damper in series with a tilting pad bearing is to optimize the damping in the rotor dynamic system. This can both increase the stability margin and decrease the amplification factor of the rotor and the vibration amplitude at the midspan. The rotor-support system is softened by the squeeze-film damper resulting in less shaft bending at the first natural frequency, which may even become a critically damped mode. A procedure for using the damper to optimize the rotor dynamic stability and response characteristics will be described.

The end user was concerned with the consequences if the squeeze-film damper lost its effectiveness because of damage or wear. Two different hypothetical conditions were considered, over-damping, caused by an overly eccentric or locked up damper (i.e. acting as a non-damper bearing), and under-damping, caused by loss of sealing at the O-rings. Analytical predictions were made for those two conditions, besides predictions made for the range of design conditions for the squeeze-film damper. Factory tests were performed for the squeeze-film damper within the range of the design conditions and with the squeeze-film damper made inoperable. The factory test was a standard mechanical run which identified the rotor's first lateral critical speed and it's amplification factor. The test results compared well with the predictions and proved the acceptability of the design. Stability parameters were compared analytically. Again, with the damper inoperable or under-damped, the rotor stability predictions indicated acceptable performance.

2. DESCRIPTION OF THE COMPRESSOR

The compressor is a recycle gas compressor installed in a refinery. It has an operating speed of 14338 RPM and is driven by a constant speed synchronous motor through a gear. Figure 1 is a sketch of the train. The gas has a Mole Weight of 3.2. The inlet pressure is 84 BAR (1219 PSIA) and the discharge pressure is 113 BAR (1641 PSIA). The compressor is a nine-stage, straight-through centrifugal compressor. The compressor is driven on the discharge end. The thrust disc is outboard of the journal bearing on the intake end. The casing shaft end seals are dry gas seals. The 5 shoe L/D = 0.44 tilting pad bearings have load between pads, a center pivot on each pad, a crowned pivot in each direction, and direct lubrication. As shipped there

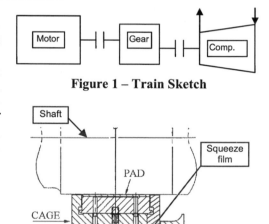

Figure 1 – Train Sketch

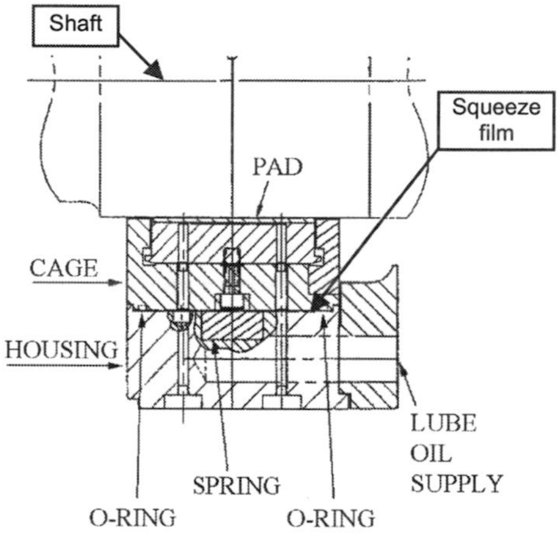

Figure 2 – Damper Bearing

are squeeze-film dampers in series with each journal bearing. Figure 2 shows a sketch of the type of tilting pad damper bearing as used in this compressor. It may appear to be novel, but as seen in Section 4.3 there is a long history of its application. The impeller eye and interstage labyrinths are teeth on rotor with no deswirling features. The stationary material next to these rotating labyrinths is an abradable material made out of PTFE and synthetic mica. The balance piston seal is a stationary aluminum toothed labyrinth with swirl brakes. There are no shunt holes from the last stage diffuser to the balance piston.

3. DISCUSSION OF THE ROTOR DYNAMICS DESIGN AND EVALUATION TOOL AND PROGRAMS

The equipment manufacturer's rotor dynamic software used for the current analysis [1] is a unique state-of-the-art program linking the aerodynamic selection tools and rotor dynamic programs into one cohesive engineering tool. This software takes the aerodynamic output (pressure, temperature, and gas properties at inlet and exit of all the impellers) and sizes the rotor dynamic components such as the couplings, balance piston, thrust bearings, journal bearings and seals. After a satisfactory set of components is obtained, the program then builds the rotor-bearing system based on the rules, logic and data used by configuration tools. The undamped critical speed analysis, the journal bearing analysis, the unbalance response analyses and the stability analysis are then performed. The program has the ability to model damper bearings and include the effects of toothed labyrinths and damper hole pattern or honeycomb seals on rotordynamic calculations. Those calculations are done in accordance with API 617 [2].

References for the individual programs used for this paper are; tilt pad bearing [3], done by the finite element and pad assembly method, rotor response [4] and stability [5], both done by the transfer matrix method, and toothed labyrinth [6], done by the bulk flow method. Usage of the first three programs, especially as related to the requirements of API 617[2], is described in the paper [7]. The undamped critical speed analysis is done by the Myklestad-Prohl transfer matrix technique.

4. DAMPER BEARING CONSIDERATIONS

4.1 Damper bearing description

The bearing cage is free to move in the housing and there is an annular gap between the cage and the housing. As shown in Figure 2, a mechanical spring supports the cage in the vertical direction. The O-rings seal the ends of the annular gap between the cage and the housing. The O-rings also support the cage in the horizontal direction and contribute to the support in the vertical direction. Pressurized lube oil is supplied into the annular gap between the cage and housing through a feed groove in the housing at the midpoint between the O-rings. The oil supply to the tilting pads of the bearings is through a feed groove in the cage that is coincident with the feed groove in the housing. The oil is free to move in an out of the annular gap at the feed grooves but not at the O-rings. The oil in the damper annulus works in all directions of motion to provide additional damping to the compressor rotor/bearing system.

Before installation in the machine, each damper bearing is given a factory bench test to determine that the damper is free to move, that it is centered within the design tolerance range, and that the clearance is to the drawings. The metallic centering spring is designed with infinite life.

4.2 Damper bearing modeling

The formulas used for the damping for the damper are shown in [8]. With the central feed grooves and the O-rings on the ends, if each land is of length L/2, the total damping of the damper is the same as the damping of a short damper of length L with leakage on both sides. If there were no O-rings on the sides, then the damping would be one-quarter of that with the O-rings on the sides. The cage is centered in the housing within design tolerances. If the damper is eccentric, then the damping is larger than if it were centered. Early testing of dampers sealed with O-rings showed that for normal bearing inlet pressures, a non-cavitated film is a good assumption [9, 10]. Test experience where the oil pressure has been varied has confirmed this. See [9-11] for studies on the effects of oil supply and sealing arrangement on squeeze- film dampers. A survey of squeeze-film dampers is given in [12]. See [13] for theory and testing of a damper bearing similar to the one described in this paper.

4.3 Damper bearing history

The equipment manufacturer has built more than 800 compressors that use damper bearings. Of these, more than 460 use the sealing O-rings to center the dampers. The O-ring centered dampers have been used since 1973, primarily in smaller barrel type compressors [14, 15]. See [16] for a description of the application of O-ring-centered damper bearings when oil-film ring seals and taper land sleeve bearings were replaced by gas seals and tilting pad bearings.

In the late 1980s, a metallic spring-centered damper bearing was developed to support the gravity load from large-frame compressors. Since the mid-1990s, the equipment manufacturer has applied this design to almost every new compressor that uses damper bearings. O-rings are still used for sealing. Metallic spring-centered dampers have been used in more than 365 compressors. The compressor that is the subject of this paper uses a metallic spring centered damper that supplies an assured centering, as shown in Figure 2. This feature was important to the robustness desired by the end user. Rig testing of the metallic spring-centered damper is described in [17]. Applications of this metallic spring-centered damper are described in the papers [18-20].

5. ROTOR DYNAMIC EXPERIENCE PLOTS

Three experience charts will be described that are used for the preliminary evaluation of the stability of centrifugal compressors. Discussions on these charts are given in [20]. This compressor application has been plotted on each of these charts with a representative sample of the compressors from the equipment manufacturer and is well within the range of experience. There is considerable experience with both damper and non-damper applications on these charts. Figure 3A is a plot of bearing span/impeller bore *vs.* the average density of the gas. Figure 3B is a plot of flexibility ratio, the ratio of the maximum continuous speed (Nmc) divided by the rigid bearing first critical speed, *vs.* the average gas density, as is used in Specification 2.6.5 of Chapter 1 of API 617 [2]. Figure 3C is a plot of discharge pressure times differential pressure across the case *vs.* flexibility ratio.

Compressors in Regions B should be subject to more strict analytical criteria than those in Regions A. See [21] for the original version of the plot as in Figure 3C. See [22] for the original version of the plot as in Figure 3B and a plot as in Figure 3C. The experience points and experience lines in [21-22] are different from the ones shown in [20] and below.

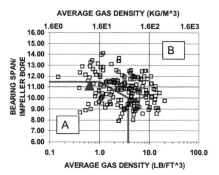

Figure 3A – Bearing Span/Impeller Bore *vs.* Average Gas Density

6. ANALYTICAL RESULTS

6.1 Bearing Analytical Models

Various tilting pad bearing models were analyzed, all with 5 pads and L/D = 0.44, both for damper (DB) and non-damper (ND). Load between the pads with center pivot bearings and dampers provided the optimum rotor dynamic characteristics. The journal bearing models considered were:

- Load between the pads and center pivot bearings (LBP, CP)
- Load on the pads and center pivot bearings (LOP, CP)
- Load between the pads and offset (0.55) pivot bearings (LBP, OP)
- Load on the pads and offset (0.55) pivot bearings (LOP, OP).

Calculations were made for minimum and maximum drawing clearances. With minimum clearances, the oil inlet temperature was assumed to be the minimum allowed and the squeeze-film damper was assumed to be at the maximum allowable eccentricity. This defines the upper stiffness limit. With maximum clearances the oil inlet temperature was assumed to be the maximum allowed and the squeeze-film damper was assumed to be centered. This defines the lower stiffness limit.

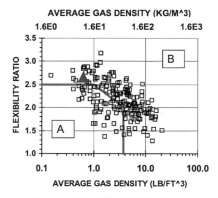

Figure 3B – Flexibility Ratio *vs.* Average Gas Density

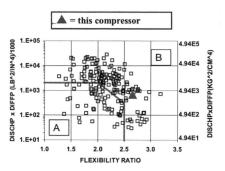

Figure 3C - Discharge Pressure x Case Differential Pressure vs. Flexibility Ratio

6.2 Stability Analysis

Figures 4A and 4B show the results of the Level I and Level II stability analysis, as required by Specifications 2.6.5 and 2.6.6 of Chapter 1 of API 617 [2]. Figure 4A is with the contract journal and damper bearings. Figure 4B is with the with the contract journal bearings and no dampers. The log decs are higher with the damper bearings. These figures show that both configurations meet the stability requirements of API 617 [2]. Each figure shows the calculated log dec *vs.* aerodynamic excitation at the midspan of the compressor. Curves are shown without the labyrinths, as is done for the Level I analysis, and with the labyrinths, as is done for the Level II analysis. There is not much difference between the Level I and Level II log decs, because the balance piston seal is toothed and has swirl brakes.

Curves are shown for minimum and maximum clearance at the journal and damper bearings. Shown as a vertical line on both plots is the anticipated cross-coupling = Q_A, the arithmetic sum of the calculated empirical aerodynamic excitations at each impeller, as in [18-20], and as was adopted in API 617 [2]. Also shown on both plots as a vertical line is Q_M = the modal sum of the excitations calculated at each impeller. As discussed in [18-20], Q_M needs to be included in the Level II analysis.

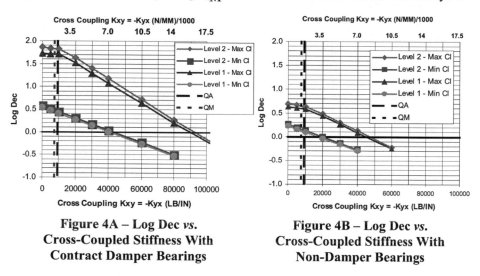

**Figure 4A – Log Dec *vs.*
Cross-Coupled Stiffness With
Contract Damper Bearings**

**Figure 4B – Log Dec *vs.*
Cross-Coupled Stiffness With
Non-Damper Bearings**

Per the Level I criteria, if the log dec at Q_A < 0.10 without the labyrinths, then a Level II analysis must be done. The log dec is >=0.42 with the damper bearings. With the non-damper bearings it is >= 0.11. The Level I criteria also uses the stability threshold and the safety factor. The stability threshold is defined as the place where the log dec crosses over to zero. The safety factor is defined as the ratio of the stability threshold to Q_A. The lowest safety factor is 4.4 for minimum clearance with dampers and 1.9 with non-dampers. Per the Level I criteria, because it is in Region B in Figure 3B and the safety factor is < 10, then a Level II analysis must be done. For the Level II analysis, this equipment manufacturer calculates the log dec with the labyrinths and Q_M included at the same time. The log dec at Q_M is >= 0.46 with the dampers and >= 0.16 with the non-dampers, therefore, in either configuration the compressor meets the stability requirement of the Level II analysis that the log dec be > 0.10.

Figure 5 shows the calculated log dec at Q_M in the Level II analysis vs. the various bearing types. The log dec is higher with damper bearings (DB), higher with center pivot bearings (CP), and about the same with load on the pad bearings (LOP) as with load between pads (LBP). With non-damper bearings (ND) and offset pivot bearings (OP) the Level II stability criteria would not be met for minimum clearance. The chart for the log dec at Q_A in the Level I analysis was similar.

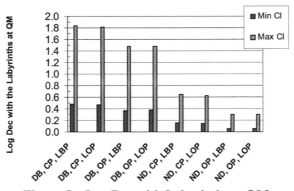

Figure 5 – Log Dec with Labyrinths at QM

Figures 6A and 6B are plots, with the contract journal bearing (CP, LBP), of the calculated log dec (at zero cross-coupled stiffness at the midspan) and the stability threshold *vs.* damper bearing damping. The labyrinth effects are not included in those figures. The plots are used to design the damper bearings. The horizontal axis extends above and below the design range of the damper bearing damping. The range of the calculated design damper bearing damping is shown on the plots. The damper was not "tuned" to be at the peak for stability, in order to lower the response to unbalance. If there were no O-rings the damping would be one quarter of that shown on the plots. If the damper bearing were completely locked up, it would be all the way to the right on the plot. Figure 6A is the same kind of plot as is shown in [5 and 7]. Figure 6B is the same kind of plot as is shown in [7, 18, and 19].

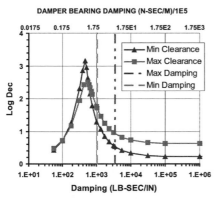

Figure 6A – Log Dec
***vs.* Damper Damping**

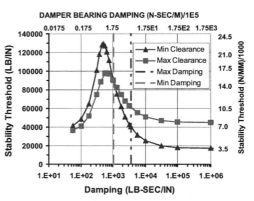

Figure 6B – Stability Threshold
***vs.* Damper Damping**

6.3 Rotor Unbalanced Response Analysis

Unbalanced rotor response analyses were made, for the various bearing types, in accordance with Specification 2.6.2 of Chapter 1 of API 617 [2] to calculate:

- The amplification factor of the first critical speed
- The midspan response at the first critical speed for a fixed midspan unbalance
- The amplitude at the bearing probes at the maximum continuous speed for fixed midspan, quarterspan, and coupling unbalances.

Figure 7A shows the calculated amplification factor of the first critical speed. For minimum design clearances, the maximum amplification factor is shown, and for maximum design clearances, the minimum amplification factor is shown. The amplification factor is calculated from several positions on the rotor - the bearing probes, the midspan, and the coupling location, thus creating a spread in values. The lowest amplification factor is with the damper bearings (DB) and center pivot (CP) journal bearings, while the highest amplification factor is with the non-damper bearings (ND) and offset pivot (OP) journal bearings. For this smaller size rotor, load between the pads (LBP) or load on the pad (LOP) does not make much difference. If this were a large rotor, it is likely that load between the pads would have shown a lower amplification factor than load on the pad.

Figure 7B shows the calculated midspan response at the first critical speed for a fixed amount of unbalance, four times the API residual unbalance, as required by Specifications 2.6.2.7 and 2.6.2.8 of Chapter 1 of API 617 [2]. The response is shown for minimum and maximum clearance journals and dampers. The lowest amplitude at midspan is with the damper bearings (DB) and center pivot (CP), while the highest amplitude is with the non-damper bearings (ND) and offset pivot (OP). Damper bearings are preferred, center pivot bearings are preferred, and again there is not much difference between load between (LBP) and load on the pad (LOP) bearings. With dampers more unbalance can be handled before touching the labyrinths.

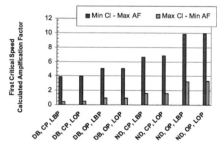

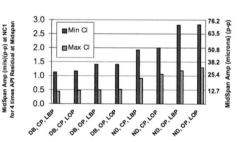

| **Figure 7A – Amplification Factor of First Critical Speed** | **Figure 7B – Amplitude at Midspan at First Critical Speed For four times API Residual Unbalance at Midspan** |

Calculations were made of the maximum amplitude seen at the bearing probes at the maximum continuous speed for fixed amounts of midspan, quarter span, and coupling unbalance. The amplitudes at the bearing probes with the midspan unbalance were slightly less with the non-damper offset pivot bearings, but not enough to offset the strong advantage seen in Figures 4-7 to use damper bearings.

For the various bearing models: The calculated first critical speed was 59 to 65 percent below the operating speed. The calculated second critical speed ranged from 34 to 46 percent more than the operating speed. There was a calculated intermediate response between the first and second critical speeds, which was close to the operating speed, and had an amplification factor under 2.5, so by Specification 2.6.6.2.10.a of Chapter 1 of API 617 [2] no separation margin was needed. This happened for damper and non-damper bearings. There was no indication of this in the mechanical test results.

To simulate a worst case for a soft system, a rotor dynamics analysis was made for maximum clearance bearings and dampers with no O-rings. Even in this configuration, the rotor dynamic requirements of API 617 [2] were met. The amplitude at the bearing probes at the maximum continuous speed with the midspan unbalance was only 12 percent higher without the O-rings than with the O-rings.

7. MECHANICAL TEST RESULTS

Factory mechanical testing was performed with and without damper bearings. The test without damper bearings was conducted with shims between the damper cage and the bearing housing, which effectively locks the dampers. Table 1 shows the predicted and tested results of the first critical speed for both location and amplification factor. Per 2.6.3.2 of Chapter 1 of API 617 [2] the actual critical should deviate by no more than 5% from the predicted value. The deviation ranges from 0 to 4%. There is no requirement in API 617[2] for a correlation between predicted and tested for the amplification factor. The correlation in this case is judged to be very good, especially as the prediction was with a pure midspan unbalance and for the test no unbalance was added at the midspan. The results confirmed that the damper bearing configuration produces lower amplification factors. Figure 8A is with damper bearings and Figure 8B is without damper bearings. They show 1X compensated (for slow roll data) from the final deceleration for each mechanical test at the thrust end vertical probe. The amplitude seen at the maximum continuous speed of about 5 microns (0.2 mils) for either configuration was well below the API 617 [2] mechanical test limit of $25.4(12000/Nmc)^{0.5} = 23$ microns (0.91 mils), Nmc = maximum continuous speed (RPM) for Nmc > 12000 rpm. If Nmc < 12000 RPM then the API 617 [2] limit is 25 microns (1 mil).

**Table 1 – First Critical Speed - Prediction and Test Data -
With and Without Bearing Dampers**

	With Bearing Dampers		Without Bearing Dampers	
	First Critical (rpm)	Amplification Factor	First Critical (rpm)	Amplification Factor
Prediction	4900 to 5500	0.4 to 3.9	4600 to 5110	1.5 to 6.6
Test	5000 to 5730	1.1 to 3.2	4700 to 5100	2.8 to 6.0

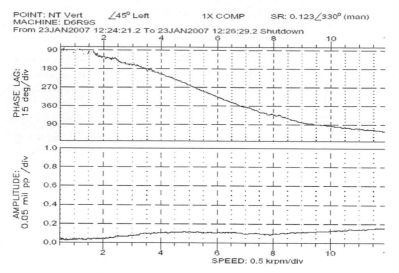

Figure 8A – Final Deceleration With Dampers

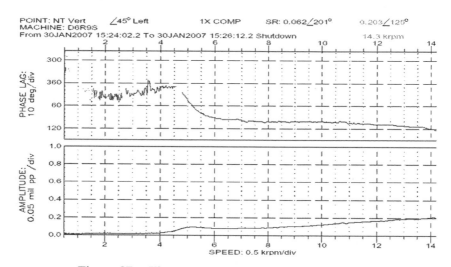

Figure 8B – Final Deceleration Without Dampers

8.   CONCLUSIONS

- The damper bearing configuration showed its superior mechanical performance for this compressor, both analytically and on test.
- The application of both damper and non-damper bearings for this particular compressor was successful and provided the robustness desired by the end user.
- An analytical methodology for designing damper or non-damper bearing applications with respect to journal bearing geometry considerations has been outlined.

9. ACKNOWLEDGEMENTS

The authors would like to thank the management of Dresser-Rand for its support and permission to publish this paper. At Dresser-Rand, Roger Vincent, Melanie Wemmell, Glenn Grosso, Cy Borer, Randy Jacobs, Ernie Pead, Thomas Soulas, Brian Stumper, Don Wehlage, and Rick Antle were among those involved in this project. The authors especially want to acknowledge the contributions of Michael Drosjack, Van Wilkinson, and others from the end user, Shell Global Solutions.

The information contained in this paper includes factual data, technical interpretations and opinions which, while believed to be accurate, are offered solely for information purposes. No representation, guarantee or warranty of any kind is made concerning such data, interpretations and opinions, including the accuracy thereof.

10. REFERENCES

[1] Ramesh K., 2002, "A State-of-the-art Rotor Dynamic Analysis Program," The 9th International Symposium on Transport Phenomena and Dynamics of Rotating Machinery, Honolulu, Hawaii, February 10-14.

[2] API Standard 617, 2002, "Axial and Centrifugal Compressors for Petroleum, Chemical, and Gas Industry Services," 7th Edition, July.

[3] Nicholas, J. C., Gunter, E. J., and Allaire, P. E., 1979, "Stiffness and Damping Coefficients for the Five-Pad Tilting-Pad Bearing," *ASLE Transactions*, Vol. 22, 2, pp. 113-124, April.

[4] Lund, J. W., 1965, "Rotor-Bearing Dynamics Design Technology, Part V: Computer Program Manual for Rotor Response and Stability," *Technical Report AFAPL-TR-65-45, Part V,* Air Force Aero Propulsion Laboratory, Wright-Patterson Air Force Base, Dayton, OH, May.

[5] Lund, J. W., 1974, "Stability and Damped Critical Speeds of a Flexible Rotor in Fluid-Film Bearings," Trans. ASME, *Journal of Engineering for Industry*, pp. 509-517, May.

[6] Kirk, R. G., 1990, "Users Manual for the Program DYNPC28 -- A Program for the Analysis of Labyrinth Seals," Negavib Research & Consulting Group, Virginia Tech, Blacksburg, VA, Jan.

[7] Memmott, E. A., 2003, "Usage of the Lund Rotordynamic Programs in the Analysis of Centrifugal Compressors," *Jorgen Lund Special Issue of the ASME Journal of Vibration and Acoustics*, Vol. 125, October.

[8] Gunter, E J, Jr., Barrett, L. E., and Allaire, P. E., 1975, "Design and Application of Squeeze Film Dampers for Turbomachinery Stabilization," *Proceedings of the 4th Turbomachinery Symposium*, Turbomachinery Laboratory, Department of Mechanical Engineering, Texas A&M University, College Station, Texas, pp.31-45, Oct.

[9] Tonnesen, J., 1976, "Experimental Parametric Study of a Squeeze Film Bearing," Transactions of the ASME, Journal of Lubrication Technology, pp. 206-213, April.

[10] Sharma, R. K. and Botman, M., 1978, "An Experimental Study of the Steady-State Response of Oil-Film Dampers," Transactions of the ASME, Journal of Mechanical Design, Vol. 100, pp. 216-221, April.

[11] Levesley, M. C. and Holmes, R, 1996, "The Effect of Oil Supply and Sealing Arrangements on the Performance of Squeeze-Film Dampers; An Experimental Study," Proc IMechE, Part J, 210, pp. 221-232.

[12] Dogan, M. and Holmes, R., 1983, "Squeeze-Film Damping of Rotor Dynamic Systems," *Shock and Vibration Digest*, Vol. 15, No. 9, pp. 3-9, September.

[13] Lund, J. W., Myullerup, C. M., and Hartmann, H., 2003, "Inertia Effects in Squeeze-Film Damper Bearings Generated by Circumferential Oil Supply Groove," *Jorgen Lund Special Issue of the ASME Journal of Vibration and Acoustics*, Vol. 125, October.

[14] Memmott, E. A., 1990, "Tilt Pad Seal and Damper Bearing Applications to High Speed and High Density Centrifugal Compressors," IFToMM, *Proceedings of the 3rd International Conference on Rotordynamics*, Lyon, pp. 585-590, Sept. 10-12.

[15] Memmott, E. A., 1992, "Stability of Centrifugal Compressors by Applications of Tilt Pad Seals, Damper Bearings, and Shunt Holes," IMechE, *5th International Conference on Vibrations in Rotating Machinery*, Bath, pp. 99-106, Sept. 7-10.

[16] Memmott, E. A., 2007, "Rotordynamic Considerations Of A High-Pressure Hydrogen Recycle Compressor," CMVA, *Proceedings of the 25th Machinery Dynamics Seminar*, St. John, NB, October 24-26.

[17] Kuzdzal, M. J. and Hustak, J. F., 1996, "Squeeze Film Damper Bearing Experimental vs. Analytical Results for Various Damper Configurations," *Proceedings of the 25th Turbomachinery Symposium*, Turbomachinery Laboratory, Department of Mechanical Engineering, Texas A&M University, College Station, Texas, September 17-19.

[18] Memmott, E. A., 2000, "Empirical Estimation of a Load Related Cross-Coupled Stiffness and The Lateral Stability Of Centrifugal Compressors," CMVA, *Proceedings of the 18th Machinery Dynamics Seminar*, Halifax, pp. 9-20, April 26-28.

[19] Memmott, E. A., 2000, "The Lateral Stability Analysis of a Large Centrifugal Compressor in Propane Service at an LNG Plant," IMechE, *Proceedings of the 7th International Conference on Vibrations in Rotating Machinery*, Nottingham, England, pp. 187-198, September 12-14.

[20] Memmott, E. A., 2002, "Lateral Rotordynamic Stability Criteria for Centrifugal Compressors," CMVA, *Proceedings of the 20th Machinery Dynamics Seminar*, Quebec City, pp. 6.23-6.32, October 21-23.

[21] Kirk, R. G. and Donald, G. H., 1983, "Design Criteria for Improved Stability of Centrifugal Compressors," AMD Vol. 55, ASME, pp. 59-71, June

[22] Fulton, J. W., 1984b, "Full Load Testing in the Platform Module Prior to Tow Out: A Case History of Subsynchronous Instability," *Rotordynamic Instability Problems in High Performance Turbomachinery,* NASA Conference Publication 2338, Texas A&M University, pp. 1-16, May 28-30.

Time-domain simulation of rotors with hysteretic damping

G Genta
Mechanics Department, Politecnico di Torino, Italy

ABSTRACT

The hysteretic damping model cannot be applied to time domain dynamic simulation and the constant equivalent damping often introduced to overcome this problem cannot simulate correctly hysteretic damping in rotordynamic analysis, in particular for what rotating damping is concerned. An alternative model based on the concepts introduced by Voigt and Biot, but with a limited number of additional degrees of freedom is proposed and the relevant equations are derived. Some examples show applications to rotordynamics.

Keywords: hysteretic damping, equivalent damping, time-domain simulation, rotordynamics.

1. INTRODUCTION

The structural or hysteretic damping model [1, 2] is based on the observation that the energy losses in engineering materials undergoing cyclic loading are proportional to the square of the amplitude and almost independent of frequency, at least in a wide frequency range. These assumptions have been thoroughly discussed in the last 60 years, but the hysteretic damping model has two basic advantages: dependence of the energy losses on the square of the amplitude, leading to linear equations of motion, and independency of frequency (i.e. constant phase lag between stresses and strains), allowing to define a constant complex stiffness or a complex Young modulus when working at the level of material properties.

These advantages are however strictly linked with two severe limitations. First, the assumption of constant phase lag cannot hold when the frequency at which the hysteresis cycle is gone through tends to zero. Second, as it is well known, the hysteretic damping model is defined specifically referring to harmonic time histories [2], i.e. it can be used for equations of motion referred to the frequency domain but not to the time domain [3].

The first point is not very severe when dealing with vibrating structures, except in particular cases of systems vibrating at very low frequencies. Actually, hysteretic damping may overestimate the damping properties of materials when the vibration

---

frequency is very low, yielding inconsistent results when the frequency tends to zero.

In rotordynamics, however, things may be different when dealing with the damping of rotating elements whirling in nearly synchronous conditions, since synchronous whirling is seen by the material constituting the rotor as a vibration at vanishing frequency.

Although this didn't prevent from using the hysteretic damping model in rotordynamics [4, 5], it caused misunderstandings and incorrect interpretations [6].

The impossibility of using the hysteretic damping model in time domain formulations is a severe limitation, preventing from performing numerical simulation, based on the numerical integration in time of the equations of motion, nowadays a basic tool in the dynamic analysis of systems of all types.

2. EQUIVALENT DAMPING

A structural element with stiffness k and loss factor η (or with imaginary part of the complex stiffness k'') has the same damping effect of an equivalent viscous damper with damping coefficient

$$c_{eq} = \frac{k\eta}{\omega} = \frac{k''}{\omega}.$$

(1)

Since c_{eq} is necessarily a positive quantity, ω must be positive. As pointed out in [6], failure to recognize this resulted in many incorrect statements in the past.

Equation (1) does not solve our problem: the value of the equivalent damping coefficient depends on the frequency at which the material goes through its hysteresis cycle, and thus cannot be introduced into time domain equations. Moreover, when the frequency tends to zero the equivalent damping tends to infinity, as discussed in detail in [3]. The two problems caused by the hysteretic model are thus not solved.

A common approximation is that of replacing the natural frequency of the system ($\omega_n = \sqrt{k/m}$ in single degrees of freedom systems), for the generic frequency ω, obtaining

$$c_{eq} = \frac{k\eta}{\omega_n} = \eta\sqrt{km} .$$

(2)

By introducing the damping ratio $\zeta = c / 2\sqrt{km}$, it follows

$$\zeta_{eq} = \frac{\eta}{2} .$$

(3)

The rationale behind this approximation is the consideration that, when damping is small, it influences the response of the system only when it works close to its resonant frequency. Thus Eq. (2) yields a fair approximation when damping is important; on the contrary it yields a poor approximation when damping has at any rate little effect on the behavior of the system.

However this is not the case for rotating damping in rotordynamics, where the variability of the frequency at which the hysteresis cycle is gone through is greater, tending to zero when synchronous whirling conditions are approached.

A model that could be applied in the time domain was studied by Biot [7] and Caughey [8], who showed it holds both in structural dynamics and in rotordynamics. It consists of a spring, with in parallel a large number of spring- damper systems (Fig. 1). Biot refers to this model as the Voigt model of viscoelastiocity; here it will be referred to as the Voigt-Biot model to avoid confusion with the Voigt law, or the Kelvin-Voigt model [9].

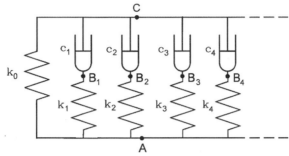

Fig. 1 Voigt model for a spring with hysteretic damping.

If the number of spring-damper series tends to infinity, the dependence of the complex stiffness of the system is almost coincident with that of a spring with hysteretic damping above an arbitrarily chosen frequency [7, 8, 10]. Below that frequency the phase angle does not remain constant, and tends to zero for $\omega \to 0$. The last point is not a drawback at all, since it eliminates the inconsistency of hysteretic damping at low frequency.

However, also this approach is not free of problems: each spring-damper series introduces a new (internal) degree of freedom, the displacement of point B_i. The lack of a mass in points B_i allows to reduce the number of states, with each damper introducing just one additional state. Each damper thus introduces a new pole, which is approximately

$$s_i = \frac{k_i}{c_i} \ . \tag{4}$$

A small number of spring-damper branches is sufficient to simulate hysteretic damping in a wide range of frequencies. As a general rule, it may be said that n dampers are enough to obtain a constant imaginary part of the complex frequency in a range of frequency spanning for more than of $n-1$ decades [10].

The nondimensional real and imaginary parts of the complex frequency of a system with hysteretic damping are reported as functions of the frequency in Fig. 2.

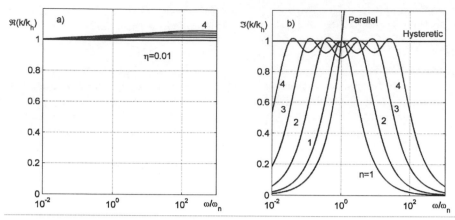

Fig. 2. Equivalent damping computed using Eq. (3) (curve labeled 'parallel'), and that computed using a Voigt-Biot damper model with 1, 2, 3, and 4 internal degrees of freedom. The real part of the complex stiffness has been computed assuming $\eta = 0.01$.

3. EQUIVALENT DAMPING OF A MULTI-D.O.F. SYSTEM

In the case of systems with many degrees of freedom, the conversion from hysteretic to viscous damping can be performed by resorting to modal decomposition. Consider a multi-degrees of freedom system with hysteretic damping, whose frequency domain dynamic stiffness matrix is

$$\mathbf{K}_{dyn} = -\omega^2 \mathbf{M} + \mathbf{K} + i\mathbf{K}'' \ . \tag{5}$$

The eigenvector matrix $\mathbf{\Phi}$ of the corresponding MK system allows computing the modal mass and stiffness matrices $\mathbf{M}$ and $\mathbf{K}$. In a similar way, an imaginary part of the modal stiffness matrix can be obtained

$$\overline{\mathbf{K}}'' = \mathbf{\Phi}^T \mathbf{K}'' \mathbf{\Phi} \tag{6}$$

If $\mathbf{K}'' = \eta\mathbf{K}$ or the usual relationships for generalized proportional damping hold, $\overline{\mathbf{K}}''$ is diagonal. If not, but the system is lightly damped as usually occurs for structural damping, an approximate uncoupling of the modal equations of motion can be performed by extracting the terms of $\overline{\mathbf{K}}''$ lying on the main diagonal.

Applying Eq. (3), an equivalent viscous damping matrix

$$\overline{\mathbf{C}}_{eq} = \mathbf{diag}\left(\frac{\overline{\mathbf{K}}''_{ii}}{\omega_{ni}} \right) \tag{7}$$

is readily obtained and back-transformed

$$\mathbf{C}_{eq} = \mathbf{\Phi}^{-T} \overline{\mathbf{C}}_{eq} \mathbf{\Phi}^{-1} \ , \tag{8}$$

where symbol $-T$ stands for inverse of the transpose.

The damping matrix $\mathbf{C}_{eq}$, perhaps added to other damping matrices that may exist, together with matrices $\mathbf{M}$ and $\mathbf{K}$, allows to write a time-domain model of the system.

An alternative approach based on Voigt damping is described in [10]. It is based on the same modal transformation, but instead of using Eq. (3) for introducing an equivalent damper for each one of the modes considered, a number of spring-damper branches of the type shown in Fig. 1 are added to each mode. Operating in this way, a number of internal degrees of freedom is added: if n modes are supplied with m dampers each, the number of degrees of freedom added to the original ones (and the number of states) is $n \times m$.

The modal equation of motion is thus

$$\begin{bmatrix} \overline{\mathbf{M}} & 0 \\ 0 & 0 \end{bmatrix} \begin{Bmatrix} \ddot{\boldsymbol{\eta}} \\ \ddot{\mathbf{x}}_B \end{Bmatrix} + \begin{bmatrix} \overline{\mathbf{C}} + \mathbf{C}_{11} & \mathbf{C}_{12} \\ \mathbf{C}_{21} & \mathbf{C}_{22} \end{bmatrix} \begin{Bmatrix} \dot{\boldsymbol{\eta}} \\ \dot{\mathbf{x}}_B \end{Bmatrix} + \begin{bmatrix} \overline{\mathbf{K}} + \mathbf{K}_{11} & \mathbf{K}_{12} \\ \mathbf{K}_{21} & \mathbf{K}_{22} \end{bmatrix} \begin{Bmatrix} \boldsymbol{\eta} \\ \mathbf{x}_B \end{Bmatrix} = \begin{Bmatrix} \overline{\mathbf{F}} \\ 0 \end{Bmatrix} . \tag{9}$$

where another set of coordinates $\mathbf{x}_B$ has been added to the modal coordinates $\boldsymbol{\eta}$. Matrices $\mathbf{C}$ and $\mathbf{K}$ are reported in [10].

Since a solution in the original coordinates (plus obviously coordinates $\mathbf{x}_B$) is required, it is possible to resort to the obvious equation

$$\begin{Bmatrix} \boldsymbol{\eta} \\ \mathbf{x}_B \end{Bmatrix} = \begin{bmatrix} \boldsymbol{\Phi} & 0 \\ 0 & \mathbf{I} \end{bmatrix} \begin{Bmatrix} \boldsymbol{\eta} \\ \mathbf{x}_B \end{Bmatrix} , \tag{10}$$

to back-transform the matrices.

Note that the back-transformation is still possible even if a reduced number of modes has been considered and hence the eigenvector matrix $\boldsymbol{\Phi}$ is not square. In this case

$$\begin{Bmatrix} \boldsymbol{\eta} \\ \mathbf{x}_B \end{Bmatrix} = \begin{bmatrix} \overline{\mathbf{M}}^{-1} \boldsymbol{\Phi}^T \mathbf{M} & 0 \\ 0 & \mathbf{I} \end{bmatrix} \begin{Bmatrix} \boldsymbol{\eta} \\ \mathbf{x}_B \end{Bmatrix} , \tag{11}$$

4. APPLICATION TO ROTORDYNAMICS

4.1. Constant equivalent damping
While the simple definition of a constant equivalent damping is adequate in most cases studied in structural dynamics, rotordynamics has, as already stated, different requirements.

It is well known that rotating hysteretic damping is stabilizing in subcritical conditions and destabilizing in supercritical ones [11, 6] and also that, while viscous rotating damping causes a gradual decrease of stability with speed, in case of hysteretic rotating damping the decrease is abrupt when crossing a relevant critical speed [12]. It is thus expected that the values of the threshold of instability computed using hysteretic and equivalent damping are different.

Consider for instance the small turbine rotor used as Example 4.3 in [3]. The loss factor of the rotor is assumed $\eta = 0.04$, while the two bearings are assumed to be hysteretic damped springs with $k = 20$ MN/m and $\eta = 0.06$. The model was reduced through Guyan reduction, obtaining a model with 8 complex degrees of freedom.

The Campbell diagram and the decay rate plot, computed using both the hysteretic and the equivalent damping models, are plotted in Fig. 3. Full lines characterize hysteretic damping, dashed lines constant viscous equivalent damping.

The modal computation was performed both by retaining all modes and only the 4 modes with the lowest natural frequencies, obtaining identical results: the relevant curves in Fig. 3 are completely superimposed.

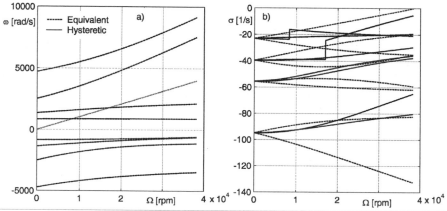

Fig. 3. Campbell diagram (a) and decay rate plot (b) for the two rotor models with hysteretic and equivalent damping.

It is clear that the two models yield the same Campbell diagram, but the decay rate plots are quite different. In particular, as expected, the two forward branches that have a critical speed in the speed range studied, show an abrupt decrease of the decay rate (in absolute value). In this case the stability of the first forward mode decreases when the equivalent model is used, while that of the second mode increases.

An important point is to note that at vanishing speed, i.e. when the rotor reduces to a structure, the equivalent model yields correct results, which become increasingly inaccurate with increasing speed. This confirms what said above: the standard equivalent model, while generally adequate for structural dynamics, is not so in rotordynamics.

4.2 Voigt-Biot damping

The procedure is similar to that seen for the application of Voigt-Biot dampers to multi-degrees of freedom vibrating system, with the difference that now two modal dampers, a rotating and a nonrotating one, must be associated to each mode.

The equation of motion of a rotor with viscous damping can be written with reference to a set of complex coordinates $\mathbf{q}$ in the form

$$\mathbf{M}\ddot{\mathbf{q}} + \left(-i\omega\mathbf{G} + \mathbf{C}_n + \mathbf{C}_r\right)\dot{\mathbf{q}} + \left(\mathbf{K} - i\omega\mathbf{C}_r\right)\mathbf{q} = F(t) \ . \tag{12}$$

This equation is based on the assumption that both the rotor and the stator are axially symmetrical. Note that, owing to the use of complex coordinates, the gyroscopic matrix is symmetrical.

The eigenvector matrix of the corresponding MK system allows computing all the modal matrices, including the modal gyroscopic matrix.

After performing the modal transformation, two Voigt-Biot dampers, a rotating and a nonrotating one, can be added to each modal system to simulate both rotating and nonrotating damping. By doing so, a number $2n$ of internal degrees of freedom must be added for each mode considered, if n is the number of branches of each one of the Voigt-Biot damper. Note that not all modal systems need to be damped, and that it is not necessary to add the same number of dampers to all modes.

By partitioning the matrices as seen for the non rotating system, and noting that several submatrices vanish, the modal equation of motion can be shown to be

$$\begin{bmatrix} \bar{\mathbf{M}} & 0 \\ 0 & 0 \end{bmatrix}\begin{Bmatrix} \ddot{\boldsymbol{\eta}} \\ \ddot{\mathbf{x}}_B \end{Bmatrix} + \begin{bmatrix} -i\omega\bar{\mathbf{G}} + \bar{\mathbf{C}}_n + \bar{\mathbf{C}}_r & 0 \\ 0 & \mathbf{C}_{22n} + \mathbf{C}_{22n} \end{bmatrix}\begin{Bmatrix} \dot{\boldsymbol{\eta}} \\ \dot{\mathbf{x}}_B \end{Bmatrix} + \ . \tag{13}$$

$$+ \begin{bmatrix} \bar{\mathbf{K}} + \mathbf{K}_{11} - i\omega\bar{\mathbf{C}}_r & \mathbf{K}_{12} \\ \mathbf{K}_{21} & \mathbf{K}_{22} - i\omega\mathbf{C}_{22n} \end{bmatrix}\begin{Bmatrix} \boldsymbol{\eta} \\ \mathbf{x}_B \end{Bmatrix} = \begin{Bmatrix} \bar{\mathbf{F}} \\ 0 \end{Bmatrix} \ .$$

where in this case also the internal coordinates $\mathbf{x}_B$ are complex.

The overlined matrices are the modal matrices of the system with viscous damping, while the non-overlined ones are the matrices added to simulate hysteretic damping. Generally speaking , $\bar{\mathbf{G}}$, $\bar{\mathbf{C}}_n$ and $\bar{\mathbf{C}}_r$ are not diagonal.

5. EXAMPLES

5.1 Rotating beam on elastic supports
Consider a beam with annular cross section (inner and outer diameters 60 and 50 mm), with a length of 1.5 m, constrained at the ends by two visco-elastic supports with a stiffness of 2 MN/m. The beam is made from steel ($E = 211$ GPa, $\rho = 7810$ kg/m³, $v = 0.3$). A shock-like excitation due to a force growing linearly to 10,000 N in 11 ms is applied to a point at 500 mm from one end when the shaft rotates at 1400 rad/s. The material of the beam has a loss factor $\eta = 0.02$, while the supports have a loss factor $\eta = 0.06$.

The nonrotating beam, with slightly different supports, was studied in [13].

By modelling the beam with 3 Timoshenko beam elements and eliminating the rotational degrees of freedom through Guyan reduction, a system with only 4 degrees of freedom is obtained.

The Campbell diagram is little influenced by damping and, owing to the negligible gyroscopic effect, is almost completely flat. It is not reported here. The decay rate plot is reported in Fig. 4a.

The lines labelled (H) describe the behavior of the system with hysteretic damping. At the critical speed the decay rate of the first forward mode has an abrupt increase. It however remains negative and the rotor is stable at all speeds.

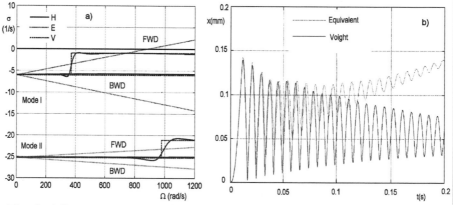

Fig. 4. a) Decay rate plot for the rotor described in example 1. The curves for hysteretic (H), equivalent (E) and Voigt-Biot damping (V) are reported. (b) Response of the same rotor to a shock. Comparison between equivalent and Voigt-Biot damping.

The lines labelled (E) are related to the simplified equivalent damping. The behavior is that typical of viscous damping: a gradual increase of the decay rate in forward whirling, with a threshold of instability in the supercritical regime. While at standstill the equivalent viscous damping is actually 'equivalent', when the system rotates the two forms of damping yield quite different results. This is true not only in synchronous whirling, where the applicability of hysteretic damping is questionable since the frequency at which the hysteresis cycle is gone through is close to zero, but at all speeds. In particular, at high speeds the two solutions diverge. The equivalent model leads to a finite value of the threshold of instability, contrary to what happens with the hysteretic model.

The curves labelled as (V) were obtained by applying 8 Voigt-Biot dampers (4 rotating and 4 non rotating) to the 4 modal systems. Each of them has three spring-damper branches, tuned at the modal frequency and at frequencies 1/10 and 10 times it. The total number of complex states of the system is 32 instead of 8. The behavior is much more similar to that related with hysteretic damping. The step of the decay rate is present, although less abrupt and the threshold of instability is not present.

The response to the shock must be computed through numerical integration in time of the equations of motion. The results, in term of the time history of the amplitude of the displacement in the rotation plane of the point where the shock is given, is plotted in Fig. 4b. In case of the Voigt-Biot damping the shock triggers a circular synchronous whirl, that however stabilizes in time to a constant amplitude and then slowly decreases asymptotically to zero (this is not shown in the figure since it takes a time of several seconds). The oscillation slowly dampens out too. In the case of the equivalent damping, since the rotor operates above its threshold of instability, the synchronous whirling grows without bounds, while the oscillations die out quicker.

5.2 Rotor with non-negligible gyroscopic effect

Consider the rotor of a small turbine described in section 4.1.

The decay rate plots obtained with hysteretic and equivalent damping are compared in Figures 3b. The computations were performed using the DYNROT code and the transformation of hysteretic into equivalent damping was performed using the option 'Dyntrans' built in the code.

The rotor with hysteretic damping is always stable, while that with equivalent viscous damping has a threshold of instability at about 38,000 rpm.

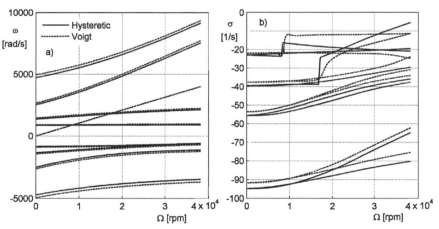

Fig. 5. Campbell diagram and decay rate plot of the rotor of Example 2. The plots obtained using hysteretic damping are compared with those computed using the Voigt-Biot damper.

The computation was repeated using 3 rotating and 3 nonrotating Voigt-Biot dampers for each mode. The results are plotted in Fig. 5. Since 4 modes were considered, the number of internal degrees of freedom or states of the system is 24, which must be added to the 4 degrees of freedom if the computation is performed in modal coordinates, for a total of 32 states. If the equations are back-transformed, there are 8 degrees of freedom for the original system, for a total of 40 states. The computation was performed in both ways, obtaining the same results, for the 4 forward and 4 backward modes shown in the plot.

The decay rate is much closer to the one for hysteretic damping. In particular, it is characterized by an abrupt decrease (in absolute value) of the decay rate of the first two modes when they cross the relevant critical speeds.

Although not being exactly equal to the plot obtained for the hysteretic damping model, Voigt damping at least corrects the largest errors introduced by the conventional equivalent damping model, in particular at high speed.

The Campbell diagrams are very similar to each other, with small differences due to the additional springs present in the Voigt-Biot model.

6. CONCLUSIONS

The Voigt-Biot model with a finite number of viscous dampers here generalized to rotordynamics allows to write the equations of motion in the time domain starting from the hysteretic damping formulation. They are a very good approximation in a wide frequency range, at the cost of a number of additional (usually referred to as internal or hidden) degrees of freedom.

Obviously it is impossible to compare the results of the proposed model with those obtained through the hysteretic model in conditions other than harmonic motion, since the latter cannot be used. However, when the system performs harmonic motions the two models yield reasonably close results.

One limitation is that hysteretic damping must be small, particularly when the system has many degrees of freedom and the Voigt-Biot dampers are applied to uncoupled modal systems. However, it is possible to perform the modal transformation only for what the hysteretic damping matrices are concerned and, if there are other forms of damping, they can be added later, after back-transforming the model to the 'physical coordinates', plus the internal ones. In this way no small damping assumptions is required for the other forms of damping.

The other limitation of hysteretic damping, linked with its poor performance at low frequency, is circumvented. In this sense, the model here suggested is better than the original hysteretic damping when the system performs low frequency motions.

Since, when dealing with rotating machinery, rotating damping must be dealt with separately from nonrotating damping, this approach requires the introduction of a larger number of internal degrees of freedom: for each mode, the number of spring-damper branches, and then of internal degrees of freedom, is doubled, if the rotating and nonrotating modal damping is modeled in the same way.

This can be mitigated by using a smaller number of internal degrees of freedom for nonrotating damping, since the problems linked with hysteretic damping are more serious in the case of rotating damping.

While the inconsistency of hysteretic damping in low frequency motion may be marginally important in structural dynamics, it constitutes a serious drawback in rotordynamics. In situations in which the whirl frequency is close to the spin speed (almost-synchronous whirling) the frequency at which the hysteresis cycle is gone through is close to zero, and this leads to a questionable applicability of the hysteretic damping concept. The fact that hysteretic rotating damping may cause a threshold of instability equal to one of the critical speeds is due to this, and may be regarded as an oversimplification linked with the type of model.

For rotating damping, the present model is thus actually better than the original model even in harmonic motion, apart from being needed when solving time-domain equations, like those encountered in non-steady-state motion, for instance in the acceleration through a critical speed or in the blade-loss problem.

The present model behaves better than the usual constant equivalent damping model not only at the critical speed crossing, but also at high speed. In particular, it retains the property of hysteretic damping of either locating the threshold of instability close to a critical speed or granting stability at all speeds, while viscous damping always causes instability at a speed large enough (even if in many cases this occurs at a speed well in excess to the actual one).

It is well known that the prediction of high speed stability of rotors is still problematic mostly owing to the uncertainties about how modeling damping. The present model offers a theoretical alternative to the simpler viscous damping model. Only experimentation can allow to validate the results obtained through the two approaches.

7.   REFERENCES

[1]   Myklestadt N.O., *The Concept of Complex Damping*, Journ. Appl. Mechanics, Vol. 19, (1952), p. 284-286.
[2]   Crandall S.H., *The Role of Damping in Vibration Theory*, Journal of Sound and Vibration, (1970), 11(1), 3-18.
[3]   Genta G., *Vibration of structures and machines*, III ed., Springer, New York (1998).
[4]   Crandall S.H., *Rotordynamics*, in: Kliemann W. and Sri Namachchivaya N. (Editors), *Nonlinear Dynamics and Stochastic Mechanics*, Boca Raton: CRC Press (1995).
[5]   Ramanujam G., Bert C.W., *Whirling and Stability of Flywheel Systems, Part 1: Derivation of Combined and Lumped Parameter Models*; *Part II: Comparison of Numerical Results Obtained with Combined and Lumped Parameter Models*, Journal of Sound and Vibration (1983), vol. 88 (3), 369-420.

[6] Genta G., *On a Persisting Misunderstanding on the Role of Hysteretic Damping in Rotordynamics*, Journal of Vibration and Acoustics, Vol. 126, July 2004, 469-471.

[7] M.A. Biot, *Linear Thermodynamics and the Mechanics of solids*, Proc. Third U.S. National Congress of Applied Mechanics (1958), pp 1-18.

[8] T.K. Caughey, *Vibration of Dynamic Systems with Linear Hysteretic Damping, Linear Theory*, Proc. of the Fourth National Congress of Applied Mechanics (1962), pp 87-97.

[9] Banks H.T., Pinter G.A., *Hysteretic Damping*, in S. Braun (ed.), *Encyclopedia of Vibration*, Academic Press, London, (2001).

[10] G. Genta, N. Amati, *On the Equivalent Viscous Damping for Systems with Hysteresis*, Atti dell'Accademia delle Scienze di Torino, (2008).

[11] Dimentberg M., *Flexural Vibrations of Rotating Shafts*, Butterworth, London, England (1961).

[12] Genta G., *Dynamics of rotating systems*, Springer, New York (2005).

MISALIGNMENT

Influence of machine alignment and load on steam turbine vibrations

P Pennacchi, A Vania
Politecnico di Milano, Dipartimento di Meccanica, Italy

ABSTRACT

Steam turbines that include a control stage often show significant changes of the vibration levels correlated with changes of megawatt load. This can be due to changes of magnitude and direction of the journal bearing loads which can cause significant variations of oil-film geometry and dynamic stiffness coefficients. A suitable choice of the bearing geometry can reduce the severity of these phenomena. Anyhow, load rises and decreases can generate additional dynamic forces, transmitted to the turbine shaft through the blades of the first stage, which cause further changes of the machine vibrations. This paper shows a diagnostic strategy used to study the influence of bearing loads and operating conditions on high pressure steam turbine vibrations. Specific techniques for the analysis of monitoring data have been integrated with model-based methods aimed to estimate the correlations between megawatt load and machine vibrations. The successful results provided by this diagnostic strategy within the analysis of the dynamic behaviour of a power unit steam turbine are shown.

1. INTRODUCTION

Changes of the shaft-train alignment caused by changes of the machine thermal state as well as changes of the forces transmitted to the shaft of steam turbines, that include a control stage, in consequence of megawatt loadings can cause significant changes of magnitude and direction of the load acting on the journal bearings of the turbine shaft. Owing to this, the oil-film geometry and the corresponding dynamic stiffness can be affected by important changes. This can cause significant changes of the machine vibrations. Moreover, the static and dynamic forces transmitted to the rotor by the steam that flows through the blades of the first stage as well as possible further forces e.g. induced by excessive thermal expansions of pipes linked to the machine case can be additional causes for changes of the machine vibrations in operating conditions. Tilting-pad journal bearings can reduce the risk of the occurrence of high vibrations of steam turbines induced by considerable changes of magnitude and direction of the bearing loads. Nevertheless, elliptical journal bearings are often mounted instead of tilting-pad journal bearings owing to their minor cost and the underestimation of the consequence of the rotor system excitations generated by some critical operating conditions. The results of detailed analyses of monitoring data integrated by those obtained by means of model-based methods can provide significant diagnostic information that can explain the reasons of the occurrence of abnormal vibrations that sometimes affect the steam turbines.

The diagnostic information provided by the proposed strategy can be very useful in fault symptom analysis to distinguish severe failures from machine malfunctions whose effects can be reduced by adjusting some process parameters. Moreover, the proposed method can be used to study corrective actions that can reduce the sensitivity of the turbine vibrations to the excitations generated by the phenomena correlated with the load control.

This paper shows the successful results provided by the proposed diagnostic strategy within investigations performed on the dynamic behaviour of the steam turbine of a power unit.

2.   CASE STUDY

An uncommon diagnostic strategy that included the use of model-based methods was developed and applied to analyse the experimental vibrations of the steam turbine of a 100 MW cogeneration power unit which showed an abnormal dynamic behaviour in operating condition. The machine-train was composed of a gas turbine (GT), a generator and a steam turbine (ST) whose rated power was 20 MW. A gearbox was mounted between gas turbine and generator while a slim auxiliary shaft and a clutch were mounted between generator and steam turbine. Figure 1 shows the machine-train diagram along with the bearing numbers hereafter used in the paper. Owing to the small flexural stiffness of the auxiliary shaft and the clutch the ST vibrations were lightly affected by those of the generator rotor.

The operating speeds of GT and ST were 5230 r/min and 3000 r/min, respectively. Each journal bearing was equipped with a couple of XY proximity probes mounted at 45° Right and 45° Left as shown in Figure 1. The regulation of the steam turbine load was obtained by means of a control stage, that is by operating on the stroke of the inlet valves that governed the steam flow.

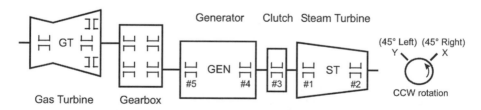

Figure 1 Machine-train diagram, bearing numbers and probe arrangement

In operating condition the ST vibrations showed to be influenced by the load. In particular the level of the synchronous (1X) vibrations significantly increased during load rises apart from the initial thermal state of the ST. This behaviour showed to be repetitive during many load rises and decreases. Figure 2 shows the historic trend of amplitude and phase of the 1X vibrations measured on bearings #1 and #2 during a load decrease occurred before a planned machine coastdown. In the same figure the trend of the load, whose values can be read on the right-hand scale of the lower diagram, is shown.

Correlations between machine response and load are not unusual in steam turbines, however in this case the changes of the vibration levels were considerable so that in the most critical conditions they approached the alarm limit suggested by the machine manufacturer (120 μm pp). Therefore, the causes of this abnormal behaviour required to be investigated in order to set up suitable corrective actions.

It was supposed that the dynamic stiffness of the oil-film journal bearings #1 and #2 of the steam turbine could play a significant role in the machine vibrations experienced in operating conditions. Figure 3 shows the main characteristics and geometrical parameters of the two elliptical bearings #1 and #2 mounted at the coupling end and the non-coupling end of the turbine shaft. The values of the main geometrical parameters of both bearings are reported in Table 1.

The machined and assembled radial clearances, C_p and C_b (Figures 3, 4), can be expressed as:

$$C_p = R_p - R \qquad\qquad C_b = R_b - R \qquad\qquad (1)$$

where R is the journal radius. The bearing pre-load, m, is expressed as:

$$m = 1 - C_b / C_p \qquad\qquad (2)$$

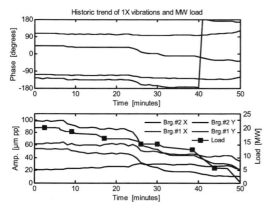

Figure 2 Historic trend of amplitude and phase of the 1X vibrations measured on bearings #1 and #2 during a load decrease. Trend of megawatt load (right-hand scale of the lower diagram).

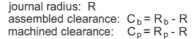

journal radius: R
assembled clearance: $C_b = R_b - R$
machined clearance: $C_p = R_p - R$

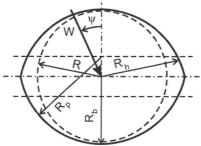

Figure 3 Geometry of the elliptical bearings #1 and #2 (exaggerated radial clearance)

Table 1 Main geometrical parameters of the oil-film journal bearings #1 and #2 of the steam turbine

		BEARING N. #1	BEARING N. #2
Journal radius [mm]	R	160.0	125.0
Length	L / R	1.0	1.0
Machined Clearance	C_p / R	0.0038	0.0040
Assembled Clearance	C_b / R	0.0017	0.0018
Pre-load Factor	m	0.5455	0.5556

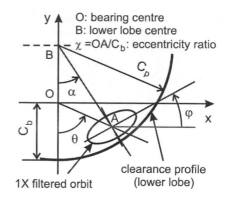

Figure 4 Journal bearing parameters

The nominal bearing loads acting on supports #1 and #2 were evaluated considering the weight of only the steam turbine shaft and the clutch. Table 2 shows the dimensionless stiffness and damping coefficients of the oil-film journal bearings evaluated by integrating the Reynolds equation considering the operating speed [1, 2].

Table 2 Dimensionless stiffness and damping coefficients of the oil-film journal bearings #1 and #2 evaluated at 3000 r/min considering the nominal bearing loads

BEARING	STIFFNESS COEFFICIENTS k_{xx}	k_{xy}	k_{yx}	k_{yy}	DAMPING COEFFICIENTS r_{xx}	r_{xy}	r_{yx}	r_{yy}
#1	1.5725	-0.6991	-4.6750	9.5010	0.8980	-1.7078	-1.7558	8.3401
#2	1.9074	-1.2601	-5.1322	9.4951	1.0657	-2.0975	-2.1214	8.0072

Owing to the considerable value of the pre-load factor, m, the direct coefficients k_{xx} and r_{xx} evaluated in the horizontal direction x are not very high. Moreover, for both bearings, they are significantly lower than the respective direct coefficients k_{yy} and r_{yy} evaluated in the vertical direction, y. That is, in operating condition, as well as at lower rotational speeds, the anisotropy of the oil-film journal bearings #1 and #2 was noticeable.

This study was integrated with the analysis of the average journal position. The gap voltage data collected at bearings #1 and #2 during the load decrease shown in Figure 2 and the subsequent machine coastdown were used to determine the journal centreline curve. The average journal displacements as well as the journal eccentricity ratio, χ, and the attitude angle, θ, were evaluated accordingly with the reference systems illustrated in Figure 4.

Figure 5 shows the dimensionless journal centreline curves of bearings #1 and #2 obtained by dividing the dimensional eccentricity values by the corresponding assembled clearance C_b of the bearing. In these figures the clearance profile of the lower lobe of the elliptical bearing is shown.

On both supports the load decrease from 20 MW to the off-load condition caused significant changes of the average journal position in comparison to the assembled clearance C_b. On bearings #1 and #2 these average journal displacements occurred with an opposite versus and similar magnitude also during load rises. This behaviour showed to be highly repetitive in occasion of several rises and decreases of the steam turbine load.

In particular, on the outboard support the load rises caused an increase of the horizontal and vertical components of the average journal position. Owing to the bearing pre-load the horizontal component of the centreline curve experienced in the shaft rotational speed range up to 3000 r/min was rather large, therefore, the additional displacement due to the load rises that occurred after the runups caused a considerable total displacement of the journal in the horizontal direction. On the contrary, on the inboard bearing #1 the megawatt load rises caused a significant decrease of the horizontal and vertical components of the average position. Owing to this, at the highest loads the journal position got near the vertical axis passing through the bearing centre.

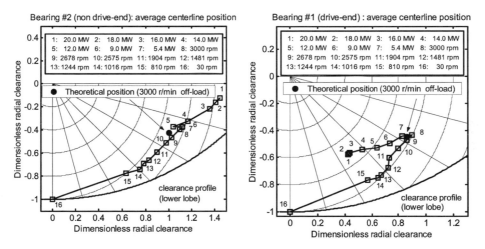

Figure 5 Average journal position at bearings #2 (ST non-coupling end) and #1 (ST coupling end) during a load decrease and a coastdown

In order to validate some of the results obtained by means of the models of bearings #1 and #2 the theoretical values of the average journal positions determined considering the nominal bearing loads and the operating speed have been shown in Figure 5. These numerical results are in good accordance with the respective experimental data collected in the off-load operating condition.

However, when the steam turbine load exceeded 12 MW the journal eccentricity ratio and attitude angle within bearings #1 and #2 became very different although the dimensionless geometrical parameters of these bearings were rather similar. This behaviour implies that the bearing loads were significantly influenced by the megawatt load. This is not unusual in steam turbines that include a control stage which causes a partial arc admission of the steam flow. In this case magnitude and direction of the resultant vector of the forces transmitted to the shaft by the steam flow depend on the opening degree of the inlet valves of the steam which is governed by the load value.

The models of bearings #1 and #2 along with the experimental average journal positions measured in operating condition were used to estimate the static horizontal and vertical components of the oil-film forces associated with different values of the load ranging from 0 MW to 20 MW. This study allowed the dependence on megawatt load of magnitude and direction of the bearing loads to be investigated.

The results of this study are shown in Figure 6. The phase, ψ, of the bearing load vector was evaluated with respect to the vertical axis as shown in Figure 3. In the upper range of the megawatt load the magnitude of the horizontal component of the static load acting on bearings #1 and #2 became rather important.

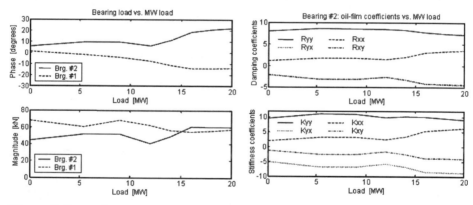

Figure 6 Dependence on megawatt load of magnitude and phase of the bearing load acting on supports #1 and #2 of the ST

Figure 7 Stiffness and damping coefficients of the oil-film journal bearing #2 (ST non-coupling-end) vs. megawatt load

The curves illustrated in Figures 5 and 6 show the presence of a deterministic relationship between megawatt load and bearing loads. This behaviour can be the consequence of the forces transmitted by the steam flow to the rotor blades of the first stage. In fact, during load rises and decreases the steam flow passes through only a partial arc of the bladed disk. Anyhow, it is not possible to exclude that the alignment of the shaft-train that included the steam turbine was affected also by changes caused by an excessive thermal expansions of the pipe that convoys the steam to the inlet valves. Although flexible joints are mounted between pipe and turbine case it is possible that under the effect of undesired forces the turbine

818

supports, which were included in the machine casing, were affected by unexpected displacements that depended on megawatt load and machine thermal state.

The significant changes of the average journal position due to the load rises and decreases caused changes of the oil-film geometry and dynamic stiffness coefficients. Figures 7 shows the dependence on the megawatt load of the stiffness and damping coefficients evaluated by means of the model of bearing #2 considering the values of eccentricity ratio and attitude angle that corresponded to the experimental average journal positions illustrated in Figure 5. Owing to the changes of the bearing load caused by the megawatt load rises, the stiffness coefficients k_{xx} and k_{yx} are affected by significant changes. Anyhow, further coefficients of both bearings #1 and #2 showed to be fairly influenced by the unit operating conditions. It is possible to suppose that in some operating conditions the anisotropy of the oil-film reached exasperated values. This could affect the levels of the turbine vibrations. With regard to this it is well known that tilting-pad journal bearings can provide better performances since they are less sensitive to changes of magnitude and direction of the bearing load.

In order to better investigate the relationship between the machine dynamic behaviour and the operating conditions of the unit the journal orbits measured on bearings #1 and #2 were analysed. The polar plots illustrated in Figure 8 show the dimensionless unfiltered orbits measured with four different values of the ST load spaced in the range up to 20 MW.

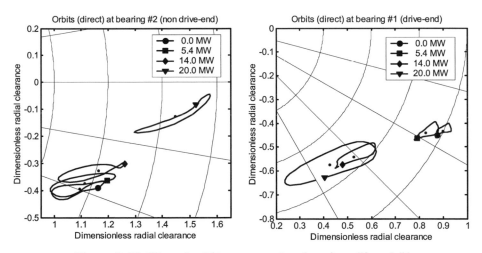

**Figure 8 Unfiltered orbits measured on bearings #2 and #1
in on-load operating conditions**

These dimensionless orbits were obtained by dividing the vibration amplitude by the assembled radial clearance C_b of the respective journal bearing. The centre of each orbit coincides with the corresponding average journal position illustrated in Figure 5. The positions of the key-phasor mark along the orbits measured on bearings #1 and #2 in operating condition were out of phase since the operating speed was about 500 r/min higher than the second flexural critical speed of the turbine shaft.

Accordingly with the vibration data shown in Figure 2 shape and dimension of the orbits were significantly influenced by the megawatt load. It is possible to note that with a nominal load of 20 MW a classic "banana shaped" orbit occurred on the outboard bearing #2. With the same load the orbit measured on the inboard bearing of the ST was rather large and mainly oriented towards the horizontal axis. This behaviour is in accordance with the low value of the direct coefficients k_{xx} and r_{xx} of the oil-film dynamic stiffness evaluated in the horizontal direction.

In addition to this the inclination angle, φ, of the major axis of the elliptic 1X filtered orbits measured on bearings #1 and #2 with different megawatt loads was studied. Owing to the significant changes of the average journal position that occurred in operating condition both magnitude and orientation of the major axis of the 1X filtered orbits were significantly influenced by the load. The angle α between the vertical axis, y, and the straight line passing thorough the centre of the bearing lower lobe and the centre of the 1X orbit, as shown in Figure 4, was considered. The angle γ obtained subtracting α from the inclination angle φ of the major axis of the 1X orbit was evaluated for different megawatt loads. The results of this investigation are shown in Table 3 along with the respective values of the journal attitude angle θ. It is possible to note that for both bearings #1 and #2 the angle γ shows only small changes in spite of the larger changes of the attitude angle. Therefore, the orientation of the 1X filtered orbits was mainly influenced by the average journal position and the shape of the clearance profile of the bearing lower lobe. The results of these diagnostic investigations pointed out that the changes of the load of the steam turbine caused significant changes of the bearing loads and oil-film stiffness and damping coefficients. It was possible to suppose that this significantly affected the vibrations of the steam turbine apart from the effects due to changes of the machine thermal state correlated with the operating conditions of the unit.

On the basis of the results of this study it was decided to substitute the two elliptical journal bearings on which the steam turbine shaft is mounted with tilting-pad journal bearings. In fact, in general, the dynamic stiffness coefficients of this kind of bearings are less sensitive to changes of magnitude and direction of the bearing load.

Table 3 Journal attitude angle, θ, and angle γ of the 1X filtered orbits measured at bearings #1 and #2 with different megawatt loads of the steam turbine

BEARING	ANGLE	STEAM TURBINE MEGAWATT LOAD				
		0 [MW]	5 [MW]	9 [MW]	14 [MW]	20 [MW]
#1	γ [degree]	- 7.5°	- 6.3°	- 9.0°	- 9.1°	- 7.5°
	θ [degree]	70.1°	71.4°	71.6°	74.3°	85.0°
#2	γ [degree]	19.8°	19.5°	22.0°	23.8°	19.7°
	θ [degree]	64.3°	61.9°	55.9°	44.2°	36.6°

3.   IDENTIFICATION OF THE EQUIVALENT EXCITATIONS

Further investigations aimed to identify the causes of the high vibration levels of the steam turbine that occurred in operating condition required to simulate the shaft-train response by means of a mathematical model of the fully assembled machine. A suitable set of equivalent excitations that allowed the machine vibrations at different measurement points to be simulated was estimated. This task was performed by means of a model-based identification method, developed in the past by the authors, which minimises the error between experimental vibrations and numerical results. A detailed description of this identification technique is shown in previous papers [3, 4, 5]. Hereafter, for the sake of brevity, only the main results of the identification of the equivalent excitations performed within this case study are shown.

The machine model used to study the dynamic behaviour of the unit, as well as to identify the equivalent excitations, is based on a well known theory [6, 7]. Figure 9 shows the Finite Element Model (FEM) of the shaft-train composed of a portion of the generator rotor, the auxiliary shaft, the clutch and the steam turbine shaft.

The dynamic effects of the oil-film journal bearings on the machine response were considered by means of dynamic stiffness coefficients that depended on shaft rotational speed and megawatt load. A simple rigid foundation was included in the model of the machine-train since no significant dynamic effect that could be ascribed to resonances of the foundation structure and supports was pointed out by the experimental vibration data collected during runups and coastdowns.

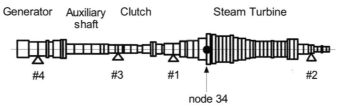

Figure 9 Finite Element Model of a portion of the slow shaft-train of the power unit

The changes of the average journal position within the ST bearings #1 and #2 that occurred in on-load operating conditions, with respect to the reference position measured in off-load operating condition, were caused by changes of the static component of forces, likely due to the control stage, which in addition generated a couple of moments M_x and M_y acting in the vertical and horizontal plane, respectively, as shown in Figure 10. This is confirmed by the data shown in Figure 5.

It is possible to suppose that the 1X harmonic component of the same set of forces, which depended on the megawatt load, generated equivalent 1X moments M_x and M_y whose magnitude and phase were affected by the operating conditions. The mathematical model of the fully assembled machine was used to evaluate these equivalent excitations whose effect was to increase the turbine vibrations during the load rises.

A single couple of 1X bending moments M_x and M_y, like those illustrated in Figure 10, was applied to different nodes of the FEM of the shaft-train illustrated in Figure 9. The influence coefficients of the shafts associated with 1X moments applied to the nodes located at the cross-sections considered within this investigation were evaluated.

A Weighted Least Squares Error Method was used to estimate the magnitude and phase of the two moments M_x and M_y that minimized

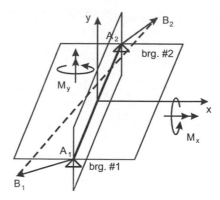

Figure 10 Vectors B_1-A_1 and B_2-A_2: changes of the average journal position, experienced with a 20 MW load with respect to the reference journal position measured in off-load condition. Equivalent moments M_x and M_y.

the error between numerical results and experimental 1X vibrations measured on bearings #1 and #2 in the X and Y directions (45° R and 45° L). These excitations were identified considering the oil-film coefficients and the experimental vibrations associated with different megawatt loads. At each measurement point the 1X vibrations that could be ascribed only to the load rises were estimated by subtracting the 1X vibration vectors measured in the off-load conditions from the corresponding 1X vibration vectors measured at different loads. Hereafter these 1X vibrations will be called additional vibrations. This common approach requires to consider unimportant the non-linear effects in the machine dynamic behaviour. Often this supposition does not significantly affect the reliability of the results provided by this investigation procedure.

These results, here not reported in detail for the sake of brevity, showed that the best simulation of the experimental additional vibrations were given by a couple of moments M_x and M_y applied to node #34 (Figure 9). With regard to this it is necessary to consider that the main aim of this study was to identify a set of equivalent excitations that were able to simulate the experimental vibrations rather than to evaluate the actual set of excitations caused by the load rises.

Figure 11 shows the changes of magnitude and phase of the identified moments, acting in the horizontal and vertical plane, associated with different values of the load. The regular changes of these equivalent excitations seem to be in accordance with a likely evolution of the phenomena correlated with the load rises.

Figure 12 shows the amplitude and phase curves of the 1X vibrations, generated on bearings #1 by the above mentioned identified moments, as a function of the load. In this figure the numerical results are compared with the corresponding experimental 1X additional vibrations in order to show the small error between monitoring data and vibration predictions.

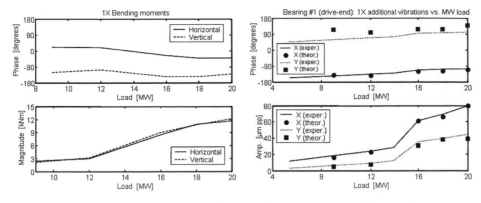

Figure 11 Magnitude and phase of the identified moments M_x and M_y vs. megawatt load

Figure 12 1X additional vibrations on bearing #1 vs. load: comparison between experimental data and numerical results

On the basis of the satisfactory results of this study it is possible to suppose that the identified equivalent moments M_x and M_y are able to successfully simulate the dynamic effects caused by the actual excitations generated by the load rises.

In general, the anisotropy of tilting-pad journal bearings is negligible or rather low, depending on their geometrical characteristics. Moreover, the stiffness coefficients of the oil-film can be increased by properly adjusting the pad pre-load. Therefore, it is possible to suppose that the effects due to the actual excitations, simulated by means of the identified equivalent bending moments, can be reduced by substituting the original elliptical journal bearings of the steam turbine with suitable tilting-pad journal bearings which can give better performances.

4.   CONCLUSIONS

The vibrations of steam turbines that include a control stage can be affected by significant changes when the unit is operated in on-load condition since magnitude and direction of the bearing loads can be highly influenced by the megawatt load value. This can cause significant changes of the dynamic stiffness coefficients of the oil-film journal bearings of the turbine shaft and then changes of the vibration levels.

In this paper a diagnostic strategy aimed to study the sensitivity of the dynamic behaviour of these steam turbines to changes of load is shown. The proposed method combines the use of suitable techniques for the analysis of monitoring data with investigation techniques based on models of oil-film journal bearings and fully assembled machine. This strategy allows the influence of machine hot-alignment and megawatt load on bearing loads and steam turbine vibrations to be successfully investigated.

The satisfactory results provided by the proposed diagnostic strategy within the analysis of the dynamic behaviour of a power unit steam turbine are shown. Moreover, a suitable set of equivalent excitations that allow the turbine additional vibrations caused by load rises to be simulated has been estimated by means of a model-based identification method. This further study provided interesting information about the effects of load on machine vibrations.

On the basis of the results obtained by means of this diagnostic strategy suitable corrective actions can be set up in order to reduce the sensitivity of the dynamic behaviour of steam turbines that include a control stage to changes of load.

REFERENCES

1. Szeri, A.Z., Fluid Film Lubrication - Theory and Design, Cambridge University Press, 1998.
2. Kirk, R.G., Gunter, E.J., Short bearing analysis applied to rotor dynamics. Part. 1: Theory, *ASME Journal of Lubrication Technology,* Vol.98, 1976.
3. Bachschmid, N., Pennacchi, P., Vania, A., (2002), Identification of multiple faults in rotor systems, *JSV Journal of Sound and Vibration*, **254**(2), pp. 327-366, 2002.
4. Bachschmid, N., Pennacchi, P., Vania, A., Experimental Results in Simultaneous Identification of Multiple Faults in Rotor Systems, *"COMADEM"*, edited by Starr, A.G. and Rao, B.K.N, Elsevier Science Ltd., Oxford, pp. 663-671, 2001.
5. Markert, R., Platz, R., Seidler, M., Model Based Fault Identification in Rotor Systems by Least Squares Fitting, in *Proc. ISROMAC-8 Int. Symposium on Transport Phenomena and Dynamics of Rotating Machinery*, Texas A&M University, **2**, pp. 901-907, 2000.
6. Adams, M.L. jr., Rotating Machinery Vibration - From Analysis to Troubleshooting, Case Western University Cleveland, Ohio, Marcel Dekker Ed., New York, USA, 2001, 347 p., ISBN 0 8247 0258 1.
7. Lalanne, M., Ferraris, G., Rotordynamics prediction in Engineering, 2nd Edition, John Wiley & Sons Ltd., Baffins Lane, Chichester, West Sussex PO19 1 UD, England, 252 p., ISBN 0 471 92633 7, 1998.

The development of harmonics in rotor misalignment

A W Lees
Swansea University, UK

J E T Penny
Aston University, UK

ABSTRACT

After rotor unbalance, misalignment is the most prevalent fault in rotor systems. Yet it remains incompletely understood. This state of affairs is not surprising since the harmonic processes occurring in a misaligned system are somewhat complex, particularly in machines mounted on oil-film journal bearings. These bearings are non-linear and hence since misalignment will cause the equilibrium shaft position to change this leads to a change in bearing stiffness, but in addition there will be changes in the damping and forcing terms at harmonics of shaft speed. Observing a completely different mechanism, Lees has recently shown the development of shaft harmonics even with a fully linear system.

The current paper seek to place these two distinct mechanisms in context and compare the generation of harmonic terms for a four bearing shaft system with non-linear bearings using short bearing theory (which is considered representative rather than accurate). The calculation is shown as both a full time integration and a harmonic balance calculation. For steady running it is shown the coupling of the torsional modes takes on a dominant effect within some of the frequency range, whilst bearing non-linearity is dominant at other speeds. A further influence of importance is shown to be the variation of bearing damping with bearing load. Results are shown for representative geometries.

1. INTRODUCTION

Misalignment in rigidly coupled multi bearing machines is a very important practical problem which remains incompletely understood. This is in fact rather an understatement as there appears to be considerable confusion in the literature. Whilst it is generally recognised that after rotor unbalance, misalignment is the second most prevalent fault in rotor systems, it is perhaps surprising that the literature on the topic is sparse. Whilst there are very many hundreds of papers dealing with unbalance, a researcher would struggle to find 20 on misalignment.

Most authors have sought to explain the harmonic excitation in terms of the non-linearity of either the bearings and/or flexible couplings in the system, and indeed it is to be expected that these elements will play a part in the dynamics of a real system. Dewell and Mitchell [1] gave a discussion of the harmonics arising in flexible couplings. Al-Hussain and Redmond [2] developed a set of non-linear equations describing the motion of a misaligned system. However, they report no twice per rev component of vibration. Ref [3] also reports a full non-linear analysis which does show non-synchronous excitation under the appropriate conditions. But the source of these harmonics is the non-linearities of the system.

In a pair of papers Xu and Maragona [4, 5] have given an analysis of a system including a flexible coupling and have backed up their predictions with laboratory experiments on a rig. Here again, however, the source of the super harmonic components in the vibration signal emanates from the non-linear behaviour of the flexible coupling.

In this paper, the harmonics generated by two quite separate mechanisms are compared. The first mechanism is simply the non-linear behaviour of the oil-film bearings, following the analysis of Adiletta, Guido and Rossi [6]. For computational convenience, a single bearing in the model was considered as non-linear. In contrast, the mechanism of coupling misalignment, causing torsional and lateral motion to be coupled, as presented be Lees [7], has been evaluated separately. It is shown that under some circumstances the two sources produce effects of a similar order of magnitude. This may well reflect the complexity of misalignment which has for so long been hindered by the lack of a comprehensive explanation of observed behaviour.

2. BEARING NON-LINEARITY

A non-linear bearing model for a short, oil-film journal bearing was presented by Adiletta, Guido and Rossi [6]. They assumed laminar and isothermal fluid flow and derived expressions for the fluid film forces in terms of the bearing displacements u and v thus

$$\begin{Bmatrix} F_x \\ F_y \end{Bmatrix} = \eta \Omega R L \left(\frac{R}{c} \right)^2 \left(\frac{L}{D} \right)^2 \begin{Bmatrix} f_x \\ f_y \end{Bmatrix} \tag{1}$$

$$\begin{Bmatrix} f_x \\ f_y \end{Bmatrix} = -\frac{c\sqrt{(u-2v')^2 + (v+2u')^2}}{c^2 - u^2 - v^2} \begin{Bmatrix} 3V(u/c) - G\sin\alpha - 2S\cos\alpha \\ 3V(v/c) + G\cos\alpha - 2S\sin\alpha \end{Bmatrix} \tag{2}$$

where u and v are the displacements of the rotor in the x and y directions respectively, η is the oil viscosity, Ω is the speed of rotation of the shaft, L, R and c are the bearing length, radius and radial clearance, and $u' = \dot{u}/\Omega$, $u'' = \ddot{u}/\Omega^2$ etc. Also in the above equation,

826

$$\alpha = \tan^{-1}\left(\frac{v+2u'}{u-2v'}\right) - \frac{\pi}{2}\operatorname{sgn}\left(\frac{v+2u'}{u-2v'}\right) - \frac{\pi}{2c}\operatorname{sgn}\left(v+2u'\right)$$

$$G = \frac{2c}{\sqrt{c^2-u^2-v^2}}\left\{\frac{\pi}{2} + \tan^{-1}\left(\frac{v\cos\alpha - u\sin\alpha}{\sqrt{c^2-u^2-v^2}}\right)\right\} \tag{3}$$

$$V = \frac{2c^2 + c\left(v\cos\alpha - u\sin\alpha\right)G}{c^2-u^2-v^2}$$

$$S = \frac{c\left(u\cos\alpha + v\sin\alpha\right)}{c^2 - \left(u\cos\alpha + v\sin\alpha\right)^2}$$

Thus for a rotor-bearing system supported by a non-linear bearing the equation of motion may be written as

$$\mathbf{M}\ddot{\mathbf{q}} + \Omega\mathbf{G}\dot{\mathbf{q}} + \mathbf{K}\mathbf{q} = \mathbf{F}_b\left(\mathbf{q},\dot{\mathbf{q}},t\right) + \mathbf{F}_{ub}\left(\Omega t\right) - \mathbf{F}_g \tag{4}$$

where $\mathbf{F}_b$ is a vector of bearing forces, $\mathbf{F}_{ub}$ is a vector of out-of-balance forces and $\mathbf{F}_g$ is a vector of gravitational forces. Note that the stiffness matrix $\mathbf{K}$ only describes the stiffness properties of the rotor. Thus the rotor is modelled as if it was a free-free rotor and the constraints due to the bearings are accounted for by the forces $\mathbf{F}_b$. The vectors $\mathbf{F}_b$, $\mathbf{F}_{ub}$ and $\mathbf{F}_g$ contain a large number of zero terms.

To solve (4) we write these equations in the form

$$\begin{Bmatrix} \dot{\mathbf{q}} \\ \dot{\mathbf{v}} \end{Bmatrix} = \begin{bmatrix} \mathbf{0} & \mathbf{I} \\ -\mathbf{M}^{-1}\mathbf{K} & -\Omega\mathbf{M}^{-1}\mathbf{G} \end{bmatrix} \begin{Bmatrix} \mathbf{q} \\ \mathbf{v} \end{Bmatrix} + \begin{Bmatrix} \mathbf{0} \\ \mathbf{M}^{-1}\left(\mathbf{F}_b\left(\mathbf{q},\mathbf{v},t\right) + \mathbf{F}_{ub}\left(\Omega t\right) - \mathbf{F}_g\right) \end{Bmatrix} \tag{5}$$

These $2n$ first order non-linear equations can be solved using an appropriate numerical procedure such as the 4th order Runge Kutta procedure. Sometimes the equations are stiff and either the equations must be reduced or a stiff equation solver was used in MATLAB in order to give reasonable run times.

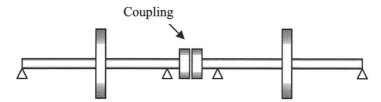

Coupling

Figure 1: Model Layout

As an example, the simple system shown in figure 1 was used to study the dynamic behaviour with a modest unbalance on the two discs. Two rotors, each 1 metre in length, are supported in oil film bearings with a clearance of 0.5mm. The shaft diameters are both 0.35m and the two discs are 1.6m diameter and thickness 0.28m.

The coupling has 12 bolts whose stiffness properties are chosen to give the appropriate torsional resonance.

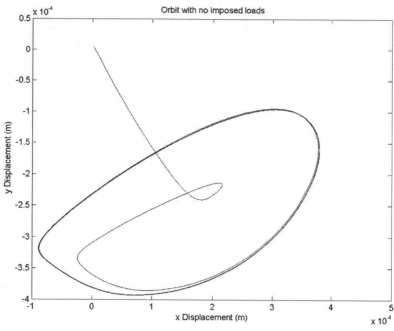

Figure 2

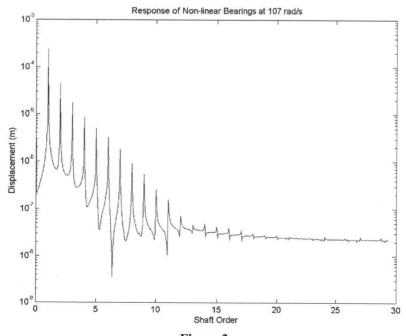

Figure 3

The rotors in this model are rigidly coupled, but this point will be discussed more fully in the next section. A nominal unbalance was simulated on each of the discs and the development of the motion was studied using a stiff equation solver in MATLAB. Typical orbits are shown in Figure 2, with a non-linear bearing but no coupling error whilst at a speed of 107 rad/s, this corresponding to the torsional natural frequency, a point which will be discussed further in section 3. Opposite vertical loadings were applied to bearings 2 and 3, and the steady state motion of the rotor was Fourier Transformed yielding the graph shown in Figure 3. The static loads applied correspond to a misalignment value of 0.5mm multiplied be coupling shear stiffness as discussed below. Figure 3 shows the generation of a range of harmonics. This is, of course, all based on short bearing theory and whilst this is known to be not accurate, it may be considered representative.

3. PARAMETRIC EXCITATION OF COUPLINGS

Some of possible problems with rigid couplings have been discussed recently by Lees [7] who showed that parallel misalignment can lead to the torsional and lateral motions becoming coupled. A similar conclusion was reached by Redmond [8] using a very different formulation of the problem.

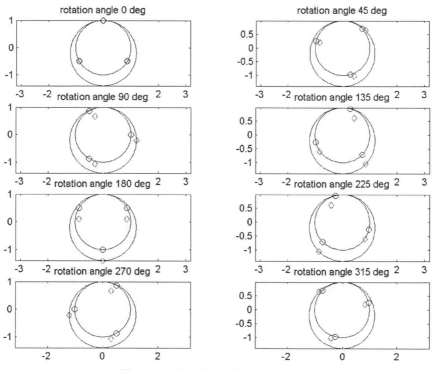

Figure 4: Idealised Coupling Motion

Figure 4 illustrates the situation for the case of a coupling with three bolts. The two shafts are assumed to rotate at the same rate but about axes which are displaced.

The location points attached to the larger coupling are denoted by diamonds whilst those on the opposing shaft are shown as circles. The figure shows the positions of the bolt locations on both shafts at a number of instants during one full cycle. It can be seen that points move relative to each other generating both torques and moments which vary in time. Reference [7] examines the dynamic behaviour of an idealised psuedo-rigid coupling. It is considered that the bolted flanges have some offset which distorts the motion. On the flange of rotor 1, the coupling bolts will be arranged about the centre. On the other rotor, the holes are not distributed around the centre, but offset. Detailed analysis of the equations are given in [7] and will not be repeated here, suffice to emphasize that this will imply that the coupling bolts have some strain energy at all except one angular position as the shafts rotate.

It is assumed that the bolt holes on the flange of the first rotor are arranged around a circle centred on the centre of the shaft cross section, but on the second rotor the bolts holes are again positioned on a circle although the centre of this circle is displaced by δ from the centre of the rotor. Hence, at zero angle, the bolts joining the two flanges have no strain, but at all other angles the bolts have strain energy.

The analysis of the motion commences with an evaluation of the energy stored within the coupling bolts. Recalling that the first shaft rotates at constant speed Ω, whilst that of the second shaft can vary, the potential energy of the bolts is given by

$$U = \frac{K_b}{2} \sum_{j=1}^{N} \left(R_j \cos(\varphi_j + \theta) + X_2 - r\cos(\alpha_j + \phi) - X_1 \right)^2$$

$$+ \frac{K_b}{2} \sum_{j=1}^{N} \left(R_j \sin(\varphi_j + \theta) + Y_2 - \delta - r\sin(\alpha_j + \phi) - Y_1 \right)^2$$

(6)

Now let

$$\theta = \Omega t + \varepsilon$$

(7)

so that, taking first order small quantities $\sin\theta = \varepsilon\cos\Omega t + \sin\Omega t$ and $\cos\theta = \cos\Omega t - \varepsilon\sin\Omega t$. Since the bolt positioning error is on rotor 1, it is clear that

$$\sum_{j=1}^{N} R_j \cos\varphi_j = 0 \qquad\qquad \sum_{j=1}^{N} R_j \sin\varphi_j = \frac{N\delta}{2}$$

(8)

We now apply Lagrange's equation to the six degrees of freedom and after a little manipulation, the equations of motion can be expressed in the form

$$M\ddot{z} + K_s z + K_c z + \Delta K(t)z = F(t)$$

(9)

The displacement vector takes the form $z = \{X_1 \quad Y_1 \quad \theta \quad X_2 \quad Y_2 \quad \phi\}^T$.

But the stiffness matrices are in three components.

Whilst the steady contribution of the coupling is described by

$$\mathbf{K}_c = \begin{bmatrix} NK_b & 0 & 0 & -NK_b & 0 & 0 \\ 0 & NK_b & 0 & 0 & -NK_b & 0 \\ 0 & 0 & NK_b r^2 & 0 & 0 & -NK_b r^2 \\ -NK_b & 0 & 0 & NK_b & 0 & 0 \\ 0 & -NK_b & 0 & 0 & NK_b & 0 \\ 0 & 0 & -NK_b r^2 & 0 & 0 & NK_b r^2 \end{bmatrix} \tag{10}$$

There is also, however, a fluctuating component arising from the coupling which is given by

$$\Delta\mathbf{K}(t) = \frac{NK_b\delta}{2} \begin{bmatrix} 0 & 0 & \sin\Omega t & 0 & 0 & -\sin\Omega t \\ 0 & 0 & \cos\Omega t & 0 & 0 & -\cos\Omega t \\ \sin\Omega t & \cos\Omega t & 0 & -\sin\Omega t & -\cos\Omega t & 0 \\ 0 & 0 & -\sin\Omega t & 0 & 0 & \sin\Omega t \\ 0 & 0 & -\cos\Omega t & 0 & 0 & \cos\Omega t \\ -\sin\Omega t & -\cos\Omega t & 0 & \sin\Omega t & \cos\Omega t & 0 \end{bmatrix} \tag{11}$$

The forcing term in equation (9) comprises both internal and external components. For the case in which there are no external forces and the excitation arises solely from the geometry of the coupling,

$$F = \frac{NK_b\delta}{2}\{\cos\Omega t \quad \sin\Omega t \quad \delta \quad -\cos\Omega t \quad -\sin\Omega t \quad -\delta\}^T \tag{12}$$

Recognising that for the complete rotor (i.e. coupled), the dynamic behaviour is described by the equation of motion

$$\mathbf{M}\ddot{z} + \mathbf{K}z + \Delta\mathbf{K}(t)z = F(t) \tag{13}$$

where $\mathbf{K} = \mathbf{K_s} + \mathbf{K_c}$.

Of course, if the coupling has a fault such that one of the bolts is defective, this changes equation (12).

4. SAMPLE RESULTS

A model was analysed which is similar to that chosen previously, but with a coupling having 12 bolts whose stiffness was selected to give a torsional natural frequency at half of the flexural critical speed. One of the bolts in the coupling is taken to be defective and this bolt is considered to take no load. As before the system is operated at the torsional resonance of 107 rad/s. Again the equations of motion were integrated with a Runge-Kutta approach, and the steady state motion was Fourier Transformed. Figure 5 shows the response for a system with linearised bearing properties but a coupling error.

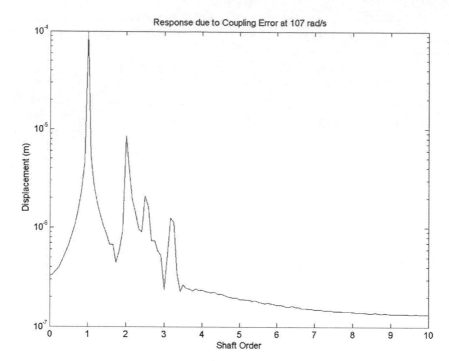

Figure 5

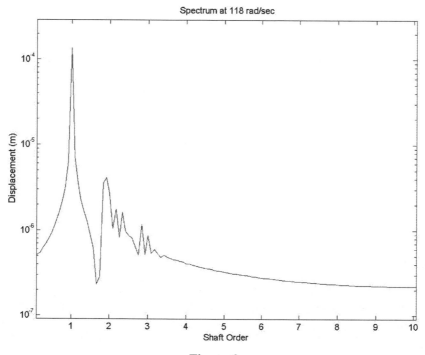

Figure 6

This spectrum is markedly different from that shown in figure 3. but it should be noticed that the twice per rev component is only marginally lower. The damping of the torsional mode was approximately 0.5% which may be too high is some cases, but the lateral damping has probably been underestimated. Figure 6 shows the spectrum when this same system was run at a speed 10% above the torsional critical speed. It is seen that the second harmonic has diminished but only by a factor of less than 2.

5. DISCUSSION

The results show a very case for computational convenience. Time integration for a realistic rotor rapidly becomes extremely demanding in terms of computing time. For such cases a harmonic balance approach may be more convenient. In the simple cases considered the twice per revolution component arising from the two mechanisms is of a similar order of magnitude. At this point several points should be considered. The relationship of torsional and flexural critical speeds has been taken as a special ratio, but it a real rotor line there may be many torsional frequencies making the situation even more complicated. Unbalance may also influence the situation and this has not, as yet, bee included in the defective coupling study. Perhaps more importantly, other torque fluctuations arising in the machine will be reflected in the translational behaviour.

It is probable that bearing non-linearity is the prime source of second harmonic vibration over a wide range of conditions. However, coupling between torsional and flexural motion emerges as a significant source in some circumstances.

6. CONCLUSIONS

a) The generation of harmonics in a misaligned system is a very complex subject with many contributory factors.
b) The non-linearities of oil film journal bearings is one important contributor.
c) Faulty couplings can give rise to the torsional and flexural equations becoming coupled.
d) Even systems with linear bearings may, under some circumstances, give rise to excitations at harmonics of shaft speed. This is due to time dependent stiffness terms rather than non-linearity.
e) These mechanisms should be studied for a realistic system.

7. REFERENCES

[1] D.L.Dewell and L.D.Mitchell, (1984), Detection of misaligned disk coupling using spectrum analysis, *Journal of Vibration, Acoustics, Stress and Reliability in Design*, 106, 9-16.

[2] K.M.Al-Hussain and I.Redmond, (2002), Dynamic response of two rotors connected by rigid   mechanical coupling with parallel misalignment, *Journal of Sound and Vibration*, 249(3), 483-498.

[3] J M Krodkiewski and J Ding, (1993), Theory and experiment on a method for on site identification of configurations of multi-bearing systems, Journal *of Sound and Vibration*, 164, 281-293.

[4] M.Xu and R.D.Marangoni, (1994), Vibration analysis of a motor-flexible coupling rotor system subject to misalignment and unbalance – Part 1: theoretical model ana0lysis, *Journal of Sound and   Vibration*, 176, 663-679.

[5] M.Xu and R.D.Marangoni, (1994), Vibration analysis of a motor-flexible coupling rotor system subject to misalignment and unbalance – Part 2: experimental validation, *Journal of Sound and Vibration*, 176, 681-691.

[6] G Adiletta, AR Guildo and G Rossi, Chaotic motions of a rigid rotor in short journal bearings, *Nonlinear Dynamics*, 10, 251-269, 1996.

[7] A W Lees, Misalignment in rigidly coupled rotors, *Journal of Sound & Vibration*, 305, pp261-371, 2007

[8] I Redmond – Shaft misalignment and vibration – a model, *IMAC, Conference on Structural Dynamics*, Orlando, FL, February 2007.

FLUID-FILM

The procedure for numerical investigation of vibration of a rotor with fluid film bearings and a disc submerged in inwettable liquid

J Zapoměl
CISS - branch of the Institute of Thermomechanics at the VSB - Technical University of Ostrava, Czech Republic

ABSTRACT

Lateral vibration of rotors is significantly influenced by their supports and by their interaction with the medium in the surrounding space. A computer modelling method is an important tool for investigation of their behaviour. In the proposed mathematical model the shaft is represented by a beam-like body and the discs are considered to be absolutely rigid. The hydrodynamic bearings are modelled by means of nonlinear force couplings. The pressure distribution in the oil film is obtained by solving the Reynolds' equation and components of the bearing force are calculated by its integration around the circumference and along the length of the bearing. The disc is circular and is placed in a cylindrical vessel of general cross section filled with a liquid. The liquid is assumed to be incompressible, inviscous, and inwettable. Because of this it does not cling to the disc surface and acts on it only by the pressure. Vibration of the liquid in the vessel is treated as 2D. The pressure in the liquid depends on the disc and vessel accelerations and on the liquid density. As only small displacements and velocities are assumed, its distribution can be described with enough accuracy by a Laplace equation. For its solution, the finite element method has been chosen and formulations for the finite elements have been derived. The developed procedure represents a new algorithm which enables to analyze a mutual interaction between the rotor, the hydrodynamic bearings, and the liquid in which the disc of the rotor is submerged.

1. INTRODUCTION

In a lot of technological applications the rotors are supported by fluid film bearings and have the discs partly or fully submerged in various liquids. Lateral vibration of such rotors is significantly influenced by their interaction with the medium in the surrounding space. A computer modelling method represents an important tool for investigation of their behaviour.

The hydrodynamic bearings are usually implemented into the computational models by means of nonlinear force couplings. The calculation of the bearing forces starts from solving the Reynolds' equation. More details can be found e.g. in [3], [5].

The movement of the disc submerged in a liquid is significantly influenced by its interaction with the neighbourhood. The forces acting on the disc are produced by the pressure induced due to oscillation of the disc and of the vessel. In general the pressure distribution together with the velocity field is described by the Navier-Stokes equations and the equation of continuity. On certain conditions (small amplitudes of oscillations) the governing equations can be reduced to the Laplace' one [1], [4]. Components of the resulting force are then obtained by integration of the pressure distribution over the total surface of the submerged part of the disc.

2. HYDRAULIC FORCES ACTING ON THE DISC SUBMERGED IN A LIQUID

The investigated system is a vessel filled with a liquid in which a disc is submerged. If the disc vibrates or if the vessel moves, then the pressure field is induced and the liquid acts on the wall of the disc by inertia forces. To determine their magnitudes it is assumed that

- the disc is circular and its surface is absolutely smooth,
- the inner wall of the vessel is cylindical of general cross section (e.g. circular, elliptical, etc.),
- the surface lines of the interior surface of the vessel and of the outer surface of the disc are vertical and parallel,
- the disc and the vessel are absolutely rigid bodies,
- the disc and the vessel can move only in the radial direction and their displacements are small,
- the liquid is incompressible, inviscous, and inwettable,
- vibration of the liquid in the vessel is treated as 2D.

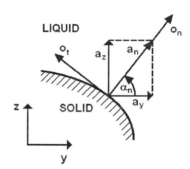

Fig.1 Acceleration of the liquid particle touching the surface of the solid body

As the disc performs oscillations only with small amplitudes, the pressure field in the liquid can be described by a Laplace's equation [1], [4]

$$\frac{\partial^2 p}{\partial y^2} + \frac{\partial^2 p}{\partial z^2} = 0 \tag{1}$$

p- - pressure,
y, z - cartesian coordinates.

As the liquid is inwettable, it does not cling to the solid body. Then the motion of the liquid particle touching the body surface is not constrained in the tangential direction but in the normal direction it moves together with the body. The boundary conditions needed to solve the Laplace equation (1) result from the Navier Stokes equation related to the direction normal to the boundary taking into account that due to the small amplitudes of the vibration the convective terms can be neglected and that the viscous forces are considerably smaller than the inertia ones [4]. After these simplifications the boundary condition takes the form

$$\frac{\partial p}{\partial n} = -\rho\, a_n \qquad (2)$$

n - coordinate in the direction of the outer normal to the boundary,

a_n - acceleration of the liquid particle on the boundary in the direction of its normal,

ρ - density of the liquid.

Consequently equation (2) can be rewritten in the following manner

$$\frac{\partial p}{\partial y}\cos\alpha_n + \frac{\partial p}{\partial z}\sin\alpha_n = -\rho\, a_n \qquad (3)$$

where

$$a_n = a_y \cos\alpha_n + a_z \sin\alpha_n \qquad (4)$$

a_y, a_z - accelerations of the sliding motion of the point on the boundary in the y, z directions,

α_n - directional angle of the outer normal of the boundary (orientation into the liquid).

For solving the Laplace's equation a finite element method can be used. As shown in [1], [4], solution of equation (1) with the boundary condition (2) is equivalent to minimizing the functional Ψ

$$\Psi = \frac{1}{2}\int_\Omega \left[\left(\frac{\partial p}{\partial y}\right)^2 + \left(\frac{\partial p}{\partial z}\right)^2\right] dy\,dz - \int_{\Gamma_D}\rho\, a_n p\, ds - \int_{\Gamma_V}\rho\, a_n p\, ds \qquad (5)$$

Ψ - functional,

Ω - investigated region (area filled with liquid),

Γ_D - interior boundary of the investigated region (outer surface of the disc),

Γ_V - outer boundary of the investigated region (interior surface of the vessel).

Functional Ψ can be expressed as a sum of three components

$$\Psi = \Psi_\Omega + \Psi_{\Gamma D} + \Psi_{\Gamma V} \qquad (6)$$

where

$$\Psi_\Omega = \frac{1}{2} \int_\Omega \left[\left(\frac{\partial p}{\partial y} \right)^2 + \left(\frac{\partial p}{\partial z} \right)^2 \right] dy \, dz \tag{7}$$

$$\Psi_{\Gamma D} = - \int_{\Gamma_D} \rho \, a_n p \, ds \tag{8}$$

$$\Psi_{\Gamma V} = - \int_{\Gamma_V} \rho \, a_n p \, ds \tag{9}$$

Let the region Ω be discretized into triangular elements. It is assumed that within each of them the pressure varies linearly. Then the functional Ψ is equal to the sum of the functionals related to all individual elements and to their sides forming the interior and outer boundaries of region Ω

$$\Psi = \sum_{i=1}^{N_E} \Psi_{\Omega Ei} + \sum_{i=1}^{N_{\Gamma D}} \Psi_{\Gamma Di} + \sum_{i=1}^{N_{\Gamma V}} \Psi_{\Gamma Vi} \tag{10}$$

The procedure described in details in [4] gives the relations for the functional of each element

$$\Psi_{\Omega Ei} = \frac{1}{2} \mathbf{p}_{Ei}^T \, \mathbf{D}_{Ei}^T \, \mathbf{D}_{Ei} \, \mathbf{p}_{Ei} \, S_{Ei} \tag{11}$$

$$\Psi_{\Gamma Di} = -\frac{1}{2} \rho \left(a_{n\Gamma DN1i} \, p_{\Gamma DN1i} + a_{n\Gamma DN2i} \, p_{\Gamma DN2i} \right) L_{\Gamma Di} \tag{12}$$

$$\Psi_{\Gamma Vi} = -\frac{1}{2} \rho \left(a_{n\Gamma VN1i} \, p_{\Gamma VN1i} + a_{n\Gamma VN2i} \, p_{\Gamma VN2i} \right) L_{\Gamma Vi} \tag{13}$$

$\mathbf{p}_{Ei}$	-	vector of pressures at nodes of the triangular element,
S_{Ei}	-	area of the triangular element,
$a_{n\Gamma DN1i}, a_{n\Gamma DN2i}$	-	acceleration of nodes N1, N2 on the interior boundary (surface of the disc) in the direction of the outer normal,
$a_{n\Gamma VN1i}, a_{n\Gamma VN2i}$	-	acceleration of nodes N1, N2 on the outer boundary (surface of the vessel) in the direction of the outer normal,
$p_{\Gamma DN1i}, p_{\Gamma DN2i}$	-	pressure at nodes N1, N2 on the interior boundary (surface of the disc),
$p_{\Gamma VN1i}, p_{\Gamma VN2i}$	-	pressure at nodes N1, N2 on the outer boundary (surface of the vessel),
$L_{\Gamma Di}$	-	length of the side of the triangular element between nodes N1 and N2 forming the interior boundary (the surface of the disc),
$L_{\Gamma Vi}$	-	length of the side of the triangular element between nodes N1 and N2 forming the outer boundary (the surface of the vessel).

The coefficient matrix $\mathbf{D}_{Ei}$ depends on coordinates of nodes of the individual elements [4]

$$\mathbf{D}_{\text{Ei}} = \begin{bmatrix} w_{21} & w_{22} & w_{23} \\ w_{31} & w_{32} & w_{33} \end{bmatrix} \tag{14}$$

where

$$\mathbf{W}_{\text{Ei}}^{-1} = \begin{bmatrix} w_{11} & w_{12} & w_{13} \\ w_{21} & w_{22} & w_{23} \\ w_{31} & w_{32} & w_{33} \end{bmatrix} \tag{15}$$

and

$$\mathbf{W}_{\text{Ei}} = \begin{bmatrix} 1 & y_{N1} & z_{N1} \\ 1 & y_{N2} & z_{N2} \\ 1 & y_{N3} & z_{N3} \end{bmatrix} \tag{16}$$

y_{N1}, z_{N1} - coordinates of node 1 of the triangular element,
y_{N2}, z_{N2} - coordinates of node 2 of the triangular element,
y_{N3}, z_{N3} - coordinates of node 3 of the triangular element.

It is considered that both the disc and the vessel are absolutely rigid bodies and that the liquid does not cling to their surfaces. Therefore accelerations of the individual liquid particles on the boundaries are functions only of components of the accelerations of the centres of the disc and of the vessel. Then substitution of (11), (12) and (13) into (10) for all elements and making use of (4) enables to express the functional Ψ in the form

$$\Psi = \frac{1}{2}\mathbf{p}^{\text{T}}\mathbf{H}\mathbf{p} - \mathbf{p}^{\text{T}}\left(a_{Dy}\,\mathbf{g}_{Dy} + a_{Dz}\,\mathbf{g}_{Dz} + a_{Vy}\,\mathbf{g}_{Vy} + a_{Vz}\,\mathbf{g}_{Vz} \right) \tag{17}$$

$\mathbf{H}$ - coefficient matrix,
$\mathbf{g}_{Dy}, \mathbf{g}_{Dz}$ - coefficient vectors,
$\mathbf{g}_{Vy}, \mathbf{g}_{Vz}$ - coefficient vectors,
a_{Dy}, a_{Dz} - y, z components of the disc centre acceleration,
a_{Vy}, a_{Vz} - y, z components of the vessel centre acceleration.

To achieve the minimum of Ψ it must hold

$$\left[\frac{\partial \Psi}{\partial \mathbf{p}} \right] = \mathbf{0} \tag{18}$$

$\mathbf{0}$ - zero vector.

After performing this manipulation calculation of the pressure distribution arrives at solving a set of linear algebraic equations

$$\mathbf{H}\mathbf{p} = a_{Dy}\,\mathbf{g}_{Dy} + a_{Dz}\,\mathbf{g}_{Dz} + a_{Vy}\,\mathbf{g}_{Vy} + a_{Vz}\,\mathbf{g}_{Vz} \tag{19}$$

The pressure at all nodes is then given as a sum of four components

$$\mathbf{p}=\mathbf{p}_{Dy}+\mathbf{p}_{Dz}+\mathbf{p}_{Vy}+\mathbf{p}_{Vz} \tag{20}$$

where

$$\mathbf{p}_{Dy}=a_{Dy}\,\mathbf{H}^{-1}\mathbf{g}_{Dy} \tag{21}$$

$$\mathbf{p}_{Dz}=a_{Dz}\,\mathbf{H}^{-1}\mathbf{g}_{Dz} \tag{22}$$

$$\mathbf{p}_{Vy}=a_{Vy}\,\mathbf{H}^{-1}\mathbf{g}_{Vy} \tag{23}$$

$$\mathbf{p}_{Vz}=a_{Vz}\,\mathbf{H}^{-1}\mathbf{g}_{Vz} \tag{24}$$

$\mathbf{p}_{Dy}$ - vector of pressures at all nodes produced by acceleration of the disc in y-direction,

$\mathbf{p}_{Dz}$ - vector of pressures at all nodes produced by acceleration of the disc in z-direction,

$\mathbf{p}_{Vy}$ - vector of pressures at all nodes produced by acceleration of the vessel in y-direction,

$\mathbf{p}_{Vz}$ - vector of pressures at all nodes produced by acceleration of the vessel in z-direction.

Because the liquid is inviscous and inwettable, no tangential forces are produced and components of the total force acting on the disc are obtained by integration of the pressure distribution around the circumference and along the height (thickness) of the submerged part of the disc

$$F_{Fy}=-h_D\int_{\Gamma_D}p_{Dy}\cos\alpha_n ds-h_D\int_{\Gamma_D}p_{Dz}\cos\alpha_n ds-h_D\int_{\Gamma_D}p_{Vy}\cos\alpha_n ds-h_D\int_{\Gamma_D}p_{Vz}\cos\alpha_n ds \tag{25}$$

$$F_{Fz}=-h_D\int_{\Gamma_D}p_{Dy}\sin\alpha_n ds-h_D\int_{\Gamma_D}p_{Dz}\sin\alpha_n ds-h_D\int_{\Gamma_D}p_{Vy}\sin\alpha_n ds-h_D\int_{\Gamma_D}p_{Vy}\sin\alpha_n ds \tag{26}$$

F_{Fy}, F_{Fz} - y, z components of the hydraulic force acting on the disc submerged in the liquid,

p_{Dy} - pressure on the interior boundary (surface of the disc) produced by acceleration of the disc in y-direction,

p_{Dz} - pressure on the interior boundary (surface of the disc) produced by acceleration of the disc in z-direction,

p_{Vy} - pressure on the interior boundary (surface of the disc) produced by acceleration of the vessel in y-direction,

p_{Vz} - pressure on the interior boundary (surface of the disc) produced by acceleration of the vessel in z-direction

h_D - height (thickness) of the disc (submerged part of the disc).

As the accelerations a_{Dy}, a_{Dz} are the same for all points situated on the interior boundary of Ω (the wall of the disc), they can be factored out before the integrals in (25) and (26) and then these two relationships can be expressed in a matrix form

$$\begin{bmatrix} F_{Fy} \\ F_{Fz} \end{bmatrix} = -\mathbf{M}_{DF} \begin{bmatrix} a_{Dy} \\ a_{Dz} \end{bmatrix} + \begin{bmatrix} -h_D \int\limits_{\Gamma_D} p_{Vy} \cos\alpha_n ds - h_D \int\limits_{\Gamma_D} p_{Vz} \cos\alpha_n ds \\ -h_D \int\limits_{\Gamma_D} p_{Vy} \sin\alpha_n ds - h_D \int\limits_{\Gamma_D} p_{Vz} \sin\alpha_n ds \end{bmatrix} \tag{27}$$

$\mathbf{M}_{DF}$ can be considered as the additional mass matrix of the disc. It expresses inertia properties of the liquid by which the vibration of the disc is influenced. In general matrix $\mathbf{M}_{DF}$ is real, full and not symmetric. Its elements depend on the mutual position of the disc relative to the wall of the vessel. If the vessel does not vibrate, the second term on the right hand side of (27) takes a zero value.

3. HYDRAULIC FORCES ACTING IN HYDRODYNAMIC BEARINGS WITH A DEEP CENTRAL CIRCUMFERENTIAL GROOVE

One possibility how to lubricate the fluid film bearings is to supply the oil into a deep circumferential groove placed in the middle of their length (Fig.2). Such bearings are marked for low oil heating but also for a reduced load capacity because the groove is deep and therefore no hydrodynamic effect is developed at its location.

To determine components of the bearing force by which the layer of lubricant acts on the rotor journal and bearing shell it is necessary to know a pressure distribution in the oil film. As the bearing gap is very narrow a classical theory of lubrication may be applied for this purpose. The pressure distribution in the oil layer is then described by the Reynolds' equation [2], [3], [5]

$$\frac{1}{R^2} \frac{\partial}{\partial\varphi}\left(\frac{h^3}{\eta} \frac{\partial p}{\partial\varphi} \right) + \frac{\partial}{\partial Z}\left(\frac{h^3}{\eta} \frac{\partial p}{\partial Z} \right) = \frac{6}{R} \frac{\partial}{\partial\varphi}\left[h(u_1 + u_2) \right] + 12 \frac{\partial h}{\partial t} \tag{28}$$

where

$h = c - e_H \cos(\varphi - \gamma)$

u_1 - circumferential velocity component of the points on the liner interior surface,

u_2 - circumferential velocity component of the points on the journal surface,

p - pressure,

c - width of the bearing gap,

R - journal radius,

h - thickness of the oil film ,

e_H - eccentricity of the journal centre (Fig.3),

γ - position angle of the line of centres (Fig.3),

φ, Z - circumferential, axial coordinates (Fig.3),

η - oil dynamical viscosity,

t - time.

The circumferential velocity components u_1 and u_2 are determined from the boundary conditions

$$u_1 = 0 \tag{30}$$

$$u_2 = R\omega \tag{31}$$

ω - angular speed of the rotor rotation.

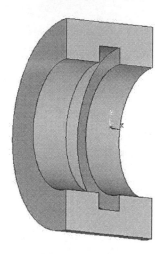

Fig.2 Bearing with a deep circumferential groove   **Fig.3 Frames of reference referred to the hydrodynamic bearing**

The Reynolds' equation holds only in the areas where the hydrodynamic effect is developed. If the pressure drops to a critical level, a cavitation occurs. The experiments carried out by Zeidan, Vance, San Andrés, and other investigators [7], [8] showed that pressure of the medium in these areas remained approximately constant. Therefore from the simplest distinguishing level the pressure distribution in the oil film can be described

$$p_d = p \qquad \text{for} \qquad p \geq p_{CAV} \tag{32}$$

$$p_d = p_{CAV} \qquad \text{for} \qquad p < p_{CAV} \tag{33}$$

p_d - pressure distribution in the bearing gap,
p_{CAV} - pressure of the medium in the cavitated area.

The pressure distribution in different parts of the bearing gap is given by the following relationships

$$p_d = p_G \qquad\qquad \text{circumferential groove} \tag{34}$$

$$p_d = p \qquad\qquad \text{full oil film} \tag{35}$$

$$p_d = p_{CAV} \qquad\qquad \text{cavitated area} \tag{36}$$

p_G - pressure of the oil supplied into the circumferential groove.

844

At the outer edges of the bearing it is assumed that the oil pressure is equal to the atmospheric one.

Components of the bearing force acting on the rotor journal are determined by integration of the pressure distribution p_d around the circumference and along the length of the bearing

$$F_{By} = -R \int_{-\frac{L}{2}}^{\frac{L}{2}} \int_{0}^{2\pi} p_d \cos\varphi \, d\varphi \, dZ \qquad (37)$$

$$F_{Bz} = -R \int_{-\frac{L}{2}}^{\frac{L}{2}} \int_{0}^{2\pi} p_d \sin\varphi \, d\varphi \, dZ \qquad (38)$$

F_{By}, F_{Bz} -   y, z components of the bearing force acting on the rotor journal,
L   -   length of the bearing.

4.   SOLUTION OF THE EQUATION OF MOTION

The hydrodynamic bearings are usually implemented into the mathematical models of rotor systems by means of force couplings. Lateral vibration of such rotors is then governed by a nonlinear equation of motion

$$\mathbf{M}\,\ddot{\mathbf{x}} + (\mathbf{B} + \eta_V\,\mathbf{K}_{SH} + \omega\,\mathbf{G})\,\dot{\mathbf{x}} + (\mathbf{K} + \omega\,\mathbf{K}_C)\,\mathbf{x} = \mathbf{f}_A + \mathbf{f}_V + \mathbf{f}_B(\mathbf{x}, \dot{\mathbf{x}}) + \mathbf{f}_F(\ddot{\mathbf{x}}) \qquad (39)$$

$\mathbf{M}, \mathbf{B}, \mathbf{K}, \mathbf{G}, \mathbf{K}_C$ -   mass, damping, stiffness, gyroscopic, circulation matrices [9],
$\mathbf{K}_{SH}$ -   stiffness matrix of the shaft [9],
$\mathbf{f}_A, \mathbf{f}_V, \mathbf{f}_B, \mathbf{f}_F$ -   vectors of applied, constrained, bearing, fluid induced forces,
$\mathbf{x}, \dot{\mathbf{x}}, \ddot{\mathbf{x}}$ -   vectors of general displacements, velocities, accelerations,
η_V -   damping coefficient of the shaft material.

Taking into consideration the relationship (27) the force by which the liquid acts on the disc can be expressed in the following manner

$$\mathbf{f}_F(\ddot{\mathbf{x}}) = -\mathbf{M}_F\,\ddot{\mathbf{x}} + \mathbf{f}_{FV} \qquad (40)$$

$\mathbf{M}_F$ - additional mass matrix,
$\mathbf{f}_{FV}$ - vector of the fluid induced forces acting on the disc due to the movement of the vessel.

After a simple manipulation the equation of motion (39) can be modified into this form

$$(\mathbf{M} + \mathbf{M}_F)\,\ddot{\mathbf{x}} + (\mathbf{B} + \eta_V\mathbf{K}_{SH} + \omega\,\mathbf{G})\,\dot{\mathbf{x}} + (\mathbf{K} + \omega\,\mathbf{K}_C)\,\mathbf{x} = \mathbf{f}_A + \mathbf{f}_V + \mathbf{f}_B(\mathbf{x}, \dot{\mathbf{x}}) + \mathbf{f}_{FV} \qquad (41)$$

(41) represents a nonlinear equation of motion. For its solution a Newmark method can be used. This method is implicit and it implies its algorithm starts from the equation of motion related to time t+Δt. At each integration step the solution arrives at solving a set of algebraic equations that are nonlinear in this case due to the bearing forces. To avoid this manipulation vector $\mathbf{f}_B$ related to the point of time t+Δt can be approximately expressed by means of its expansion into a Taylor series in the neighbourhood of time t

$$\mathbf{f}_{B,t+\Delta t} = \mathbf{f}_{B,t} + \mathbf{D}_{B,t}\,(\dot{\mathbf{x}}_{t+\Delta t} - \dot{\mathbf{x}}_t) + \mathbf{D}_{K,t}\,(\mathbf{x}_{t+\Delta t} - \mathbf{x}_t) + ... \tag{42}$$

$$\mathbf{D}_{K,t} = \left[\frac{\partial \mathbf{f}_B(\mathbf{x},\dot{\mathbf{x}})}{\partial \mathbf{x}}\right]_{\mathbf{x}=\mathbf{x}_t,\,\dot{\mathbf{x}}=\dot{\mathbf{x}}_t} \tag{43}$$

$$\mathbf{D}_{B,t} = \left[\frac{\partial \mathbf{f}_B(\mathbf{x},\dot{\mathbf{x}})}{\partial \dot{\mathbf{x}}}\right]_{\mathbf{x}=\mathbf{x}_t,\,\dot{\mathbf{x}}=\dot{\mathbf{x}}_t} \tag{44}$$

Then substitution of only the linear portion of the Taylor series (42) and carrying out several simple operations give the equation of motion related to time t+Δt that has the form which requires to solve only a set of linear algebraic equations at the current integration step

$$(\mathbf{M}+\mathbf{M}_F)\,\ddot{\mathbf{x}}_{t+\Delta t} + (\mathbf{B}+\eta_V\,\mathbf{K}_{SH}+\omega\,\mathbf{G}-\mathbf{D}_{B,t})\,\dot{\mathbf{x}}_{t+\Delta t} + (\mathbf{K}+\omega\,\mathbf{K}_C-\mathbf{D}_{K,t})\,\mathbf{x}_{t+\Delta t} =$$
$$= \mathbf{f}_{A,t+\Delta t} + \mathbf{f}_{V,t+\Delta t} + \mathbf{f}_{B,t} - \mathbf{D}_{B,t}\dot{\mathbf{x}}_t - \mathbf{D}_{K,t}\mathbf{x}_t + \mathbf{f}_{FV,t+\Delta t} \tag{45}$$

More details on application of this approach can be found e.g. in [6].

5.   EXAMPLE

Rotor of the investigated rotor system (Fig.4) consists of a shaft (SH) and of two discs (D1, D2). The shaft is coupled with a rigid frame (FR) by two hydrodynamic bearings (B1, B2). Each of them is equipped with one deep central circumferential groove into which the lubricating oil is supplied. Disc D1 mounted on the overhung end of the shaft is placed in a vessel (VS) filled with liquid and is totally submerged. Disc D2 is coupled with a rigid rotor of an electric motor by a prestressed flexible belt. The rotor rotates at constant angular speed (300 rad/s) and is loaded by centrifugal forces caused by unbalances of both discs. The task was to analyze the motion of the rotor after the initial transient component of its vibration is died out.

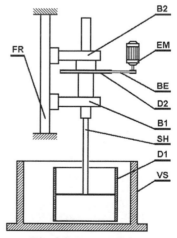

Fig.4 Scheme of the investigated rotor system

In the computational model the shaft was represented by a beam like body, both discs were considered as absolutely rigid, and the liquid as incompressible, inviscous and inwettable. For the purpose of the calculation the shaft was discretized into finite elements.

Some of the results are summarized in the following figures. The equilibrium positions of the rotor journal centres in bearings B1 and B2 are drawn in Fig.5. Trajectories of the disc D1 centre and of the rotor journal centre in bearing B1 after the initial transient component of the vibration is died out are shown in Fig.6 and 7. The Fourier transform of the time history of y-displacement of the disc centre is drawn in Fig.8. It is evident that in the resulting response except the vibration component having the principal frequency (300 rad/s) a number of subharmonic and ultraharmonic ones are excited too.

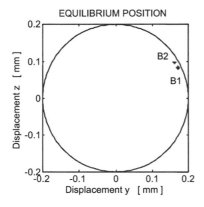

Fig.5 Equilibrium position of the journals

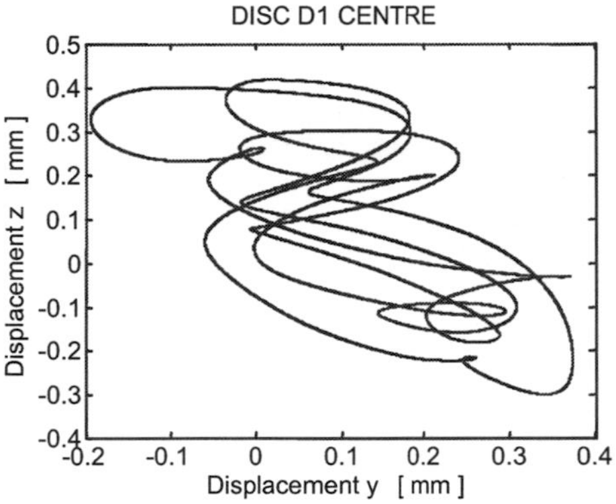

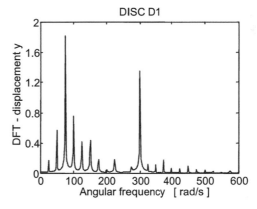

Fig.6 Trajectory of the disc D1 centre

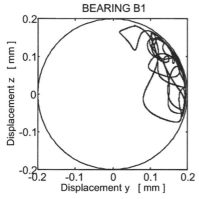

Fig.7 Trajectory of the journal centre in B1

Fig.8 Fourier transform of the disc y-displacement

6. CONCLUSIONS

The described method represents a new and complex approach to investigation of a mutual interaction of the rotor, of the hydrodynamic bearings, and of the liquid in which the disc of the rotor is submerged.

The developed procedure is intended for preparation of the input data for the investigation of the characteristics of the induced vibration of the rotor system that is excited by imbalance of the discs and by hydraulic forces induced by oscillation of the vessel in which the disc is submerged.

The results of the computer simulations show that the procedure proposed for solving the equation of motion is numerically stable. It enables the application of a reasonably short integration step, also for the cases when the vibration of the rotor becomes chaotic or considerably nonlinear.

ACKNOWLEDGEMENT

This research work has been supported by the grant projects GA101/06/0063 and AVO Z20760514. The support is gratefully acknowledged.

REFERENCES

[1] Bathe K.-J.: Finite Element Procedures in Engineering Analysis, Prentice-Hall, Inc.,Englewood Cliffs, New Jersey
[2] Cameron A.: The Principles of Lubrication, Longmans Green and Co. Ltd., 1966, London
[3] Krämer, E.: *Dynamics of rotors and foundations*, Springer-Verlag. Berlin, Heidelberg, 1993
[4] Levy S., Wilkinson J.P.D.: The Component Element Method in Dynamics with Application to Earthquake and Vehicle Engineering, McGraw-Hill, Inc., 1976
[5] Vance J.M.: *Rotordynamics of Turbomachinary*, John Wiley & Sons, Inc., 1988
[6] Zapoměl J. and Malenovský E.: Approaches to numerical investigation of the character and stability of forced and and self-excited vibration of flexible rotors with non-linear supports, *7th International Conference on Rotating Machinery* (IMechE Conference Transactions), University of Nottingham, 2000, pp. 691-699.
[7] Zeidan F.Y., Vance J.M.: Cavitation regimes in squeeze film dampers and their effect on the pressure distribution, *STLE Tribology Transactions*, vol.33, 1990, pp. 447-453
[8] Zeidan F.Y., Vance J.M.: Cavitation leading to a two phase fluid in a squeeze film damper, *STLE Tribology Transactions*, vol.32, no.1, 1989, pp. 100-104
[9] Zorzi E.S., Nelson H.D.: Finite element simulation of rotor-bearing systems with internal damping, Transactions of the ASME, *Journal of Engineering for Power*, January 1977, pp.71-76

Shape optimization of a labyrinth seal: leakage minimization and sensitivity of rotordynamic coefficients

A O Pugachev, M Deckner
Institute of Energy Systems, Technische Universität München, Germany

ABSTRACT

The present paper deals with the shape optimization and rotordynamic analysis of a short staggered labyrinth seal. A leakage minimization problem is considered by applying a two-dimensional CFD analysis. Design variables are axial positions of the teeth and dimensions and position of the rotor step. A non-gradient-based method is applied. An optimized seal has about 8 percent leakage decrease over the baseline configuration. A full three-dimensional eccentric CFD model along with a whirling rotor method is used for the prediction of rotordynamic coefficients. The optimized sealing configuration demonstrates an increase in direct stiffness and damping. Leakage and rotordynamic coefficients for the baseline configuration are compared to the experimental results.

NOTATION

e = rotor eccentricity, [m]
F = seal force, [N]
f = objective function
HS = step height, [m]
K = direct stiffness, [N/m]
k = cross-coupled stiffness, [N/m]
LS = step length, [m]
m = number of initial feasible points
n = number of design variables
PS = step position, [m]
PT = tooth position, [m]
q = magnetic excitation coefficient, [N/m]
r = magnetic stiffness, [N/m]
x = design variable

Ω = whirl velocity, [rad/s]
ω = angular velocity, [rad/s]

Subscripts
0 = test case without flow through the seal
1 = test case with flow through the seal
l = lower bound
r = radial
t = tangential
u = upper bound

Abbreviations
CFD = computational fluid dynamics

1 INTRODUCTION

In spite of development of new sealing technologies labyrinth seals still remain the most common sealing elements in turbomachinery. Advantages such as wide range of operating pressures, temperatures and rotating speeds, ability to withstand rotor excursions compensate a relatively high leakage in comparison to novel sealing concepts. Though labyrinth seals are very common and well-studied, there is a potential to minimize their leakage characteristics. A straightforward solution is the reduction of radial clearances. This is not always possible, because the minimal value of radial clearance is limited by service conditions and is normally set on the design phase. The improvement of leakage performance of seals could also be achieved by varying positions of sealing teeth and dimensions of cavities, using inclined teeth etc. Generally, a staggered labyrinth seal with inclined teeth demonstrate better leakage reduction than other types of labyrinth seals (e.g. (1)). However, an optimization of a specific labyrinth sealing configuration based only on engineering experience could not be effective, so numerical techniques for shape optimization should be used.

The problem of labyrinth seal optimization has been studied so far mostly by manual theoretical and experimental analysis of particular designs. The work by Rhode et al. (2) studies the influence of step height, cavity length, tooth width and clearance on the leakage of a stepped labyrinth seal. The measurements on the improved sealing configuration showed a gain of 60% - 79% over the baseline configuration in leakage reduction for different pressure differentials. A more recent work (3) investigated two staggered labyrinth seals with straight and inclined teeth. A seal with long teeth, inclined at the angle of 60° to the stator surface, demonstrated about 19% of leakage reduction compared to the seal with straight teeth.

Schramm et al. (4) performed the shape optimization study on a two-tooth-on-rotor stepped labyrinth seal. A two-dimensional CFD model was applied to predict the seal leakage. For optimization the stochastic non-gradient-based method of simulated annealing was used. This method is generally more computationally intensive due to the high number of trial solutions in comparison to the gradient-based method or other direct methods. It can provide a global optimum point though. The distance from the stator step to the tooth and the step height were selected as design variables. Optimization process for the pressure ratio ~1.08 needed about 900 evaluations of an objective function. The optimal design showed a leakage reduction of about 10%.

An important factor that should be considered in design of labyrinth seals is their impact on the rotordynamic stability. Under certain conditions (small clearances, high velocities, high pressure differentials etc.) a destabilizing force generated in the labyrinth seal could reach large values and bring the rotor to instability. In the conventional approach gas seals are represented by a set of stiffness and damping coefficients (see for example (5)). A linear rotordynamic model of a labyrinth seal for small motion around the stator centre could be expressed as follows:

$$F_r/e = -K - c\Omega$$
$$F_t/e = k - C\Omega$$

(1)

In this equations, F_r and F_t are the components of the aerodynamic force acting in the seal; e is the rotor eccentricity; Ω is the rotor whirl speed; K, k and C, c are direct and cross-coupled, stiffness and damping coefficients respectively.

The influence of the labyrinth configuration on the rotordynamic coefficients was also studied in many works (an extensive analysis can be found in (5)). Baumann (1) reported that the staggered labyrinth seals have unfavourable rotordynamic characteristics compared with other types. It was also reported that an axially shifted rotor step from its centred position under the tooth provides positive direct stiffness, while in the centred position direct stiffness is negative. In the recent work of Kwanka (6) experimental results on rotordynamic coefficients for different labyrinth seals were summarized. The influence of operating parameters as well as the number of cavities, length and height of the cavity were analyzed.

The aim of this work is to study the influence of geometrical parameters on leakage and dynamical characteristics of a short staggered labyrinth seal with three teeth on stator. The analysis is performed using optimization technique along with CFD calculations. The theoretical study consists of the two parts. First, a quasi two-dimensional CFD model is used for the seal shape optimization in respect of leakage minimization at one operating point. At the second step, a full three-dimensional eccentric CFD model is applied for the baseline and optimized sealing configurations for the rotordynamic analysis. Predictions for rotordynamic coefficients at three pressure differentials are obtained using rotating frame of reference (a whirling rotor method). CFD modelling is carried out using commercial software ANSYS CFX and ICEM CFD. To validate the models, theoretical results for the baseline sealing configuration are compared with measurements.

2 SHORT STAGGERED LABYRINTH SEAL

2.1 Experimental Analysis
The experimental investigations on the short staggered labyrinth seal have been carried out on the seal test rig at the Institute of Energy Systems, Technische Universität München. Fig 1 shows the geometric configuration of the studied seal. The rotor has a diameter of 180.05 mm. The seals teeth are equally spaced with a pitch of 14 mm and a radial clearance of 0.27 mm. The complete seal length is 60 mm. The test rig is a double-flow assembly. Depending on pressure drop, a part of compressed air injected at the centre of the assembly flows through two identical seals. A bypass channel withdraws the rest of the air. The test rig allows generating an inlet swirl which is almost independent from the inlet static pressure. Detailed description of the test rig can be found in (7), (8).

The experimental approach to identify the rotordynamic coefficients of gas seals used in this work was originally presented by Kwanka and Mair (9). The identification procedure is based on the alternation of rotor vibrations due to aerodynamic forces in seals. The idea is to bring the rotor to the stability limit by means of additional excitation, first without and then with the leakage flow through test seals. The difference in the rotor behaviour determines the seal influence on rotordynamics, i.e. seal stiffness and damping coefficients. A magnetic bearing is used as exciter (10). It is positioned close to the test seal, however not interfering with the flow at the seal outlet. Two catcher bearings keep the rotor from destructive oscillations. The magnetic bearing generates tangential and normal forces during the rotor excitation toward the stability limit. A value of tangential force is corresponded to a magnetic excitation coefficient q. A normal exciting force changes a rotor whirling frequency. The compensation of this frequency shift gives a second parameter – magnetic stiffness r. During the experiment the magnetic bearing operating parameters q and r are converted via the registered values of the calibrated control box. Assuming that indexes 0 and 1 refer to the case without and with sealing flow respectively, one can write the following relationship between measured values and forces acting in the seal:

$$R_r/e = r_0 - r_1$$
$$R_t/e = q_0 - q_1 \qquad (2)$$

Equation (1) and equation (2) form a relationship between magnetic bearing operating parameters and seal rotordynamic coefficients. Performing at least two measurements with different whirl velocity Ω, one can identify the four coefficients K, k, C, c. The value of the operated rotor whirl frequency is varied by changing the distance between bearing supports. The forward and backward whirling motions in the frequency range of 27-31 Hz are used for the coefficients identification.

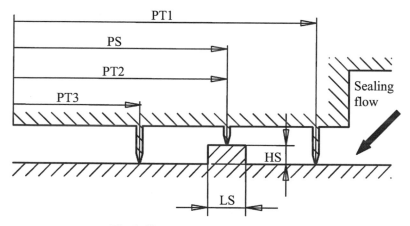

Fig 1 Short staggered labyrinth seal

2.2 Leakage and Rotordynamic Coefficients Calculations

Calculation of leakage and four rotordynamic coefficients of the studied seal is of interest to this work. The mass flow rate is calculated by applying a quasi two-dimensional CFD analysis. The computation domain includes the additional downstream region to more accurate modelling of outlet flow. A structured hexahedral grid with one cell in circumferential direction is used. One requirement for the optimization is an automatically grid generation which is realised by means of a command script. Compressible Reynolds-averaged Navier-Stokes equations are closed by the shear stress transport (SST) turbulence model with automatic wall functions. Pressure at inlet and outlet boundaries and wall velocity on the rotor surface are set. Periodic boundary conditions on the two-dimensional slice faces are applied.

To determine stiffness and damping coefficients a whirling rotor method is used (11). This method is based on a full three-dimensional eccentric CFD model of the seal. Under the same assumptions of the linear model in equation (1) the circular whirling orbit is treated quasi-steady by switching to the rotating frame of reference. In this case, the dynamic behaviour of the seal could be modelled by static calculations. As with experimental identification, the procedure for calculation of four coefficients involves at least two CFD solutions at different rotor whirl speeds. The normal and tangential components of the aerodynamic force in the seal are determined by integrating the pressure field on the rotor surface. Finally, the rotordynamic coefficients are determined from the linear relations between the force components and the whirl speed in equation (1). A three-dimensional structured hexahedral grid is extruded from the quasi two-dimensional one. Then the grid is transformed to set the rotor at the eccentric position.

CFD analysis provides a general approach to prediction of seal performance. Generality in geometry and working conditions is especially important in optimization studies. At the same time a single calculation cycle during the optimization should be executed, where possible, fast. Therefore one have to find a "computational optimum" between sophisticated CFD models that could be computationally very costly and simplified ones without lack of validity. In this study, the inlet swirl in the two-dimensional model does not correspond entirely precise to the experimental conditions. In the three-dimensional model inlet swirl should be modelled accurate because of its strong influence on the rotordynamic coefficients (especially, on the cross-coupled stiffness). The 3D CFD model used in this work takes into account different inlet swirl speeds. However, the accurate modelling of the inflow geometry is not considered. A mesh independence study is not specially carried out; nevertheless information on the influence of the computational grid on the CFD predictions for the studied seal from the previous work (see (8)) is used.

3 OPTIMIZATION PROCEDURE

3.1 Design Optimization Problem
A problem of leakage minimization for a labyrinth seal can be formulated as follows:

$$\text{Minimize} \quad f\left(x_1, x_2, \ldots, x_n\right)$$
$$x_i^l \leq x \leq x_i^u, \quad i = 1, 2, \ldots, n \tag{3}$$

This nonlinear optimization problem in equation (3) consists of the objective function f subjected only to side constraints imposed on the design variables x. The objective function in equation (3) is the mass flow rate through the seal. Design variables describe a seal shape. In this study the set of design variables include six geometrical parameters (see Fig 1): positions of three teeth (PT1, PT2, and PT3), position and size of the rotor step (PS, LS, and HS). Design variables are chosen in such way to analyze possibilities of leakage reduction without significant modification of the baseline configuration. The variation bounds of design variables define all allowable sealing configurations. Generally, the lower and upper bounds in side constraints are set according to physical and manufacturing limitations. For the described optimization problem, in order to narrow the search range, the bounds are set on the basis of geometric considerations and experience. The design variable bounds and the baseline values are summarized in Table 1.

Table 1 Design variables bounds

	PT1, [mm]	PT2, [mm]	PT3, [mm]	PS, [mm]	LS, [mm]	HS, [mm]
Lower bound	42.9	28.9	2.0	26.9	2.0	1.0
Upper bound	49.9	38.9	24.9	40.9	6.0	3.0
Baseline value	47.9	33.9	19.9	33.9	6.0	3.0

3.2 Optimization Algorithm

The solution of the optimization problem described above is obtained using a direct search technique – a complex method, which is an extension of the simplex method on constrained optimization problems (12). The method is computationally very simple, for it does not require calculation of gradients of objective function and constraints. The drawbacks of the method are its inability to handle with equality constraints and inefficiency for large problems (i.e. with many design variables).

The method starts with the baseline configuration and $m \geq (n+1)$ feasible points, which are generated randomly within the side constraints. These feasible points form an initial complex that has to be modified (it moves and shrinks in the search region) during the iterative procedure. At each iteration, the worst point of the current complex is replaced by a new feasible point by expansion and contraction of the complex. The worst point means the sealing configuration from the current set with the largest leakage. The complex tries to expand away from the worst point. If a new point is even worse, this point is contracted back toward the complex centroid until the better objective function is found. If an improved point could not be obtained, the current worst point is removed from the complex and a new feasible point is randomly generated. Eventually, the complex should collapse into the optimum point.

4   RESULTS AND DISCUSSION

A shape optimization study for the leakage minimization applied to the two-dimensional CFD model is conducted for one operating point with an inlet pressure of 0.5 MPa. Ten feasible sealing configurations are used as an initial set. Fig 2 shows a variation of objective function during the optimization. Convergence of the complex method slows down with iterations. The algorithm stalls two times that leads to the generation of new randomly defined feasible points. At the end of the optimization the objective function of mass flow rate is decreased by about 8%. Due to the fact that clearances are being kept constant and inclined teeth are not considered, decrease in the mass flow rate is relatively small.

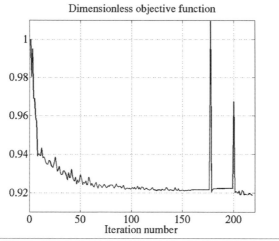

Dimensionless objective function

Fig 2 Convergence of the optimization process

Fig 3 shows the history of changes in design variables during the optimization. A zero value corresponds to the baseline configuration, ±1 refer to the design variables bounds (see Table 1). Variations of teeth axial positions are presented in Fig 3, left. The value of the third tooth position (PT3) converges to its low limit, whereas PT1 and PT2 take on their optimized values within the range space. Comparing with the baseline configuration PT1 and PT2 tend to the smaller and the greater values respectively. Design variables which describe the rotor step geometry are presented in Fig 3, right. Position of the step changes only slightly in the direction of inlet section compared to the baseline value. Height and length of the rotor step vary similar and tend to the smaller values.

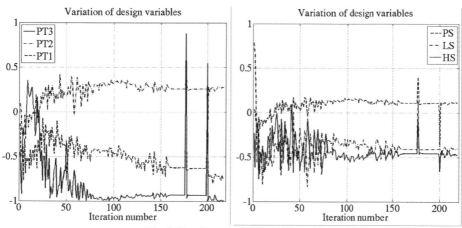

Fig 3 Design variables history

Although it is not evident from Fig 3, it should be noted that in case of the optimized configuration the rotor step stands out in front of the second tooth only at a minimal value set by bounds. This can be seen in Fig 4 where seal shapes for the initial and optimized configurations are compared. In the baseline configuration both cavities have the same dimensions. After the optimization the length of the first cavity has decreased slightly, while the length of the second cavity has increased considerably. Inlet cavity has also increased, outlet cavity has practically disappeared.

The flow patterns in the baseline and optimized sealing configurations are also shown in Fig 4. The optimized rotor step changes the flow in the cavities. There is only one stagnation point in the first cavity. In the second cavity the main vortex is developing clockwise due to the smaller step height, while in the baseline configuration the big vortex is developing counter-clockwise.

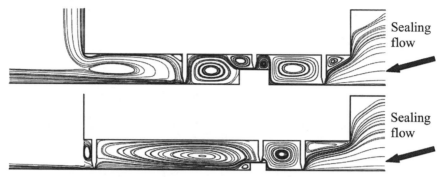

Sealing flow

Sealing flow

Fig 4 Baseline (above) and optimized (below) seal shapes

One limitation in the presented analysis is imposed on the position of the rotor step – it must be located under the second tooth and not elsewhere in the cavity. This is a disadvantage of the grid generation procedure. Although it could be managed by using more complicated command script for the grid generation, it is beyond the scope of the work. An additional study on varying downstream size of the rotor step along the second cavity of the optimized shape does not show noticeable mass flow rate reduction.

Leakage performance of the baseline and optimized shapes at different inlet pressures is demonstrated in Fig 5. Theoretical results for the baseline configuration show good agreement with the measurements. As it has been expected due to the linear character of the mass flow function, the leakage decrease of about 8% for the optimized shape remains in the whole operating range.

The second part of the work is addressed to the rotordynamic performance of the baseline and optimized labyrinth seals. Experimental values of rotordynamic coefficients for the baseline configuration are used for the validation of the three-dimensional CFD model. Fig 6 shows stiffness and damping coefficients as functions of inlet pressure. These data correspond to the rotor speed of 750 rev/min and inlet swirl of approximately 140 m/s. Cross-coupled stiffness and direct damping increase with rising inlet pressure. Direct stiffness and cross-coupled damping remain relatively constant. Theoretical values of stiffness coefficients demonstrate good agreement with the experiments. Predictions of damping coefficients correlate with the measured trends. Direct damping at high pressures is underpredicted. Generally, the results are consistent with the expectation of poor rotordynamic characteristics of staggered seals.

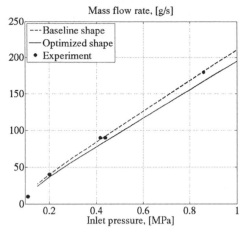

Fig 5 Mass flow rate versus inlet pressure

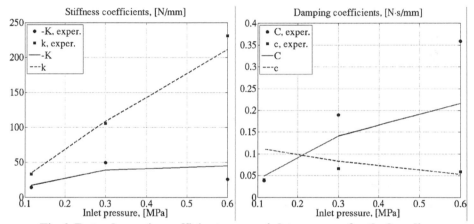

Fig 6 Rotordynamic coefficients versus inlet pressure for the baseline seal shape

Fig 7 presents a comparison of the rotordynamic coefficients for the baseline and optimized sealing configurations. The difference in cross-coupled stiffness is insignificant. The direct stiffness and direct damping are increased. A notable change occurs in the cross-coupled damping. The reason of this significant increase is not clear. Theoretical calculation just as experimental identification of cross-coupled damping is very sensitive to the level of accuracy, especially in case of labyrinth seals where damping coefficients are relatively small. As a whole, the rotordynamic performance of the optimized sealing configuration remains unsatisfactory.

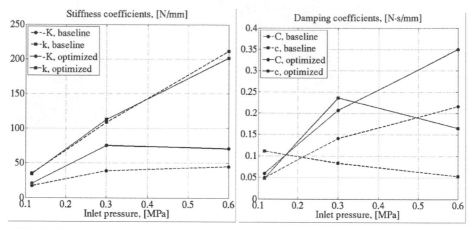

Fig 7 Rotordynamic coefficients versus inlet pressure for the both seal shapes

5   CONCLUSION

A shape optimization problem is considered for the leakage minimization of the short staggered labyrinth seal. Considerable design changes are not taken into account. The seal overall dimensions and clearances are constant. The results show that only redistribution of the three teeth and changing the rotor step dimensions of a particular seal allow reducing the leakage by ~8%. The optimization algorithm based on the complex method demonstrates a convergence slow down which could be due to relatively large number of design variables (six variables) and non-optimal dimension of the initial set (ten variants). Another point is the resolution of the computational grid in the CFD model. The small changes in the design variable coming from the optimization algorithm might not always be caught by the used computational grid. This could also lead to the slow convergence of the optimization cycle.

Both, baseline and optimized sealing configuration are analysed in respect of their rotordynamic performance at one inlet swirl speed and three pressure differentials. In case of the baseline configuration theoretical results are compared to the experimental values with reasonable agreement. Analysis of both seals shows that the seal with optimized shape has slightly improved rotordynamic characteristics over the baseline shape.

6   REFERENCE LIST

1.   Baumann, U., 1999, Rotordynamic stability tests on high-pressure radial compressors, *28th Turbomachinery Symposium*, Texas A&M University, College Station, Texas, USA, pp. 115-122.
2.   Rhode, D.L., Ko, S.H. and Morrison, G.L., 1994, Leakage optimization of labyrinth seals using a Navier-Stokes code, *Tribology Transactions*, Vol. 37, No. 1, pp. 105-110.

3.   Vakili, A.D., Meganathan, A.J., Michaud, M. and Radhakrishnan, S., 2005, An experimental and numerical study of labyrinth seal flow, *ASME Turbo Expo 2005*, Reno-Tahoe, Nevada, USA, Paper GT2005-68224.

4.   Schramm, V., Denecke, J., Kim, S. and Wittig, S., 2004, Shape optimization of a labyrinth seal applying the simulated annealing method, *International Journal of Rotating Machinery*, Vol. 50, No. 5, pp. 365-371.

5.   Childs, D., 1993, *Turbomachinery rotordynamics. Phenomena, modeling, and analysis*, Wiley, New York.

6.   Kwanka, K., 2007, Rotordynamic coefficient of short labyrinth gas seals – general dependency on geometric and physical parameters, *Tribology Transactions*, Vol. 50, No. 4, pp. 558-563.

7.   Schettel, J., Deckner, M., Kwanka, K., Lüneburg, B. and Nordmann, R., 2005, Rotordynamic coefficients of labseals for turbines – comparing CFD results with experimental data on a comb-grooved labyrinth, *ASME Turbo Expo*, Reno-Tahoe, Nevada, USA, Paper GT2005-68732.

8.   Pugachev, A.O., Deckner, M., Kwanka, K., Helm, P. and Schettel, J., 2006, Rotordynamic coefficients of advanced gas seals: measurements and CFD simulations, *Internation Scientific Symposium on Hydrodynamic Theory of Lubrication*, Orel, Russia, Vol. 1, pp. 545-555.

9.   Kwanka, K. and Mair, R., 1995, Identification of gas seal dynamic coefficients based on the stability behavior of a rotor, *1st European Conference on Turbomachinery*, Erlangen-Nürnberg, Germany, VDI-Report 1186, pp. 297-309.

10.   Ulbrich, H., 1988, New test techniques using magnetic bearings, *1st International Symposium on Magnetic Bearings*, ETH Zurich, Switzerland, pp. 281-288.

11.   Athavale, M.M., Przekwas, A.J. and Hendricks, R.C., 1992, A finite volume numerical method to calculate fluid forces and rotordynamic coefficients in seals, *28th AIAA Joint Propulsion Conference*, Nashville, USA, AIAA Paper 92-3712.

12.   Rao, S., 1996, *Engineering Optimization – Theory and Practice*, Wiley, New York.

Effect of cavitation on identification of the configuration state of statically indeterminate rotor bearing systems

N Feng, E Hahn
University of New South Wales, Australia
W Hu, N Zhang
The University of Technology, Sydney, Australia

ABSTRACT

The vibration behaviour of a statically indeterminate rotor bearing foundation system (RBFS) with hydrodynamic bearings is influenced by the system configuration state (SCS). This SCS is usually unknown in existing turbomachinery installations; and attempts to identify it generally utilise the Reynolds equation in conjunction with measurements of the relative motion between the rotor and bearings to determine the bearing reaction forces. This requires one to specify, amongst other parameters, the cavitation in the bearings and it is the aim of this paper is to examine how important the correct specification of this cavitation is for proper SCS identification. This is done via numerical experiments on a simple unbalanced statically indeterminate four bearing RBFS. Simple circular end feed journal bearings are used and cavitation is assumed to occur at atmospheric pressure, at the lubricant saturation vapour pressure, or midway between. It is shown that the accuracy of the identified SCS may be significantly influenced by the assumed cavitation model and the therefrom predicted unbalance responses may be quite at variance from the actual one. It is concluded that a more accurate specification of the onset of cavitation is necessary for any SCS identification approach which relies on the Reynolds equation to determine the bearing reaction forces.

1.   NOTATION

$\mathbf{b}_i$	mean displacement vector of i'th bearing: i = 1,...,4	**M**	mass matrix
C	damping and gyroscopic matrix	P	film pressure
C	radial clearance	P_c	cavitation pressure
$\mathbf{c}_i$	location vector of i'th unloaded bearing: i = 1,...,4	P_i	inlet pressure
$\mathbf{e}_i$	mean journal eccentricity vector at i'th bearing: i = 1,...,4	P_o	outlet pressure
F	force vector	$\mathbf{r}_i$	mean displacement vector of rotor at i'th bearing: i = 1,...,4
H	film thickness at B (Fig. 3)	R	journal radius
K	stiffness matrix	**X**	displacement vector
		y, z	displacement of O_j from O_b (Fig.3)

Z	axial variable in Reynolds equation	*Subscripts (if not otherwise defined)*	
ω	angular velocity of rotor	b	bearing
μ	lubricant viscosity	e	external
ψ	angular location of B (Fig. 3)	p	foundation or pedestal
'	differentiation with respect to time	r	rotor
		u	unbalance
		y, z	horizontal and vertical directions respectively

2.   INTRODUCTION

The vibration behaviour of a statically indeterminate rotor bearing foundation system (RBFS) with hydrodynamic bearings, such as turbomachinery in power generation, is very much influenced by the relative transverse alignment of the bearings, the so-called system configuration state (SCS). This is because the SCS determines the reaction loads at the bearings, and these reaction loads can significantly affect the stiffness and damping properties of the individual bearings, and hence, the system natural frequencies, the system stability and the unbalance response.

Important as a knowledge of the SCS might be, it is generally unknown in existing statically indeterminate installations due thermal effects and foundation settlement [1]. Hence, it is of interest to be able to identify the SCS in such installations. There appears to be no simple way for achieving this using presently available instrumentation. Most attempts to date suggest using the Reynolds equation in conjunction with rotor and bearing motion measurements to determine the pressure distributions in the bearings and therefrom the bearing reaction forces [2]. Basic to this approach for bearing force determination is the need to model lubricant cavitation (here loosely defined as the appearance of gaseous or vapour phases in the otherwise homogeneous liquid phase of the lubricant). The problem with cavitation is that its onset is unpredictable and its proper modelling is overly complex. Whether cavitation occurs at atmospheric pressure, or at the saturation vapour pressure of the lubricant, or at some pressure inbetween, is unclear. Its onset apparently depends not only on the possibility of air ingress from the bearing sides, but also on dissolved gases coming out of solution [3,4].

The specification of cavitation is not the only potential source of error in evaluating the SCS if the bearing forces are to be determined via Reynolds equation. Significant inaccuracies can also arise due to errors in other essential input parameters such as the bearing clearances, the mean lubricant viscosities, as well as the measurements of the instantaneous relative displacements between the journals and their respective bearings. These effects have already been discussed earlier [5,6,7]. Here we restrict ourselves solely to problems due to inaccurate modelling of cavitation; ie the aim of this paper is to examine how important the specification of cavitation is when using the Reynolds equation to determine the reaction forces needed to identify the SCS. This is to be done via numerical experiments on a simple unbalanced statically indeterminate four bearing RBFS, using, for ease of computation, simple circular end feed journal bearings; the approach being to guess the cavitation level without knowing what the assumed specified value happens to be. While it is recognised that present day turbomachinery uses more complex bearing profiles, it is felt that these simpler bearings are equally effective in evaluating the sensitivity of the vibration behaviour of statically indeterminate systems to cavitation modelling.

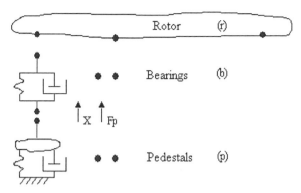

Fig 1 Schematic of a general RBFS

3.   THEORY

Figure 1 is a schematic of a general RBFS, the foundation here being simple undamped flexible pedestals connected to rigid ground. The equations of motion for the rotor and foundation are given by:

$$\mathbf{M}_r\ddot{\mathbf{X}}_r + \mathbf{C}_r\dot{\mathbf{X}}_r + \mathbf{K}_r\mathbf{X}_r = \mathbf{F}_b + \mathbf{F}_e + \mathbf{F}_u \tag{1}$$

$$\mathbf{M}_p\ddot{\mathbf{X}}_p + \mathbf{C}_p\dot{\mathbf{X}}_p + \mathbf{K}_p\mathbf{X}_p = \mathbf{F}_p \tag{2}$$

where $\mathbf{X}_r$ and $\mathbf{X}_p$ are absolute displacements measured from some fixed datum line, here taken to be the horizontal line joining the first and last bearing, as shown in Figure 2 for a statically indeterminate four bearing RBFS. The transverse displacement of the centres of bearings 2 and 3, i.e. (c_{2y}, c_{2z}) and (c_{3y}, c_{3z}) then define the SCS. Note that in this description, it is the relative transverse alignment of the bearing centres in their statically deflected location (prior to the introduction of the rotor) which defines the SCS.

The SCS affects the solution of the above equations by influencing the magnitudes of the stiffness forces in Eqn (1). The actual determination of the SCS (presuming it is unknown) is well described in earlier works (e.g. refs [5-7]). Suffice it to say that, assuming absence of error in modelling and measurements, identification of the SCS is possible if there is at least one speed in the operating speed range at which there is a stable periodical solution to the above equations and the bearing forces $\mathbf{F}_b$ at that speed are known. Note that the ability to correctly evaluate these forces can be a twofold requirement for properly evaluating the system vibration behaviour; they are needed for the solution of the above equations in their own right but are also needed for determination of the SCS, if it is not already known.

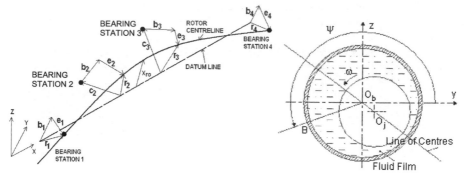

Fig 2 Location of the bearing centres with respect to the deformed rotor centreline

Fig 3 Simple circular journal bearing

These bearing forces are normally obtained from the solution of the Reynolds equation, a partial differential equation describing the pressure distribution in the bearing lubricant film. For the simple circular bearing shown in Figure 3 it is given by [8]:

$$\frac{1}{R^2}\frac{\partial}{\partial\psi}\left(h^3\frac{\partial P}{\partial\psi}\right)+\frac{\partial}{\partial Z}\left(h^3\frac{\partial P}{\partial Z}\right)=12\mu\left[\frac{\omega}{2}(y\sin\psi-z\cos\psi)-(y'\cos\psi+z'\sin\psi)\right] \quad (3)$$

where $\quad h=C-y\cos\Psi-z\sin\Psi$ $\quad\quad\quad\quad\quad\quad\quad\quad\quad\quad\quad\quad$ (4)

The boundary conditions pertaining to Eqn (3) are the side inlet pressure P_i and the side outlet pressure P_o. If it be assumed that the fluid film pressure equals the cavitation pressure P_c whenever the pressure is less than P_c, one can determine the bearing force components according to:

$$\begin{Bmatrix} F_y \\ F_z \end{Bmatrix}=-\int P\begin{Bmatrix}\cos\Psi \\ \sin\Psi\end{Bmatrix}dA \quad\quad\quad\quad\quad\quad\quad\quad\quad\quad\quad\quad (5)$$

where the integration is over the bearing surface area A. Note that the bearing forces are influenced not only by the mean viscosity of the lubricant, the bearing clearance and the relative displacement and velocity between the journal and bearing housing (all of which affect the solution for the pressure distribution and all of which can be measured in practical installations) but also by the assumed value for P_c. It is the sensitivity of the identified SCS and the corresponding unbalance response to the correct specification of this pressure that is of interest.

4.   NUMERICAL EXPERIMENTS

Rotor, bearing and pedestal details of the RBFS used for the numerical experiments are given in Table 1 and Figure 4. The dimensions for this four bearing statically indeterminate RBFS correspond approximately to an experimental rig. B1, B2, B3 and B4 are simple pedestal supported self aligning circumferentially grooved bearings, while D1, D2 and D3 are disks with unbalance at disk D2 The pedestal damping was assumed to be negligible and each pedestal was assumed to have a mass of 20 kg. The inlet and outlet pressures at all bearings were atmospheric.

864

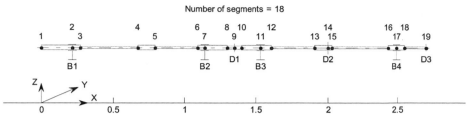

Fig 4 Rotor, bearing and pedestal details

Table 1 Rotor, bearing and pedestal details

Rotor Data					
Segment	Diameter (m)	Length (m)	Segment	Diameter (m)	Length (m)
1	0.060	0.2175	10	0.060	0.1345
2	0.060	0.0535	11	0.060	0.0755
3	0.037	0.3980	12	0.037	0.3000
4	0.061	0.1200	13	0.061	0.0890
5	0.037	0.3000	14	0.061	0.0310
6	0.060	0.5150	15	0.037	0.3980
7	0.060	0.1585	16	0.060	0.0615
8	0.042	0.0500	17	0.060	0.0595
9	0.042	0.0500	18	0.037	0.1500
Density :			7824.1 kg/m^3		
Young's Modulus :			200 Gpa		
Shear Modulus :			80 Gpa		
Unbalance :			0.0004kg m at node 14		
Disk Mass at Node 9			3.0930 kg		
Disk Mass at Node 14			3.2960 kg		
Disk Mass at Node 19			1.1223 kg		

Bearing/Journal Data				Pedestal Stiffness Data (N/m)		
		Radial Clearance(μm)			K_y	K_z
Diameter	0.060 m	Bearing 1	97	Bearing 1	1.870E6	2.963E7
Length	0.025 m	Bearing 2	113	Bearing 2	1.533E6	9.030E6
Mean Viscosity	0.014 Pa.s	Bearing 3	112	Bearing 3	1.933E6	7.306E6
		Bearing 4	110	Bearing 4	1.850E6	2.786E7

For an assumed SCS of c_{2y} = +0.65 μm, c_{2z} = -1450 μm, c_{3y} = -90.60 μm, and c_{3z} = -1600 μm, and an arbitrarily assumed P_c of 0.25 atmospheres, the equilibrium periodic (steady state) solutions to the equations of motion were evaluated using harmonic balance [9] over the speed range from 100 rad/s to 300 rad/s in steps of 10 rad/s. Each of these equilibrium solutions was tested for stability by using the harmonic balance solution at some instant of time as the initial condition input to obtain the transient solution of the equations of motion using 4th order constant step Runge Kutta integration. A solution was accepted as stable if the transient solution did not deviate from the harmonic balance response orbit. The stable response orbits at the four bearings comprised the 'measurements' available for data processing to identify the SCS and, having identified the SCS, to generate the predicted unbalance response, which could then be compared with the 'measurements'. Gyroscopic and rotary inertia effects, though they could have been included, were neglected to simplify computations.

The only essential input lacking for correct SCS identification and subsequent response predictions was the value of P_c used when generating the 'measurements'. This pressure, for want of better knowledge, was assumed to be absolute zero, or 0.5 atmospheres, or 1 atmosphere. Note that even though in practice, orbit measurements tend to be limited to relative journal to bearing displacements, and the relative journal to bearing velocities need to be calculated from these displacements, for these numerical experiments the actual velocities were also made available for data processing so as to ensure that the only source of error in the identified SCS and response predictions was the uncertainty in the assumed P_c. Actually, two sets of 'measurement' data were prepared, the second set being identical to the first, except in the case of the second set, P_c was assumed to be 0.75 atmospheres.

5. RESULTS AND DISCUSSION

Though it is sufficient in theory to utilise 'measurements' pertaining to only one rotor speed in order to identify the SCS, in practice one would utilise data pertaining to as many speeds as practical to help minimise error. Hence, a SCS was here identified from the 'measurements' pertaining to each speed that produced a stable response and the final SCS used to predict the unbalance response was obtained by averaging the many SCSs thus identified (a maximum of twenty one). Table 2 summarises the thus identified SCSs for the three assumed cavitation pressures for the two sets of 'measurements'. In this table, CAV01, CAVH1 and CAVT1 refer to the identifications obtained from the first set of 'measurements' (when the actual cavitation pressure was 0.25 atmospheres P_{ATM}) for assumed cavitation pressures of absolute zero, 0.5 P_{ATM} and 1 P_{ATM} respectively, while CAV02, CAVH2 and CAVT2 refer to the corresponding identifications obtained from the second set of 'measurements' (when the actual P_c was 0.75 P_{ATM}), again for the same three assumed cavitation pressures. The predicted values for c_{2y} range from -27.74 μm to 44.94 μm whereas its actual value was 0.65 μm in all cases. Similarly, the predicted values for c_{3y} range from -52.44 μm to 95.42 μm (actual value was -90.60 μm), the predicted values for c_{2z} range from -1526 μm to -1668 μm (actual value was -1450 μm) and the predicted values for c_{3z} range from -1676 μm to -1784 μm (actual value was -1600 μm). In all cases there was an over 20 μm error in magnitude in the identified horizontal and vertical SCS components.

Table 2 Identified SCSs

		System Configuration States [μm]					
	ACTUAL	**CAV01** Actual $P_C = 0.25\ P_{ATM}$	**CAV02** Actual $P_C = 0.75\ P_{ATM}$	**CAVT1** Actual $P_C = 0.25\ P_{ATM}$	**CAVT2** Actual $P_C = 0.75\ P_{ATM}$	**CAVH1** Actual $P_C = 0.25\ P_{ATM}$	**CAVH2** Actual $P_C = 0.75\ P_{ATM}$
		Assumed $P_C = 0$		Assumed $P_C = 1\ P_{ATM}$		Assumed $P_C = 0.5\ P_{ATM}$	
c_{2y}	0.65	-27.74	-19.96	44.94	38.06	40.58	35.72
c_{3y}	-90.60	-20.93	-52.44	95.42	68.78	88.65	60.95
c_{2z}	-1450	-1561	-1526	-1661	-1668	-1648	-1642
c_{3z}	-1600	-1706	-1676	-1768	-1784	-1760	-1764

The extent to which these errors in the identified SCSs and the assumed cavitation pressures on which the identifications were based affect the predicted unbalance responses for the first 'measurement' set (P_c = 0.25 atmospheres) is shown in Figures 5 to 7 which compare the actual and predicted maximum eccentricities (in magnitude) in the y and z directions at each of the four bearings. While there can be significant differences between actual and predicted maximum eccentricities, there appears to be little difference between actual and predicted resonance speeds. The difference between the actual and predicted unbalance responses based on the identified SCSs corresponding to the second 'measurement' set (P_c = 0.75 atmospheres) were similar. All responses (actual and predicted) were stable over the selected speed range. Figure 8 displays typical response orbit comparisons, here showing predicted and actual orbits at the four bearings at 180 rad/s and 300 rad/s for the case CAV01 (i.e. using the SCS corresponding to CAV01 in Table 2).

These results are of course a function of the chosen RBFS. However, since the cavitation pressure assumption can result in such erroneous SCSs and corresponding unbalance response predictions, and considering that in practice additional errors in the identified SCS will be present due to motion measurement errors as well as assumed radial clearance and/or assumed mean viscosity errors, it is concluded that use of Reynolds equation for determining bearing reaction forces is unlikely to result in satisfactory SCS identification and vibration behaviour predictions, though in the cases here examined there is admittedly little difference in the unbalance responses.

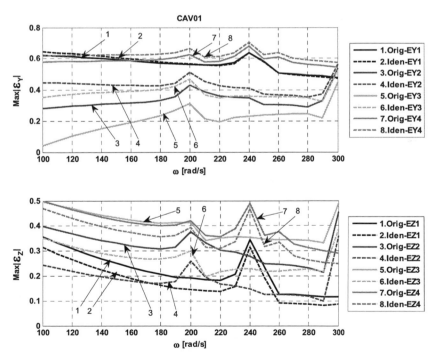

Fig 5 Actual and predicted maximum eccentricities (in magnitude) in the y and z directions at the four bearings for case CAV01

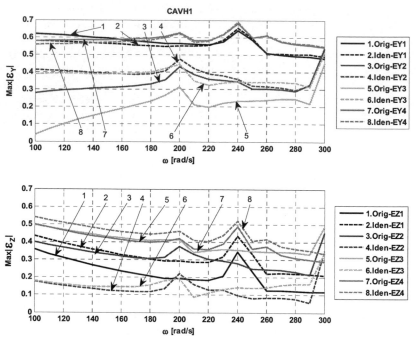

Fig 6 Actual and predicted maximum eccentricities (in magnitude) in the y and z directions at the four bearings for case CAVH1

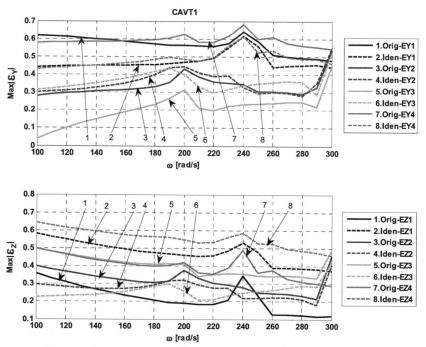

Fig 7 Actual and predicted maximum eccentricities (in magnitude) in the y and z directions at the four bearings for case CAVT1

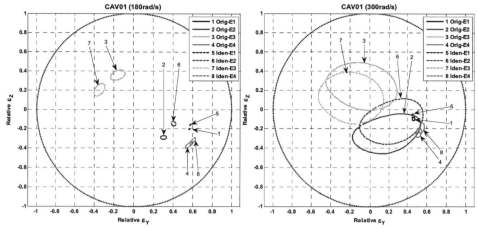

**Fig 8 Predicted and actual orbits at the four bearings at
180 rad/s and 300 rad/s for case CAV01**

6. CONCLUSIONS

In SCS identification approaches which rely on Reynolds equation for the determination of bearing reaction forces, the accuracy of the identified SCS may be significantly influenced by the assumed cavitation model; and the therefrom predicted unbalance responses may be quite at variance with the actual ones.

Considering that errors in the identified SCS in addition to those due to an assumed cavitation model will be present in practice due to motion measurement errors as well as assumed radial clearance and/or assumed mean viscosity errors, the use of Reynolds equation for determining bearing reaction forces is unlikely to result in satisfactory SCS identification and vibration behaviour predictions.

7. REFERENCES

[1]   S.Edwards, A.W.Lees and M.I.Friswell (2000), "Experimental identification of excitation and support parameters of a flexible rotor bearings foundation system from single run-down", Journal of Sound and Vibration, **332**(5), pp.963-992.

[2]   J.M.Krodkiewski and J.Ding (1993), "Theory and experiment on a method for onsite identification of configurations of multi-bearing rotor systems", Journal of Sound and Vibration, **64**, pp.281-293.

[3]   D.Dyer and B.R.Reason (1976), "A study of tensile stresses in a journal bearing oil film", Journal of Mechanical Engineering Science, **18**(1), pp.46-52

[4]   I.Etsion and L.P.Ludwig (1990), "Observation of pressure variation in the cavitation region of submerged journal bearings", Journal of Lubrication Technology, **22**(2), pp.163-170.

[5] W.Hu, N.S.Feng, and E.J.Hahn (2004), "A comparison of techniques for identifying the configuration state of statically indeterminate rotor bearing systems", Tribology International, **37**, pp.149-157.

[6] W.Hu, N.S.Feng and E.J.Hahn (2004), "Identification of the configuration state of turbomachinery using rotor deformation measurements", 8th Intl Conf. on Vibrations in Rotating Machinery, IMechE Conference Transactions, Swansea, September, pp.403-412.

[7] N.S.Feng, E.J.Hahn and W.Hu (2005), "On predicting the vibration behaviour of turbogenerators by identifying their configuration state from rotor motion measurements", The 12th International Congress on Sound and Vibration (ICSV12), Lisbon, July5, 8pp.

[8] O.Pinkus, and B.Sternlicht (1961), Theory of Hydrodynamic Lubrication, McGraw Hill.

[9] EJ.Hahn and P.Y.P.Chen (1994), "Harmonic Balance Analysis of General Squeeze Film Damped Multi-Degree of Freedom Rotor Bearing Systems". Trans. ASME, Journal of Tribology, **116**(3), pp.499-507.

A study on the oil whip of an elastic rotor supported in tilting-pad journal bearings

H Taura
Department of Mechanical Engineering,
Nagaoka University of Technology, Japan

M Tanaka
Toyama Prefectural University, Japan

1. INTRODUCTION

Tilting-pad journal bearings (TPJBs) are widely used in high speed rotating machinery because the bearings are superior to other journal bearings in the capability of suppressing oil whip that often occurs in high speed rotating machinery. However, the oil whip occurred in some actual rotating machinery equipped with the bearings (1-2). Furthermore, the oil whip was also found to take place in some test rigs (3-6).

At the eighth IMechE Vibration Conference held in Swansea in 2004, the authors (7) reported that oil whip actually occurred in a test rig of an elastic rotor supported in a TPJB at one end and in a self-aligning ball bearing at the other end. The authors found oil whip occurred when the TPJB was installed in a non-pivotable bearing seat. They also conducted the linear stability analysis of the rotor-bearing system with the oil film moment being considered, compared the theoretical threshold shaft speeds with the experimental results, and concluded that the oil whip is caused by the effect of the oil film moment (the angular stiffness and angular damping coefficients) generated by the journal tilt motion in the bearing. However, the theoretical model assumed two identical TPJBs to support the rotor at both ends, which was different from the test rig used. Furthermore neither the destabilizing mechanism nor the stability behavior of the rotor-bearing system was fully discussed,

This paper presents new theoretical rotor-bearing models exactly corresponding to the test rig used, studies the stability behavior of the rotor-bearing system comprehensively and discusses the destabilizing mechanisms in TPJBs.

2. THEORETICAL MODEL

Two theoretical models corresponding to the test rig used in Reference (7) are newly introduced. The former models given in Reference (7) assumed the rotor was supported by two identical TPJBs at both ends, which was different from the real rotor-bearing system in the test rig used. On the other hand, the new models incorporate the ball bearing at one end in accordance with the test rig used.

2.1 Non-Pivotable Bearing Model

Figure 1 shows the "non-pivotable bearing model". A rotor consists of a uniform, massless, flexible shaft with a concentrated mass at its mid span, and is supported in a five pad TPJB at one end, while the other end of the shaft is assumed to be simply supported, corresponding to the ball bearing. Since the TPJB can not tilt because of its non-pivotable seat, the model incorporates the angular spring and damping coefficients of oil film moment, in addition to the full set of translational spring and damping coefficients of oil film force (8). The mass of the rotor, Young's modulus, the length and moment of inertia of area of the shaft are M, E, $2l$ and I, respectively. Each pad of moment of inertia, I_p has a single degree of freedom of rolling on the inner surface of the pad carrier ring. The vertical and the horizontal coordinates of the rotor are x_d and y_d, and those of the journal are x_j and y_j respectively, in the shaft rotating plane. The tilt angles of the journal in the horizontal and the vertical plane are θ_x and θ_y, and the tilt angle of ith pad is δ_i.

When the journal shows small displacements x_j and y_j and small angular displacements θ_x and θ_y from its equilibrium and also the mass shows small displacements x_d and y_d from its equilibrium, additional small forces F_{jx} and F_{jy}, and also additional small moments M_{jx} and M_{jy} are applied to the journal, and they are expressed by using the linear beam theory as follows:

$$
\begin{cases}
F_{jx} = \dfrac{k'_s}{2}\left(\dfrac{22}{7}x_d - \dfrac{16}{7}x_j - \dfrac{5}{7}l\theta_y\right) \\[2mm]
F_{jy} = \dfrac{k'_s}{2}\left(\dfrac{22}{7}y_d - \dfrac{16}{7}y_j + \dfrac{5}{7}l\theta_x\right) \\[2mm]
M_{jx} = \dfrac{k'_s l}{2}\left(-\dfrac{6}{7}y_d + \dfrac{5}{7}y_j - \dfrac{2}{7}l\theta_x\right) \\[2mm]
M_{jy} = \dfrac{k'_s l}{2}\left(\dfrac{6}{7}x_d - \dfrac{5}{7}x_j - \dfrac{2}{7}l\theta_y\right)
\end{cases}
\tag{1}
$$

where $k'_s = 48EI/l^3$. Additional small forces F_{dx} and F_{dy} are applied to the mass, expressed as follows:

$$
\begin{cases}
F_{dx} = -k'_s\left(\dfrac{16}{7}x_d - \dfrac{11}{7}x_j - \dfrac{3}{7}l\theta_y\right) \\[2mm]
F_{dy} = -k'_s\left(\dfrac{16}{7}y_d - \dfrac{11}{7}y_j + \dfrac{3}{7}l\theta_x\right)
\end{cases}
\tag{2}
$$

872

The equations of motion for the rotor are formulated from Eqs. (2), as follows:

$$\begin{cases} M\ddot{x}_d + k_s'\left(\dfrac{16}{7}x_d - \dfrac{11}{7}x_j - \dfrac{3}{7}l\theta_y\right) = 0 \\[2mm] M\ddot{y}_d + k_s'\left(\dfrac{16}{7}y_d - \dfrac{11}{7}y_j + \dfrac{3}{7}l\theta_x\right) = 0 \end{cases} \qquad (3)$$

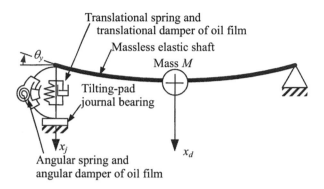

Fig. 1 Non-pivotable bearing model

The sum of oil film forces caused by the translational motion of the journal and the tilt motion of the pads balances with the restoring force of the elastic shaft:

$$\begin{cases} c_{xx}\dot{x}_j + c_{xy}\dot{y}_j + k_{xx}x_j + k_{xy}y_j + \displaystyle\sum_{i=1}^{n}\left(c_{x\delta i}\dot{\delta}_i + k_{x\delta i}\delta_i\right) = \dfrac{k_s'}{2}\left(\dfrac{22}{7}x_d - \dfrac{16}{7}x_j - \dfrac{5}{7}l\theta_y\right) \\[3mm] c_{yx}\dot{x}_j + c_{yy}\dot{y}_j + k_{yx}x_j + k_{yy}y_j + \displaystyle\sum_{i=1}^{n}\left(c_{y\delta i}\dot{\delta}_i + k_{y\delta i}\delta_i\right) = \dfrac{k_s'}{2}\left(\dfrac{22}{7}y_d - \dfrac{16}{7}y_j + \dfrac{5}{7}l\theta_x\right) \end{cases} \qquad (4)$$

Equations of motion of the pads:

$$I_p\ddot{\delta}_i + c_{\delta i x}\dot{x}_j + c_{\delta i y}\dot{y}_j + k_{\delta i x}x_j + k_{\delta i y}y_j + c_{\delta i \delta i}\dot{\delta}_i + k_{\delta i \delta i}\delta_i = 0 \quad (i{=}1...5) \qquad (5)$$

The oil film moment caused by the tilt motion of the journal balances with the restoring moment of the shaft:

$$\begin{cases} c_{\theta x \theta x}\dot{\theta}_{jx} + c_{\theta x \theta y}\dot{\theta}_{jy} + k_{\theta x \theta x}\theta_{jx} + k_{\theta x \theta y}\theta_{jy} = \dfrac{k_s'l}{2}\left(-\dfrac{6}{7}y_d + \dfrac{5}{7}y_j - \dfrac{2}{7}l\theta_x\right) \\[3mm] c_{\theta y \theta x}\dot{\theta}_{jx} + c_{\theta y \theta y}\dot{\theta}_{jy} + k_{\theta y \theta x}\theta_{jx} + k_{\theta y \theta y}\theta_{jy} = \dfrac{k_s'l}{2}\left(\dfrac{6}{7}x_d - \dfrac{5}{7}x_j - \dfrac{2}{7}l\theta_y\right) \end{cases} \qquad (6)$$

where k_{jk}, $k_{j\delta i}$, $k_{\delta ik}$, $k_{\delta i\delta i}$ ($i{=}1, \ldots, 5$, j, $k{=} x$, y) represent the full set of spring coefficients, c_{jk}, $c_{j\delta i}$, $c_{\delta ik}$, $c_{\delta i\delta i}$ the full set of damping coefficients, $k_{\theta j\theta k}$ the angular spring coefficients, and $c_{\theta j\theta k}$ the angular damping coefficients.

The linear stability of the rotor-bearing system is examined as follows: A characteristic equation of the system is derived from Eqs. (3) to (6), and numerically solved for complex eigenvalues for a specific operating condition by means of the QR method. The system is unstable if there is more than one eigenvalue with its positive real part, and stable if no such eigenvalues are obtained. Consequently the stability threshold speed is searched for numerically with the shaft speed being swept, and is obtained where the maximum real part changes its sign.

2.2　Pivotable Bearing Model

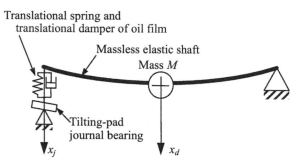

Fig. 2 Pivotable bearing model

Figure 2 shows the "pivotable bearing model". The bearing can tilt around x and y axis because of its pivotable bearing seat. If the journal tilts in the bearing clearance due to the whirling of the shaft, the bearing can follow the motion immediately and tilt due to the action of the oil film moment, resulting in no reaction of the oil film moment against the journal. The same x-y-z coordinate system is defined like Fig. 1. Equations of motion for the rotor and the pads at the equilibrium are formulated as follows:

Equations of motion of the rotor:

$$\begin{cases} M\ddot{x}_d + k_s\left(x_d - \dfrac{x_j}{2}\right) = 0 \\[2mm] M\ddot{y}_d + k_s\left(y_d - \dfrac{y_j}{2}\right) = 0 \end{cases} \tag{7}$$

Balance of forces on the journal:

$$\begin{cases} c_{xx}\dot{x}_j + c_{xy}\dot{y}_j + k_{xx}x_j + k_{xy}y_j + \displaystyle\sum_{i=1}^{5}\left(c_{x\delta i}\dot{\delta}_i + k_{x\delta i}\delta_i\right) = \dfrac{k_s}{2}\left(x_d - \dfrac{x_j}{2}\right) \\[3mm] c_{yx}\dot{x}_j + c_{yy}\dot{y}_j + k_{yx}x_j + k_{yy}y_j + \displaystyle\sum_{i=1}^{5}\left(c_{y\delta i}\dot{\delta}_i + k_{y\delta i}\delta_i\right) = \dfrac{k_s}{2}\left(y_d - \dfrac{y_j}{2}\right) \end{cases} \tag{8}$$

Equations of motion of the pads:

$$I_p\ddot{\delta}_i+c_{\delta ix}\dot{x}_j+c_{\delta iy}\dot{y}_j+k_{\delta ix}x_j+k_{\delta iy}y_j+c_{\delta i\delta i}\dot{\delta}_i+k_{\delta i\delta i}\delta_i = 0 \qquad (9)$$

The linear stability limit is calculated from Eqs. (7) to (9) in the way similar to the case for the non-pivotable bearing model.

2.3 Calculation Conditions
Stability threshold shaft speeds are calculated under the conditions shown in Table 1, corresponding to the test ring used in Reference (7).

Table 1 Bearing Dimension & Operating Conditions

Journal diameter, mm	40
L/D	0.4
Pad number	5
Clearance ratio	1.5/1000 ($2C_b/D$)
Load direction	LOP
Preload factor $m_p(=1-C_b/C_p)$	0, 0.18, 0.6
load factor κ	0.17 to 1.0
Shaft speed, r/min	< 7500
Natural frequency of 1st shaft bending mode, Hz	15

3. RESULTS AND DISCUSSIONS

The load applied to the TPJB changes between full load ($Mg/2$, κ=1) and κ=0.17.

3.1 Theoretical Results of Non-Pivotable Bearing Model
Figure 3 compares the measured stability threshold shaft speeds (Reference (7)) and the predictions obtained with the newly derived non-pivotable theoretical model. The qualitative agreement is good, but the quantitative agreement was found to show little improvement. The discrepancy can be mainly attributed to the damping action of the rubber coupling that is not included in the model. The calculated vibration frequencies were about 15Hz, which agrees with the experimental results.

Figure 4 shows variations of dimensionless angular spring coefficients (upper) and dimensionless angular damping coefficients (lower) with Sommerfeld number S ($=\mu DLN/W(R/C_p)^2$) in the case of m_p=0. The two cross-coupling terms of angular spring coefficients $K_{\theta x\theta y}$ and $K_{\theta y\theta x}$ are not negligibly small but significantly as large as the direct terms. While the sign of $K_{\theta x\theta y}$ is negative, that

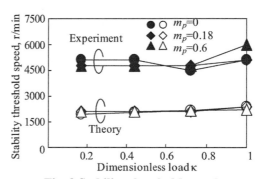

Fig. 3 Stability threshold speed (non-pivotable bearing model)

of $K_{\theta y \theta x}$ is positive. The two terms increase in magnitude at higher Sommerfeld number, becoming much larger than the direct terms. On the other hand, the cross-coupling terms of angular damping coefficients remain very small over a wide range of Sommerfeld number. Similar results were also obtained in the cases of m_p=0.18, 0.6. From these results, the oil whip found in the case of non-pivotable bearing seat can be reasonably attributed to the destabilizing effect of the cross-coupling terms of the angular spring coefficients. Consequently, the oil film moment (angular stiffness and damping of oil film) acting on the journal proves to destabilize the rotor-bearing system.

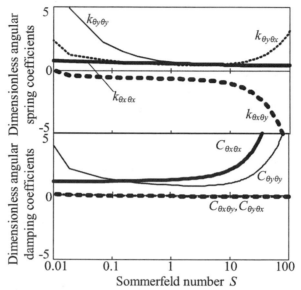

Fig. 4 Angular spring coefficients (upper) and angular damping coefficients (lower) (m_p=0)

3.2 Theoretical Results of Pivotable Bearing Model

Figure 5 shows the stability threshold shaft speeds with the newly derived pivotable bearing model for the conditions shown in Table 1, specifically in the case of m_p =0. However, the experiments given in Reference (7) showed that the rotor always kept a stable operation up to 7500 r/min (the maximum shaft speed of the test rig) regardless of load factor and preload

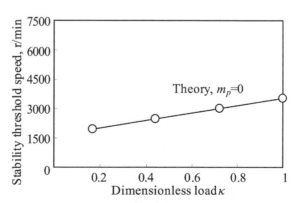

Fig. 5 Theoretical stability threshold speed (pivotable bearing model)

876

factor. The discrepancy can be again attributed to the damping action of the rubber coupling as shown in the non-pivotable bearing case. The calculated vibration frequencies were always about 15 Hz which agrees to the natural frequency.

On the other hand, in the cases of the m_p =0.18 and 0.60, the stability threshold shaft speeds increase drastically up to 8.5×10^5 r/min and 9.2×10^5 r/min respectively. These results do not disagree to the observed fact that the rotor remained stable up to 7500 r/min in the experiments (Reference (7)). Furthermore, the vibration frequencies calculated were as high as 8 kHz(m_p =0.18) and 13 kHz(m_p =0.60). The results suggest the vibration mode should be different from that of the natural frequency of the first bending mode, 15 Hz.

Consequently, the variation of the stability threshold shaft speed is investigated with m_p swept from 0 to 0.18. Figure 6 shows the result. The vertical coordinate represents the logarithmic value of the shaft speed, and the horizontal one m_p. When m_p increases, the bearing modulus λ defined by Eq. (10) can not stay constant because of the geometry.

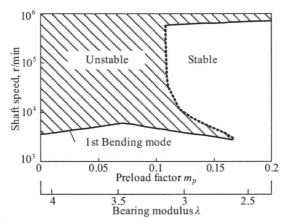

Fig. 6 Stability chart (pivotable bearing model, $\kappa = 1$)

$$\lambda = \frac{\mu DL}{W} \left(\frac{R}{C_p} \right)^2 \frac{\sqrt{g/C_p}}{2\pi}$$ (10)

Preload factor m_p varies with C_p for C_b being fixed. As a result, λ at m_p =0 is 4.08, which is the value at κ=1 in Fig. 5, and λ decreases from 4.08 to 2.33 with the increase in m_p from 0 to 0.2. The value of λ is also given on the horizontal coordinate for reference. The lines correspond to the boundaries between stable and unstable region of operation of the rotor-bearing system, and the system is unstable in the shaded area. If m_p is less than 0.17, the rotor-bearing system turns to be unstable at relatively low shaft speeds. However, in the case of m_p greater than 0.11 and lower than 0.17, the system shows stable operation for a range of shaft speed above certain values and finally turns unstable again at very high shaft speeds. In the case of m_p greater than 0.17, the rotor-bearing system turns unstable only at very high shaft speeds. Consequently, the oil whip in the case of pivotable bearing seat can be suppressed by increasing preload factor, until shaft speed becomes very high.

These theoretical analyses shown so far suggest two different modes of oil whip in the case of pivotable bearing seat, that is, the first one for low pad preload factors with relatively low threshold shaft speeds and the second one for high pad preload factor with very high threshold shaft speeds.

The destabilizing mechanism for the first mode oil whip can be examined by transforming the full-set of oil film coefficients into the reduced oil film coefficients expressed as 4 spring coefficients and 4 damping coefficients by means of the transform method proposed by Parsell et al. (9). The method needs to assume the vibration frequency with which the journal and the pads vibrate synchronously. The imaginary part of a complex root of the characteristic equation of the rotor-bearing system represents the vibration frequency of a vibration mode. The imaginary part of the eigenvalue with the largest real part is used for the method because it is always confirmed to be equal to the frequency of the first bending mode.

Figure 7 shows the reduced spring and damping coefficients in the case of $m_p=0$, $\kappa=1$. The horizontal coordinate represents the shaft speed, and the vertical coordinates the dimensionless spring coefficients (upper) or the dimensionless damping coefficients (lower). The vertical dotted line corresponds to the stability threshold shaft speed. The rotor-bearing system is stable on the left hand side of the line, while it is unstable on the right hand side. The cross-coupling terms of the reduced spring and damping coefficients are negligibly too small to destabilize this rotor-bearing system in this case, although they normally causes oil whip of a rotor supported in fixed bore bearings.

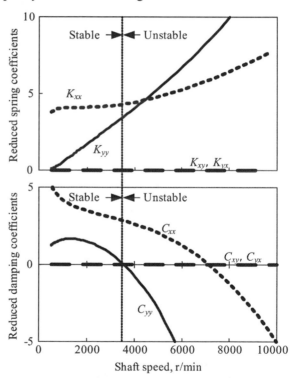

Fig. 7 Reduced spring and damping coefficients (pivotable bearing model, $m_p = 0$, $\kappa = 1$)

878

On the other hand, both of the direct terms of the reduced damping coefficients C_{xx} and C_{yy} are found to decrease with the increase in the shaft speed, eventually turning negative. C_{yy} turns to be negative first at and beyond the stability threshold shaft speed, and then C_{xx} becomes negative. Next, the effect of the negative direct damping coefficients on the stability of the rotor-bearing system is discussed. For simplicity, a symmetrical rotor is assumed to be supported in two identical bearing at both ends. With the Routh-Hurwitz's stability criterion of oil whip shown by Hori (10), both of the following two equations with respect to C_{xx} and C_{yy} must be satisfied at the same time if the rotor-bearing system is stable:

$$C_{xx}C_{yy} - C_{xy}C_{yx} > 0 \tag{11}$$

$$K_{xx}C_{yy} + K_{yy}C_{xx} - K_{xy}C_{yx} - K_{yx}C_{xy} > 0 \tag{12}$$

Since the cross-coupling terms of the reduced spring and damping coefficients can be assumed to be zero, Eqs. (11) & (12) are expressed as follows:

$$C_{xx}C_{yy} > 0 \tag{13}$$

$$K_{xx}C_{yy} + K_{yy}C_{xx} > 0 \tag{14}$$

If either C_{xx} or C_{yy} turns to be negative, Eq. (13) is not satisfied. If both of them are negative, Eq. (13) is satisfied, but Eq. (14) is not satisfied, because K_{xx} and K_{yy} are always positive. Consequently, the destabilizing mechanism of the first mode oil whip with the low frequency can be attributed to the effect of the negative values of the two direct terms of the reduced damping coefficients, C_{xx} and C_{yy}.

Next, the destabilizing mechanism for the second mode oil whip is discussed in the case of higher preload factors. Figure 8 shows the variations of the reduced spring coefficients with shaft speed in the case of m_p =0.2, κ=1. The vertical dotted line represents the very high stability threshold shaft speed calculated. While the cross-coupling terms of the reduced spring coefficients are much smaller than the direct terms at the shaft speeds lower

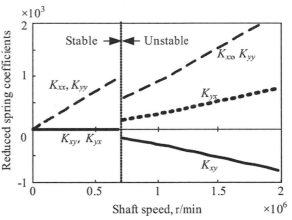

Fig. 8 Reduced spring coefficients (pivotable bearing model, $m_p = 0.2$, $\kappa = 1$)

than the stability threshold shaft speed, they are found to increase in magnitude with the shaft speed above the threshold speed. This drastic increase of the cross-coupling terms can be attributed to the moment of inertia of the pad, as Lund (11) pointed out. The pads can not follow the high frequency vibration of the journal owing to the moment of inertia of pad, and the phase delay of the pad vibration against the journal vibration increases. As a result, the cross-coupling terms of the spring coefficients increases, and the rotor-bearing system becomes unstable. Orcutt(3) pointed out that the moment of inertia of the pad used in the actual machines was too small to cause the oil whip, but this result shows that, even if the moment of inertia of the pad is small, the oil whip can occur if shaft speeds are very high.

The reduced spring coefficients are found to be discontinuous at the stability threshold shaft speed, because the vibration mode changes from the first mode of the low frequency to the second mode of the very high frequency. The spring coefficients are isotropic because the eccentricity ratio is approximately zero at very high shaft speeds.

4   CONCLUSIONS

This paper presents the theoretical results of the linear stability analysis of a Jeffcott rotor supported in a 5-pad tilting-pad journal bearing at one end and in a self-aligning ball bearing at the other end. Two theoretical models are newly derived. One of them corresponds to the case of the tilting pad journal bearing installed in a non-pivotable seat, resulting in the angular oil film coefficients as well as the translational oil film coefficients due to the relative tilt motion of the journal in the bearing. The other corresponds to the case of the tilting pad journal bearing installed in a pivotable seat, resulting in only the translational oil film coefficients without the angular oil film coefficients because of no relative tilt motion of the journal in the bearing.

The following conclusions are derived from the results with the first theoretical model for the non-pivotable bearing seat.

(1)   The cross-coupling terms of the angular stiffness coefficients increase with increasing Sommerfeld number, becoming larger than the direct terms of the angular stiffness coefficients, and eventually cause the oil whip. The stability threshold shaft speed is insensitive to the variations of pad preload factor m_p or load factor κ.   Consequently, this oil whip can be suppressed only with the bearing seat being changed into the pivotable one.

(2)   The predicted threshold shaft speed shows good qualitative agreement with the measurements, but the quantitative agreement shows little improvement, compared to the former theoretical model (Reference (7)).

The following conclusions are derived from the results with the second theoretical model for the pivotable bearing seat.

(3)   The linear stability chart finds two different modes of oil whip, that is, the low frequency mode oil whip of low threshold shaft speeds for low pad preload factors and the very high frequency mode oil whip of very high threshold shaft speeds for high pad preload factor.

(4)   The low frequency mode oil whip is caused by the negative direct terms of reduced translational damping coefficients transformed from the full set of oil film coefficients. This oil whip can be easily suppressed by increasing pad preload factor.

(5)   The very high frequency mode oil whip is caused by the increase in the cross-coupling terms of the reduced translational stiffness coefficients with Sommerfeld number, resulting from the increasing phase delay of pad vibration against journal vibration. This oil whip is the ultimate mode oil whip when shaft speed is very high, even if the moment of inertia of pad is reasonably small.

ACKNOWLEDGEMENT

The authors express their sincere gratitude to Dr. Shinobu Saito, IHI Corporation for giving them useful information and also to Daido Metal Corporation for kindly supplying test rigs.

REFERENCES

1   Data Book of v-BASE (in Japanese), the Japan Society of Mechanical Engineers, Tokyo, (1994), pp. 100-101.
2   Proceedings of v-Base Forum (in Japanese), the Japan Society of Mechanical Engineers, No. 95-8 II(1995-8), pp. 20-21.
3   Orcutt, F. K., The Steady-State and Dynamic Characteristics of the Tilting-Pad Journal Bearing in Laminar and Turbulent Flow Regimes, Transactions of the ASME, Journal of Lubrication Technology, Vol. 89(1967), pp. 392-404.
4   Saito, S., Critical Speed and Stability of Rotor Supported in Tilting-Pad Journal Bearings(in Japanese), Transactions of the Japan Society of Mechanical Engineers, Series C , Vol. 47, No. 413 (1981), pp. 30-37.
5   Flack, R. D. and Zuck, C. J., Experiments on the Stability of Two Flexible Rotors in Tilting Pad Bearings, Tribology Transactions, Vol. 31, No. 2 (1988), pp. 251-257.
6   Lie, Y., You-Bai, X., Jun, Z. and Damou, Q., Experiments on the destabilizing factors in tilting pad journal bearings, Tribology International, Vol. 22, No. 5 (1989), pp. 329-334.
7   Taura, H. and Tanaka, M., Self-Excited Vibration of Elastic Rotors in Tilting-Pad Journal Bearings, IMechE Conference Transactions, eighth International Conference on Vibration of Rotating Machinery, (2004), pp. 35-43.
8   Allaire, P. E., Parsell, J. K. and Barrett, L. E., Pad Perturbation Method for the Dynamic Coefficients of Tilting-Pad Journal Bearings, Wear, Vol.72, No. 1(1981), pp.29-44.

9   Parsell, J. K., Allaire, P. E. and Barrett, L. E., Frequency Effects in Tilting-Pad Journal Bearing Dynamic Coefficients, ASLE Transactions, Vol. 26(1982), pp. 222-227.

10 Hori, Y., A Theory of Oil Whip, Trans. of ASME, J. of Applied Mechanics, Vol. 26, No. 2(1959), pp. 189-198.

11 Lund, J. W., Spring and Damping Coefficients for the Tilting-Pad Journal Bearing, ASLE Transactions, Vol. 7 (1964), pp. 342-352.

BALANCING

Multi-objective genetic algorithm application in unbalance identification for rotating machinery

H Fiori de Castro, K Lucchesi Cavalca, L Ward Franco de Camargo
State University of Campinas, Brazil

ABSTRACT

This paper presents a non-linear model updating, applying multi-objective genetic algorithm as searching method, in order to obtain the tuning of mathematical model parameters in rotating systems. An experimental setup is used for this purpose, which consists of a rotor on a rigid foundation supported by two cylindrical journal bearings and with a central disk of considerable mass, introducing an unbalance excitation force. A finite element model is used to simulate rotor and shaft. A multi-objective genetic algorithm allows the individual analysis of each objective function, which is, in this case, based on the difference between the geometric properties of the experimental and adjusted elliptical orbits. The experimental fitting is satisfactory, but a better fitting can be obtained with a more detailed finite element model of the rotor-bearing system.

1. INTRODUCTION

The study of rotating machines occupies a special position in the context of machines and structures, in view of the significant amount of typical phenomena in the operation of those equipments. Mathematical models are developed to foresee the dynamic behaviour of those systems and, to obtain reliable results, it is necessary to use updating techniques to adjust the results of the simulations of the models to those obtained in experimental tests.

This paper presents a fitting process of a non-linear model of cylindrical journal bearings, as part of a rotating system modelled by finite element method. The unknown parameters considered in the system are unbalance and oil viscosity of each bearing. The objective functions adopted are based on the difference between the geometric properties of the experimental and simulated elliptical orbits. The fitting process involves the simultaneous consideration of multiple performance criteria and at last two objective functions are considered.

Castro et. al. (2004 and 2005) and Castro & Cavalca (2006) studied this problem considering a single objective fitting process. Weights were chosen for each one of the objective functions, representing the importance of each parameter in the adjustment, in order to describe the problem using only one objective function. In this work, each objective function will be treated separately, so a multi-objective genetic algorithm is proposed to allow the individual analysis of each function and the identification of the unknown parameters of the model. Pareto ranking is applied to evaluate the performance of each potential solution. Niching and Fitness Sharing schemes are considered to obtain a diversity of solutions.

The use of multi-objective optimization approach results in a Pareto optimal set of solutions and there is not the need of immediately identifying the best one. Using high level information, it is possible to choose among all of the solutions properly and fit the mathematical model to obtain reliable responses for the physical system studied. Once the viscosity and rotor unbalance are estimated, it is possible to approach the mathematical model, and then to obtain more reliable responses for the physical system studied.

This research also proposes a contribution for the rotating machine design area as it presents a relatively simple method of tuning and validation of computational models for machines and structures, considering the conflicting aspects of parameters used on the objective functions.

2.   MATHEMATICAL MODEL

The dynamic model of the system includes interactions between rotors, shafts and journal bearings. The finite element method is applied to the shaft and to the lumped mass in the disk. The model of the non-linear supporting forces due to the cylindrical journal bearings is based on the work of Capone (1986 and 1991). Those non-linear forces are obtained of the resolution of the Reynold's equation for short berings.

Eq. (1) describes the pressure distribution inside the cylindrical journal bearing (Fig. 1.a), based on the Reynolds' equation solution for laminar flux condition. This expression considers the oil thickness h and the axial gradient z, due to the losses of lubricating fluid in short journal bearing.

$$\frac{\partial}{\partial \upsilon}\left(h^3 \frac{\partial p}{\partial \upsilon}\right)+k^2 \frac{\partial}{\partial z}\left(\frac{h^3}{\mu}\frac{\partial p}{\partial z}\right)=\frac{\partial h}{\partial \upsilon}+2\frac{dh}{d\tau} \tag{1}$$

The pressure gradient in circumferential direction can be neglected for short journal bearing in relation to the axial gradient (Childs, 1993). Therefore, the result of the differential equation with this simplification is:

$$p(\upsilon,z)=\frac{1}{2}\left(\frac{L}{D}\right)^2\left[\frac{(x-2\dot{y})\sin(\upsilon)-(y+2\dot{x})\cos(\upsilon)}{(1-x\cos(\upsilon)-y\sin(\upsilon)^3)}\right](4z^2-1) \tag{2}$$

In order to determine the force generated by the oil film pressure distribution, the shaft contact area, $dA = R.d\upsilon.L.dz$, is considered in Eq. (3):

$$Fh = \begin{Bmatrix} Fh_x \\ Fh_y \end{Bmatrix} =$$

$$= -\mu\omega\left(\frac{R^2}{C^2}\right)\left(\frac{L^2}{D^2}\right)(RL)\frac{\left[(x-2\dot{y})^2+(y+2\dot{x})^2\right]^{\frac{1}{2}}}{(1-x^2-y^2)}\begin{Bmatrix} 3xV(x,y,\alpha)-\sin(\alpha)G(x,y,\alpha)-2\cos(\alpha)F(x,y,\alpha) \\ 3yV(x,y,\alpha)-\cos(\alpha)G(x,y,\alpha)-2\sin(\alpha)F(x,y,\alpha) \end{Bmatrix} \qquad (3)$$

Where the terms V, G and F are respectively given in Eq. (4), (5) and (6).

$$V(x,y,\alpha) = \frac{2+(y\cos(\alpha)-x\sin(\alpha))G(x,y,\alpha)}{(1-x^2-y^2)} \qquad (4)$$

$$G(x,y,\alpha) = \int_{\alpha}^{\alpha+\pi}\frac{d\upsilon}{(1-x\cos(\upsilon)-y\sin(\upsilon))} = \frac{\pi}{\sqrt{1-x^2-y^2}} - \frac{2}{\sqrt{1-x^2-y^2}}\tan^{-1}\left(\frac{y\cos(\alpha)-x\sin(\alpha)}{\sqrt{1-x^2-y^2}}\right) \qquad (5)$$

$$F(x,y,\alpha) = \frac{(x\cos(\alpha)+y\sin(\alpha))}{(1-x^2-y^2)} \qquad (6)$$

The differential equation of motion must be written in two coordinates, x and y, respectively Eq. (7) and (8) (see Castro & Cavalca, 2006).

$$[M]\frac{d^2x}{dt^2} + ([C]+[G])\frac{dx}{dt} + [K]x = Fh_x(x,y,\frac{d}{dt}x,\frac{d}{dt}y) + \omega^2 ME\cos(\omega t) \qquad (7)$$

$$[M]\frac{d^2y}{dt^2} + ([C]+[G])\frac{dy}{dt} + [K]y = Fh_y(x,y,\frac{d}{dt}x,\frac{d}{dt}y) + \omega^2 ME\sin(\omega t) - W \qquad (8)$$

The matrixes M, C, G and K are respectively the mass, damping, gyroscopic and stiffness matrixes of the shaft and the lumped mass, which are obtained by a classical finite element method. The horizontal rotor weight is represented in these equations by W.

The solution of the equation of motion can be obtained by the application of numerical methods. The Newmark integration method was considered in this study.

In order to proceed with the simulations, a finite element model is considered. Figure 1 represents the horizontal rotor (Fig. 1.b) and its finite element model (Fig. 1.c).

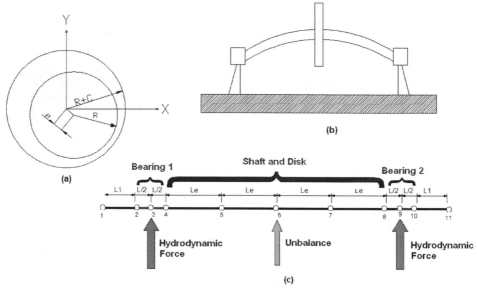

Figure 1 - (a) Journal Bearing scheme; (b) Horizontal Rotor physical model; (c) Horizontal Rotor Finite Element Model.

3.   OBJECTIVE FUNCTION

The fitting method includes the control variables that can be measured in a experimental setup (amplitude and elliptical shape of the rotor orbits at the bearings). These control variables depend on the variables that are fitted in the optimization process (oil viscosity and unbalance moment).

A diagnosis technique, based on Bachschmid et. al. (2004), is proposed to characterize the elliptical form of the orbit. This analysis includes the degree of ellipticity, the dimension of the major axis of the orbit a and its inclination with respect to the horizontal axis θ (Fig. 2).

The displacement in x e y directions will be described by Eq. (9) and (10), respectively, where xo, xs, xc, yo, ys e yc are the first order Fourier coefficients.

$$x(t) = x_0 + x_c \cos(\omega t) + x_s \sin(\omega t) \tag{9}$$

$$y(t) = y_0 + y_c \cos(\omega t) + y_s \sin(\omega t) \tag{10}$$

The inclination of the major axis θ and the major radius α are given by:

$$\theta = 0.5 \tan^{-1}\left(\frac{2(x_c y_c + x_s y_s)}{x_c^2 + x_s^2 - y_c^2 - y_s^2} \right) \tag{11}$$

$$a = \sqrt{\frac{2\left(x_s y_c - x_c y_s\right)^2}{x_c^2 + x_s^2 + y_c^2 + y_s^2 - \sqrt{\left(y_c^2 + y_s^2 - x_c^2 - x_s^2\right) + 4\left(x_s y_s + x_c y_c\right)^2}}} \tag{12}$$

The degree of ellipticity is described by the Shape and Directivity Index (SDI):

$$-1 \leq SDI = \frac{\left|r^f\right| - \left|r^b\right|}{\left|r^f\right| + \left|r^b\right|} \leq 1 \tag{13}$$

Where r^f and r^b are the forward and backward coefficients of the complex harmonic signal $p(t)$ of frequency ω.

$$p(t) = x(t) + jy(t) = r^f e^{j\omega t} + r^b e^{-j\omega t} \tag{14}$$

The difference between the experimental and the adjusted orbit can be calculated by the difference of their parameters: degree of ellipticity (*SDI*), dimension of the major axis (*a*) and the inclination of the major axis in relation to the horizontal (*θ*). The minimization process of each one of these parameters can be considered as a function objective to be minimized, constituting a multi-objective problem. Eq. (15) presents the objective functions of the proposed problem and the orbits are adjusted for a certain number of nodes. Each equation will be treated separately.

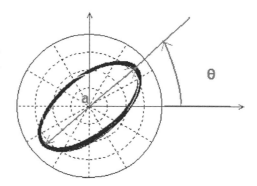

Figure 2 – Elliptic Orbit Parameters

$$f_{SDI} = \sum_{i=1}^{nodes} \left| \frac{SDI_{experimental\_i} - SDI_{adjusted\_i}}{SDI_{exp\_i}} \right|, \quad f_a = \sum_{i=1}^{nodes} \left| \frac{a_{experimental\_i} - a_{adjusted\_i}}{a_{exp\_i}} \right| \quad \text{and}$$

$$f_\theta = \sum_{i=1}^{nodes} \left| \frac{\theta_{experimental\_i} - \theta_{adjusted\_i}}{\theta_{exp\_i}} \right| \tag{15}$$

4.   MULTI-OBJECTIVE GENETIC ALGORITHM

4.1 Genetic algorithm
Genetic algorithm (GA) is a random search technique that simulates the process of genetic evolution. Each parameter is codified by a gene using an appropriate representation. The corresponding genes for all of the parameters form a chromosome capable to describe an individual solution of the problem. Each individual is evaluated in each generation based on its fitness with regard to an objective function previously defined. A group of chromosomes representing

several individual solutions of the problem includes a population in which an individual with the best characteristics (closer to the global solution) has more chance of being selected to reproduce. The evolution is executed using a series of stochastic operators to manipulate the genetic code, in other words, with the evolution of the search, the population includes fittest solutions and, eventually, converges.

Most of GA's includes operators for individuals' selection for reproduction, for new individuals' formation starting from selected ones and to determine the composition of the population in the future generation, see Fig 3. Crossover is accomplished using crossings to combine different individuals' genes (parents) to produce the offspring. The mutation operator is applied to modify the new individuals' code. The new individuals are inserted in a new population and the procedure is restarted.

The selection process is an important phase of GA because selects the most capable individuals to reproduce and participate in the next generation (iteration). The individual's possibility to be selected depends on its fitness. The objective of the reproduction operator is to make copies of good solutions and eliminate bad solutions in a population, while maintaining constant the size of the population. In this work, a proportional selection method is used, as proposed by Holland (1975).

The crossover operator is considered the most important operator of GA. Usually, two chromosomes - called parents - are selected and random portions of genetic material are exchanged among them to generate new chromosomes - called here new individuals (children). The parents are selected among the existent chromosomes in the population, in order preferably as for its fitness, for the new individuals to inherit the parents' good characteristics. Applying iteratively the crossing operator, genes (or parameter solutions) of chromosomes with better fitness will more frequently appear in the population.

The mutation operator introduces random changes in the characteristics of the chromosomes and it is, usually, applied on the genes. The mutation is a critical element of GA because, instead of guiding the population for the convergence, as the crossover operator, it reintroduces the genetic diversity in the population and allows the search process to escape from local optima.

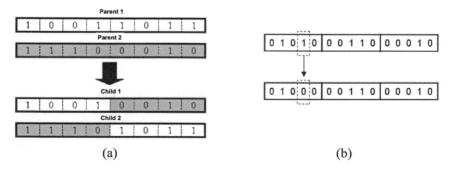

(a)   (b)

Figure 3 – Genetic Operators. (a) One point crossover; (b) One point mutation

4.2. Multi-Objective Problems

Real engineering problems, like the tuning of mathematical model parameters in rotating systems, usually involve simultaneous consideration of multiple performance criteria, in which objectives are non-commensurate and are frequently in conflict amongst themselves. Optimize the problem with regard to only one objective eventually can generate an inadmissible solution in relation to the other objectives. Once there is more than one conflicting objective, a single best solution usually does not exist and the concept of Pareto optimality can be applied.

The multi-objective genetic algorithm applied in this work was based in Fonseca and Fleming's non-elitist MOGA (1993). MOGA differs from a classic genetic algorithm because the fitness is evaluated for each solution of the population. The remaining algorithm (universal stochastic selection, crossover and mutation) is the same of the classic genetic algorithm. However, MOGA uses the concept of Pareto ranking and incorporates niching and fitness sharing to created a dispersion of solutions.

The Pareto ranking method finds the group of individuals that are not dominated by the rest of the population, according to the Pareto criterion. The best rank is attributed ($r_i = 1$) to that group. After this step, other group of non dominated solutions is obtained, without consider the first group, and an immediately worse rank is attributed. The iteration proceeds until all individuals are ranked. In general, for a solution i, a rank r_i is defined representing the number of solutions η_i that dominates the solution i, added to one, as indicated by Eq. (16). Once executed the Pareto ranking, a fitness line is defined based on the ranking.

$$r_i = 1 + \eta_i \tag{16}$$

The idea of fitness sharing was proposed by Goldberg and Richardson (1987) in the investigation of the local optima for multi-modal functions. The niching method is used to maintain the diversity of solutions, dividing the population in groups (that gather individuals with similar characteristics) to reduce the competition for resources and to create stable sub-populations, each one of them concentrated in a niche of the search space. This method encourages the search in unexplored sections of the Pareto front artificially reducing the fitness of solutions in areas of high population density.

The fitness of a solution i is calculated by Eq. (17).

$$F_i = N - \sum_{k=1}^{r_i-1} \mu(k) - 0.5(\mu(r_i) - 1) \tag{17}$$

where N is the number of individuals in the population, ri is the rank i and $\mu(ri)$ is the number of solutions at rank r_i.

The Euclidean distance between each pair of solutions i and j in the same rank is determined by Eq. (18) and compared to the niche size (sigma share), which is the key parameter of the niche mechanism. To obtain the sharing function (Eq. (19)) and the niche count (Eq. (20)).

$$d_{ij} = \sqrt{\sum_{k=1}^{M} \left(\frac{f_k^i - f_k^j}{f_k^{max} - f_k^{min}} \right)^2}$$

(18)

$$Sh(d_{ij}) = \begin{cases} 1 - \left(\dfrac{d_{ij}}{\sigma_{Share}} \right)^{\alpha} & \text{if } d_{ij} \leq \sigma_{Share} \\ 0 & \text{if } d_{ij} > \sigma_{Share} \end{cases}$$

(19)

$$nc_i = \sum_{j=1}^{\mu(r_i)} Sh(d_{ij}), \quad 0 \leq nc_i \leq 1$$

(20)

The shared fitness is obtained by dividing the fitness of each solution by its corresponding niche count, that causes a decrease in the fitness value of closer solutions. These values are scaled so that their average is the same as the original fitness value.

4.3 Algorithm

The basic algorithm applied in this paper is given below. The total number of generations was chosen as the stop criterion.

- Step 1. Generate an initial population: randomly generated. The binary representation was used for each individual of the population.
- Step 2. For each generation, do:
- Step 3. Transform Genotype in Phenotype: convert the binary string in real values;
- Step 4. Calculate the objective functions: calculate the parameters of the orbit to solve the Eq. 15, considering each individual of the population as a solution;
- Step 5. Fitness evaluation: use Pareto ranking to calculate the rank of each solution and the niche and fitness sharing methods;
- Step 6. Apply the Selection operator: use the proportional selection based on the fitness;
- Step 7. Apply the Crossover operator: choose randomly a crossing point and two individuals to be crossed;
- Step 8. Apply the Mutation operator: choose randomly an individual and the bit that will be the mutation point;
- Step 9. If the stop criterion is not satisfied, go to next generation and go back to step 3: Through the use of the genetic operators, a new population, closer to the final solution, was created and it must be evaluated and improved.

5. EXPERIMENTAL SETUP

The experimental setup consists of two hydrodynamic bearings and a lumped mass assembled in the centre of the shaft (Fig. 4), similar to a Laval rotor. The shaft of the physical system is made of steel with a 12 mm diameter. The distance between the bearings is 600 mm. The concentrated mass consists of a disk with an external diameter of 94.7 mm, a length of 47.0 mm and a mass of 2.34 kg. A pair of cylindrical hydrodynamic bearings are used to support the shaft, made of in bronze, with a

Figure 4: Experimental test rig.

radial clearance of 90 μm and L/D ratio of 0.64. The bearings are lubricated with AWS 32 oil, whose operating temperature is close to 25°C. The nominal frequency of the electric motor lays in the range of 1 to 60 Hz. The rotor support structure can be considered rigid in these operational range of frequencies.

6. EXPERIMENTAL RESULTS AND DISCUSSION

The natural frequency of the rotor was verified by the unbalance response of the rotor. The highest amplitude occurred close to 23.0 Hz. In order to determine the unbalance, as well as the oil viscosity of each bearing, the fitting process of the orbits in each bearing were accomplished at 22.5 Hz (near the resonance frequency). A stationary condition was considered in data acquisition as well as in the adjustment simulations. The GA parameters that gives good results for this problem are: total number of generations: 80; Population size: 50; Mutation probability: 0.25; Mutation rate: 0.8; Crossover probability: 0.7.

6.1 The Pareto's front

Fig.5 shows the Pareto's front with 12 points of the last iteration. Then, it is possible to analyze the non dominated solutions from the objective functions used in the searching process, given by Eq. 15. It is interesting to recall that the objective functions f_{SDI}, f_a and f_θ are dimensionless, being optimal when close to zero, which minimizes the relative error or the difference between adjusted and experimental geometric parameters of the orbits.

Points 1 and 2 are not non dominated solutions, because all objective functions f_{SDI}, f_a and f_θ present high relative errors, significantly greater than 1. These points still remain among the possible solutions due to the application of the diversity operator in the last iteration, but they do not belong to the Pareto's front. Point 3, instead, can be characterized as a non dominated solution, due to the lowest error in the objective function f_θ, although the objective functions f_{SDI} and f_a present high errors in the result. The possible solutions numbered from 4 to 12 certainly are non dominated solutions, which means they attend the conflicting aspect characteristic of multi-objective problems, i.e., any change in one of the search variables (SDI, a or θ) eventually could generate a better result for one of the objective functions, but it could also present worse solutions to the other functions (see points 4 and 6 in Table 1). The physical solution is, therefore, inside the Pareto's front (points 3 to 12). In order to determine which solutions are close to the physical result, a design analysis should be made, comparing the adjusted simulated orbit in the bearings with the experimental ones.

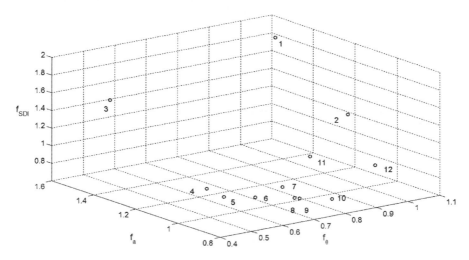

Figure 5: Pareto's front points.

6.2 The fitting process results
The adjusted parameters and the objective functions results for points 4 and 6 in Fig. 5 are indicated in Table 1. The algorithm was able to find good results for the orbits. Fig. 6 shows the experimental and adjusted orbits for a rotation of 22.5 Hz in each bearing.

Table 1 – Adjusted Parameters

	Solution A; point 6	Solution B; point 4
Unbalancing Moment [kg.m]	$1.24 \cdot 10^{-4}$	$1.24 \cdot 10^{-4}$
Viscosity in Bearing 1 [Ns/m²]	$6.03 \cdot 10^{-3}$	$7.25 \cdot 10^{-3}$
Viscosity in Bearing 2 [Ns/m²]	$9.12 \cdot 10^{-3}$	$9.40 \cdot 10^{-3}$
SDI – Bearing 1 / 2	0.57 / 0.64	0.60 / 0.63
Amplitude a [μm] – Bearing 1 / 2	28 / 23	26 / 24
Angle θ [degrees] – Bearing 1 / 2	25 / 30	27 / 32
Experimental SDI – Bearing 1 / 2	0.48 / 0.44	
Experimental Amplitude [μm] – Bearing 1 / 2	34 / 23	
Experimental Angle [°] – Bearing 1 / 2	40 / 28	

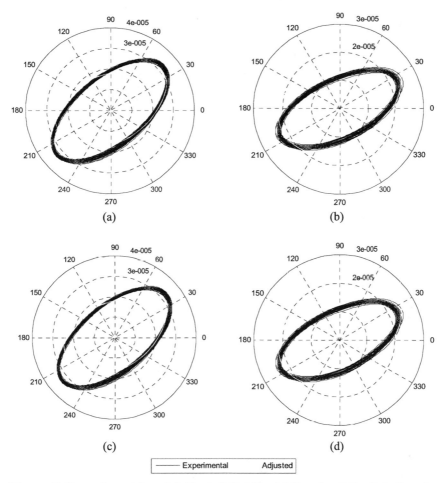

(a)   (b)

(c)   (d)

Experimental   Adjusted

**Figure 6- Experimental and Adjusted Orbits. (a) Bearing 1 for Solution A;
(b) Bearing 2 for Solution A; (c) Bearing 1 for Solution B;
(d) Bearing 2 for Solution B.**

For the selected solutions of points 4 an 6, Bearing 1 presents a good fitness for the orbit degree of ellipticity and inclination angle, while bearing 2 presents good fittness for all parameters, i.e., the orbit degree of ellipticity, the dimension of the major axis and the inclination angle. It confirms that the solutions are distributed in a Pareto front: an improvement in one (or more) objective(s) represents a degradation in other(s). As Solution A returned a smaller viscosity than Solution B for both bearings, it was obtained a slight better result for the adjust in Bearing 1 by Solution A and a slight better result in Bearing 2 by Solution B in terms of the orbits amplitudes.

The fitting process returned better results for Bearing 2, due the fact that the current analytical model is not considering the coupling stiffness near Bearing 1. Another parameter that is not considered in the model is the proportional damping of the

895

shaft. As the rotation speed is near the critical speed the effects of the structural damping parameters can be significant. As the population size is 50 individuals, the algorithm returned 50 final solutions, however their dispersion proposed by niching and fitness sharing was not confirmed. Many of the solution were identical and some changes in the algorithm have to be made to improve the search for new regions in the Pareto front.

7.   CONCLUSIONS

In this paper, a multi-objective genetic algorithm, based on Fonseca and Fleming's MOGA (1993) was applied to adjust a non-linear rotating system model. The adjustment method is useful for the analysis of rotating machines due to eventual limitations of the mathematical models in representing the real structure in an acceptable way. A very promising approach was obtained to unbalance moment using only the displacement measurements in the bearings, which makes the method particularly interesting to practical applications in real machines.

The adjusted orbits to experimental data presented some differences, that can be due some variables that are not considered in the model, like motor-shaft coupling and proportional damping in the shaft. Although the process presented good results, the diversity of solutions should be improved in future tests.

The continuity of this work will consider some of these variables and new search algorithms (also based on Genetic Algorithm) will be used for the estimation of the Pareto's front, in order to obtain a better diversity of solutions, what is primordial in multi-objective analysis.

8.   ACKNOWLEDGEMENTS

The authors thank FAPESP, CNPq and CAPES for the sponsoring of this research.

9.   REFERENCES

Bachschmid, N., Pennacchi, P., Vania, A., 2004, Diagnostic Significance of Orbit Shape Analysis and its Application to Improve Machine Fault Detection. Journal of the Brazilien Society of Mechanical Science and Engineering, v. XXVI, n. 2, pp. 200-208.

Capone, G., 1986, Orbital motions of rigid symmetric rotor supported on journal bearings., La Meccanica Italiana, pp. 37-46.

Capone, G., 1991, Descrizione analitica del campo di forze fluidodinamico nei cuscinetti cilindrici lubrificati., L.Energia Elettrica, pp. 105-110.

Castro, H. F.; Idehara, S. J. ; Cavalca, K. L.; Dias Junior, M. 2004. Updating Method Based on Genetic Algorithm Applied to Non-linear Journal Bearing

Model. In: 8th International Conference on Vibrations in Rotating Machinery, p. 433-444.

Castro, H. F., Cavalca, K. L., Mori, B. D. 2005. Journal Bearing Orbits Fitting Method with Hybrid Meta-heuristic Method, Proceedings of the COBEM 2005, 6-11 November, Ouro Preto, Brazil, pp. 1 – 10.

Castro, H. F., Cavalca, K. L. 2006. Hybrid meta-heuristic method applied to parameter estimation of a non-linear rotor-bearing system, 09/2006, IFToMM - 7th International Conference on Rotor Dynamics,Vol. 1, pp.11-20, Viena, Austria.

Cavalca, K.L, Idehara, S.J., Dedini, F. G, Pederiva, R.,2001, Experimental non-linear model updating applied in cylindrical journal bearing, Proceedings of ASME Design Engineering Technical Conference 2001, 9-12 September, Pittsburgh, USA, pp. 1 –9.

Childs, D., 1993, Turbomachinery Rotordynamics . Phenomena, Modeling and Analysis., John Wiley & Sons, New York, p.476.

Deb, K. 2001. Multi-Objective Optimization using Evolutionary Algorithms. John Wiley and Sons, Chichester, 515p.

Fonseca, C. M. and Fleming, P. J. 1993. Genetic Algorithms for multi-objective optimization: Formulation, discussion, and generalization. In Proceedings of the Fifth International Conference on Genetic Algorithms, pp. 416-423.

Goldberg, D. E. and Richadson, J. 1987. Genetic Algorithms with sharing for multimodal function optimization. In Proceedings of the First International Conference on Genetic Algorithms and their Applications, pp 41-49.

Holland, J. H., 1975. Adaptation in Natural and Artificial Systems. Ann Arbor, MI: MIT Press.

Device asymmetries and the effect of the rotor run-up in a two-plane automatic ball balancing system

D J Rodrigues, A R Champneys, M I Friswell, R E Wilson
University of Bristol, UK

ABSTRACT

We present a numerical investigation of a two-plane automatic ball balancer (ABB) for rotating machinery. This device consists of a pair of circular races that are set perpendicular to the shaft in two distinct planes. Within each race are a series of balancing balls which move to compensate any mass imbalance arising from rotor eccentricity or misalignment. We build upon an earlier bifurcation analysis of the steady state dynamics to consider device asymmetries such as non equal ball masses and we also investigate the effect of the rotor run-up.

1. INTRODUCTION

An automatic ball balancer (ABB) is a device which reduces vibrations in rotating machinery by compensating for the mass imbalance of the rotor. The ABB consists of a series of balls that are free to travel around a race which is set at a fixed distance from the shaft. During balanced operation the balls find such positions that the principal axis of inertia is repositioned onto the rotational axis. Because the imbalance does not need to be determined beforehand ABB's are ideally suited to applications where the amount of imbalance varies with the operating conditions. For example, automatic balancers are currently used in optical disk drives, machine tools and washing machines.

The first study of an ABB was carried out by Thearle in 1932 [1], and Lee and Van Moorhem demonstrated the existence of a stable balanced steady state at rotation speeds above the first critical frequency in a theoretical and experimental analysis [2]. However, in some cases the ABB may not balance the system even when operating at supercritical rotation speeds [3]. In 1977, Hedaya and Sharp [4] extended the autobalancing concept by proposing a two-plane device that can compensate for both shaft eccentricity and shaft misalignment, see Fig 1. Misalignment induces tilting vibrations and so models based on a 4DOF rotor which include gyroscopic effects must be considered [5, 6, 7]. In a previous study we used Lagrange's method and rotating coordinates to derive an autonomous set of governing equations [8]. A symmetric system was then considered and numerical continuation techniques were employed to map out the stability boundaries of the balanced state in various parameter planes. The setup on a real machine is, however, usually asymmetric, for example the balls may have different masses or the balancing planes may not be equally spaced. Here we shall extend our work by considering the effect of these asymmetries and in addition we shall investigate the influence of the rotor run-up [9].

The rest of this paper is organised as follows. In section 2 we present and discuss the equations of motion for the ABB. The steady states of the system are considered in section 3, and we focus on using numerical bifurcation theory to investigate the effect that the physical parameters have on their stability. The effects of the rotor run-up are then investigated in section 4 and finally we draw conclusions and discuss possible directions for future work in section 5.

2. EQUATIONS OF MOTION

The setup of the ABB is illustrated in Fig. 1, and is based on a rigid rotor which has been fitted with a two-plane automatic balancer. The rotor has mass M, moment of inertia tensor $diag[J_t, J_t, J_p]$, and is mounted on two compliant linear bearings which are located at S_1 and S_2. The automatic balancer consists of a pair of races that are set normal to the shaft in two separate planes. Each race contains two balancing balls of mass m_k, which move through a viscous fluid and are free to travel, at a fixed distance R_k from the shaft axis. The position of the kth ball is specified by the axial and angular displacements z_k and α_k, which are written with respect to the $C\xi\eta z$ rotor axes.

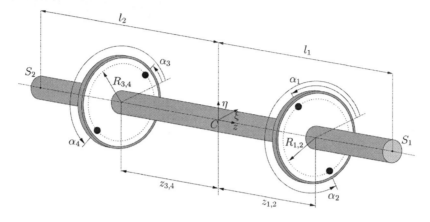

Fig. 1: Schematic diagram of a two-plane automatic balancer.

The equations of motion have been derived in [7, 8] using Lagrange's method and they can be written in the inertial space frame as

$$\mathbf{M}\ddot{\mathbf{q}} + (\mathbf{C} - i\dot{\varphi}_0\mathbf{G})\,\dot{\mathbf{q}} + \mathbf{K}\mathbf{q} = (\dot{\varphi}_0^2 - i\ddot{\varphi}_0)\mathbf{f}_I e^{i\varphi_0} + \sum_{k=1}^{4}(\dot{\varphi}_k^2 - i\ddot{\varphi}_k)\mathbf{f}_{b_k} e^{i\varphi_k}, \tag{1}$$

$$\tilde{J}_p\ddot{\varphi}_0 + c_r\dot{\varphi}_0 + \sum_{k=1}^{4} c_b(\dot{\varphi}_0 - \dot{\varphi}_k) + \Im((\ddot{\mathbf{q}} \cdot \mathbf{f}_I)e^{-i\varphi_0}) = L(\dot{\varphi}_0), \tag{2}$$

$$m_k R_k^2 \ddot{\varphi}_k + c_b(\dot{\varphi}_k - \dot{\varphi}_0) + \Im\left((\ddot{\mathbf{q}} \cdot \mathbf{f}_{b_k})\,e^{-i\varphi_k}\right) = 0, \quad k = 1,\ldots,4. \tag{3}$$

Here $\mathbf{q} = [X + iY, \phi_Y - i\phi_X]^T$ is the (complex) vector of the vibrational degrees of freedom[1], φ_0 is the rotor angle and $\varphi_k = \varphi_0 + \alpha_k$ is the angular displacement of the kth ball. The mass, gyroscopic, damping and stiffness matrices are given respectively by

$$\mathbf{M} = \begin{bmatrix} M & 0 \\ 0 & J_t \end{bmatrix} + \sum_{k=1}^{4} \begin{bmatrix} m_k & m_k z_k \\ m_k z_k & m_k z_k^2 \end{bmatrix}, \quad \mathbf{G} = \begin{bmatrix} 0 & 0 \\ 0 & J_p \end{bmatrix}, \quad \mathbf{C} = \begin{bmatrix} c_{11} & c_{12} \\ c_{12} & c_{22} \end{bmatrix}, \quad \mathbf{K} = \begin{bmatrix} k_{11} & k_{12} \\ k_{12} & k_{22} \end{bmatrix},$$

and the mass imbalance and ball vectors are given by

$$\mathbf{f}_I = \begin{bmatrix} M\epsilon e^{i\beta_1} \\ \chi\,(J_t - J_p)\,e^{i\beta_2} \end{bmatrix} \quad \text{and} \quad \mathbf{f}_{b_k} = \begin{bmatrix} m_k R_k \\ m_k R_k z_k \end{bmatrix}, \quad k = 1,\ldots 4.$$

---

[1]As defined in [10]

Here ϵ and χ are the rotor eccentricity and misalignment respectively and $\beta_{1,2}$ are the fixed phases of these imbalances with respect to the rotor ξ axis. In addition, $\tilde{J}_p = J_p + J_t\chi^2 + M\epsilon^2$ is the modified polar moment of inertia, $L(\dot{\varphi}_0)$ is the driving torque generated by the motor and c_r is the torque damping. Finally c_b is the viscous damping of the balls in the race as they move through the fluid. We note that by taking $m_k = 0$ in (1), we recover the equations of motion for a four degree of freedom rotor [10]. Also, by setting the tilt angles $\phi_x = \phi_y \equiv 0$, the system reduces to the equations of motion for the planar automatic balancer [11]. The form of the governing equations suggest that automatic balancing can be viewed as a synchronization phenomena of coupled rotors [12]. For smooth operation we require that the ball speeds $\dot{\varphi}_k$ synchronize with the rotor speed $\dot{\varphi}_0$ and furthermore that the phases of the balls $\varphi_k - \varphi_0$ are such that their forcing cancels out, or at least reduces, the forcing from the rotor imbalance. It is usual to assume that the motor can provide enough torque to realise a given angular velocity profile and so we shall impose the spin speed $\Omega(t) = \dot{\varphi}_0(t)$ instead of considering (2) with a given driving torque. Also in order to aid the stability analysis we can use the transformation $\mathbf{r} = \mathbf{q}e^{-i\varphi_0}$ to rewrite the equations of motion in the rotating frame to give

$$\mathbf{M}\ddot{\mathbf{r}} + [\mathbf{C} + i\Omega(2\mathbf{M} - \mathbf{G})]\dot{\mathbf{r}} + \left[\mathbf{K} - \Omega^2(\mathbf{M} - \mathbf{G}) + i\left(\dot{\Omega}\mathbf{M} + \Omega\mathbf{C}\right)\right]\mathbf{r}$$

$$= (\Omega^2 - i\dot{\Omega})\mathbf{f}_I + \sum_{k=1}^{4}\left[(\Omega + \dot{\alpha}_k)^2 - i\left(\dot{\Omega} + \ddot{\alpha}_k\right)\mathbf{f}_I e^{i\alpha_k}\right], \tag{4}$$

$$m_k R_k^2(\dot{\Omega} + \ddot{\alpha}_k) + c_b\dot{\alpha}_k + \Im\left\{\left[\left(\ddot{\mathbf{r}} + 2i\Omega\dot{\mathbf{r}} - \left(\Omega^2 - i\dot{\Omega}\right)\mathbf{r}\right)\cdot\mathbf{f}_{b_k}\right]e^{-i\alpha_k}\right\} = 0, \quad k = 1,\ldots,4.$$

In the remainder of this paper we shall investigate the model given by equation (4), however we note that many effects such as bearing anisotropy, geometric defects and dry friction between the race and balls are not included. We refer the interested reader to [13, 14, 15].

3. BIFURCATION ANALYSIS OF THE BALANCED STEADY STATE

Steady state solutions are obtained by setting all time derivatives in the equations of motion (4) to zero. Moreover, if we also set the vibrational coordinates $\mathbf{r} = 0$, we arrive at the following condition for a balanced steady state

$$\mathbf{f}_I + \sum_{k=1}^{4}\mathbf{f}_{b_k}e^{i\alpha_k} = 0. \tag{5}$$

Of course this equation simply states that the forces and moments acting on the rotor due to the imbalance and balancing balls must be in equilibrium. The solution is physically unique and exists provided that the balls have a mass large enough to cope with the imbalance. The ball positions $\alpha = \alpha^*$ can be determined in closed form but the equations are long and so are not presented here.

Next we shall use the continuation package AUTO [16] to compute bifurcation diagrams showing the regions of stability in various parameter planes. The boundaries of stability are formed by Hopf bifurcations which mark the onset of rotor vibrations. In order to relate our findings to systems with different parameters it will prove useful to non-dimensionalize the equations of motion. We rescale with respect to the mass of the rotor M, the race radius R, and the critical frequency associated with the cylindrical whirl $\Omega_1 = \sqrt{k_{11}/M}$, for details see the appendix. Unless otherwise stated, for the rest of this study we shall consider the non-dimensional ABB model with the following parameters

$$z_{1,2} = z_{3,4} = 2, \quad l_1 = l_2 = 3, \quad m_k = m, \quad \mathbf{K} = \begin{bmatrix} 1 & 0 \\ 0 & 9 \end{bmatrix},$$

$$c = 0.02 \quad \text{where} \quad \mathbf{C} = c\mathbf{K} \quad \text{and} \quad c_b = 0.01. \tag{6}$$

The stiffness and damping values are based on a rotor with two equal isotropic bearings where the centre of mass is located at the midspan and the shaft has a length of six times the race radius. The approximate critical frequencies for the cylindrical and conical whirls occur respectively at

$$\Omega_1 = \sqrt{\frac{k_{11}}{M}} \equiv 1, \quad \text{and} \quad \Omega_2 = \sqrt{\frac{k_{22}}{J_t - J_p}}. \tag{7}$$

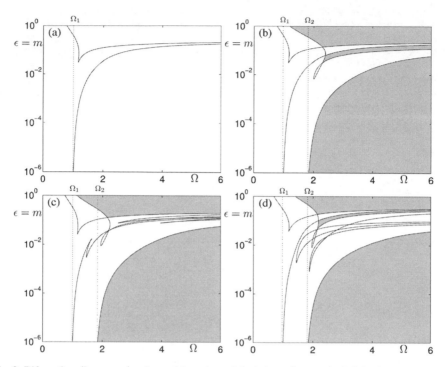

Fig. 2: Bifurcation diagrams showing stable regions of the balanced state (shaded) in the case of a static imbalance. The eccentricity ϵ is varied against Ω, whilst m is kept equal to ϵ so that α^* remains constant. Panel (a) is a bifurcation diagram for a 'disk' type rotor $J_p > J_t$ and panel (b) is for a 'long' type rotor $J_p < J_t$. Similar diagrams for the 'long' type rotor are shown in (c) where one of the balls has a mass 20% greater than the others and in (d) where $z_{1,2} = 1$ and $z_{3,4} = 3$ so that the balancing planes are not equidistant from the midspan.

Figure 2 shows stability diagrams for the static imbalance case (i.e. an imbalance with no misalignment $\chi = 0$). The eccentricity ϵ, is plotted against Ω, whilst we also vary the ball mass so that $m = \epsilon$. Thus, the mass of the balls scales with the imbalance and the balanced state α^* does not change value. Physically the condition $m = \epsilon$ means that each ball has enough mass to compensate for the rotor eccentricity, therefore as there are a total of four balls, the ABB is far from its balance correction limit. A logarithmic scale is used for the vertical axis so that a wide range of eccentricities can be considered. Panel (a) shows the situation for a 'disk' type rotor where $J_p > J_t$, here there is no stable region of balanced operation because the influence of the gyroscopic term is such that the eigenfrequency corresponding to the conical whirl is always greater than the rotor speed Ω. This means that there is no conical critical speed Ω_2 and thus no associated self-aligning process[2], hence the ABB is not stabilised with respect to conical motions. The method of direct separation of motion has been used by Sperling

---

[2]Self aligning is the phenomena whereby a rotor will tend to rotate about its principal axis of inertia at supercritical rotation speeds.

et al. [12] to derive this result and in addition they discuss how it relates to Blekhman's generalised self-balancing principle [17]. From a practical viewpoint however, the prognosis for the autobalancing of 'disk' type rotors is not as bad as it may first seem. Because the conical mode has no associated critical speed, 'disk' rotors often need only to be balanced with respect to the static unbalance and a single plane ABB can be used to provide a partial unbalance compensation [18].

Next we illustrate in panel (b) the case for a 'long' rotor where $J_p < J_t$. We have taken $J_t = 3.25$ and $J_p = 0.5$ which corresponds to a solid cylindrical rotor with a height of six times its radius. For these values the critical speed corresponding to the conical whirl occurs at $\Omega_2 \simeq 1.81$. Here there are several regions of balanced operation though for applications the main area of interest occurs where there is a large connected stable region for small eccentricities and supercritical rotation speeds. The Hopf curve which bounds this region asymptotes towards $\Omega = \Omega_2$ as $\epsilon \to 0$, hence there is no stable region in the subcritical regime. We note at this point that these results for the stability of the balanced state are only valid *locally* and there most likely exists competing dynamics in much of the stable range. We shall now investigate how asymmetries of the ABB device can effect its stability. Panel (c) shows the situation where one of the balls has a mass 20% greater than the other balls. We see that the stable regions remain largely unchanged, however extra Hopf instability curves are present. These arise because the introduction of a different ball mass breaks the symmetry of the system. Another factor which must be noted is that unequal balls cannot counterbalance each other by settling to opposite sides of the race.

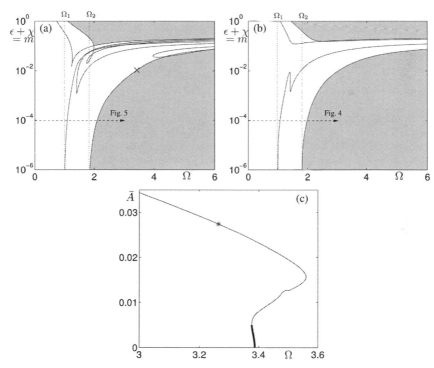

Fig. 3: Bifurcation diagrams in the case of a 'long' rotor with a dynamic imbalance. Panel (a) is for a race damping value of $c_b = 0.01$ and panel (b) is for a higher race damping value of $c_b = 0.1$. The indicated one-parameter sweeps in Ω are illustrated in Figs. 4 and 5. The limit cycle which emanates from the Hopf bifurcation marked by a × in (a) is continued in panel (c). The thick curve represents a stable limit cycle and the thin curve an unstable one.

Thus the overall capability of the ABB is reduced as unequal balls will inevitably add an imbalance to a rotor that is already well balanced [19]. Next the diagram in panel (d) shows the case where the balancing planes are not equally spaced from the midspan. We have taken $z_{1,2} = 1$ and $z_{3,4} = 3$, and again the main point to note is the robustness of the stable region for low eccentricities and supercritical rotation speeds. We shall now return to the symmetric ABB setup but consider a rotor that suffers from a dynamic imbalance (i.e. an imbalance with both an eccentricity ϵ and a misalignment χ). Similar plots to those of Fig. 2 are illustrated in Fig. 3 for the dynamic imbalance case with $m = \epsilon + \chi$, $\epsilon = \chi$ and a constant phase $\beta_1 = 1$. In panel (a) the race damping parameter is $c_b = 0.01$ whereas for panel (b) the race damping value is increased to $c_b = 0.1$. As a result much of the complicated structure in the high eccentricity regime is smoothed out although the main influence of this parameter is manifest in the transient dynamics and in the size of the basin of attraction of the balanced state. The limit cycle which is born at the marked Hopf bifurcation for $c_b = 0.01$ and $m = 0.01$ is continued in panel (c), here the measure $\bar{A}$ is the average rotor vibration at points one unit length from the midspan and is given by

$$\bar{A} = \frac{1}{2}\left(\sqrt{(x + \phi_y)^2 + (y - \phi_x)^2} + \sqrt{(x - \phi_y)^2 + (y + \phi_x)^2} \right).$$

The Hopf bifurcation is supercritical and so there is a small region, indicated by the bold curve, where the limit cycle is stable. In a controlled experiment we would expect to see small oscillations of the balls about their balanced positions before the balls would desynchronize with the rotor. For smaller values of the imbalance, say $\epsilon + \chi = 1 \times 10^{-4}$, the Hopf bifurcation is subcritical and the transition to the desynchronized state would be immediate. The ability to follow the desynchronized limit cycles with continuation software is a work in progress.

4. THE EFFECT OF THE ROTOR RUN-UP

In this section we consider the effect of the rotor run-up on the vibrations of the system. In Fig. 4 we plot the absolute ball speeds $\dot{\varphi}_k$ against the rotor speed Ω for the sweep shown in Fig. 3 (b) with a race damping value of $c_b = 0.1$ and where the other parameters are as given in the previous section. The mass of the balls is $m = 1 \times 10^{-4}$ and we *slowly* increase the rotor speed over a time scale of $t = 1 \times 10^4$ corresponding to approximately 3 hours. Here we can see that the ball speeds initially synchronize with

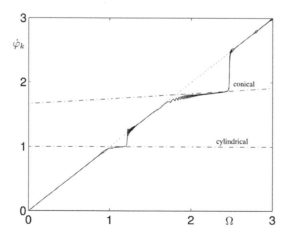

Fig. 4: Diagram showing the ball speeds $\dot{\varphi}_k$ against rotor speed Ω for the dynamic imbalance case with $c_b = 0.1$ and $m = 1 \times 10^{-4}$, see Fig. 3 (b).

the rotor, however as the rotor approaches and passes through its critical speeds the balls tend to 'stall' and synchronize with the rotor eigenfrequencies. As the rotation speed is increased still further the balls again resynchronize with the rotor. If the race damping value is too low, say $c_b = 0.01$, then the balls may never resynchronize with the rotor and they stay at the eigenfrequency speed. This behaviour was first analysed in [20] and is similar to the Sommerfeld effect in which a rotating machine with an insufficiently powerful motor has difficulty in passing through the critical speeds.

The optimization of the velocity profile for a particular application lies outside the scope of this paper, however we shall now consider a more realistic rotor run-up that we can model by the Hill function

$$\Omega(t) = \Omega_{max} \frac{t^n}{t_{1/2}^n + t^n}. \tag{8}$$

This function is plotted in Fig. 5 (a) for the parameter values $(\Omega_{max}, t_{1/2}, n) = (3, 10, 3)$, note that $\Omega \approx 0.9\Omega_{max}$ for $t = 20$. We use this rotor run-up to perform the indicated sweep through Fig. 3 (a) for

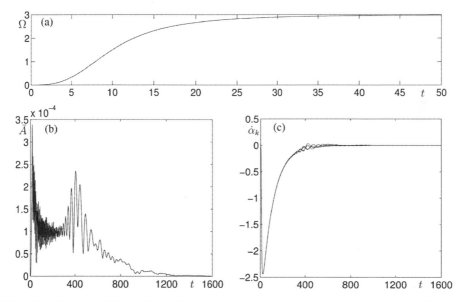

Fig. 5: Simulations which include the effect of rotor run-up, parameters are the same as that for Fig. 3 (a). The considered angular velocity profile is shown in panel (a), the amplitude of rotor vibrations $\bar{A}$ in (b) and the relative ball speeds $\dot{\alpha}_k = \dot{\varphi}_k - \dot{\varphi}_0$ in (c).

the race damping value of $c_b = 0.01$. The results are shown in Fig. 5 (b) and (c), initially the balls speeds lag behind the rotor but eventually the balls catch up synchronizing with the rotor, furthermore they have phases which compensate for the rotor imbalance. It is interesting that there is a period of increased vibrations as the balls approach the rotor speed $\dot{\varphi}_0$. This occurs because the balls desynchronize with each other before resynchronizing with the rotor. However, the resulting vibrations can be significantly reduced if the race damping parameter is increased to say $c_b = 0.1$. Ideally a value of c_b should be chosen so that the balls are close to being critically damped [21], however realising this aim in practice may prove difficult. If the balls undergo non-synchronous motions and complete many circuits of the race then energy will be dissipated in the fluid causing it to heat up and its viscosity to decrease.

5. CONCLUSION

We have used rotating complex coordinates to present a simple model for a rigid rotor with a two-plane ABB. Stability charts obtained by numerical continuation of the Hopf bifurcation curves together with simulations show that the considered device can effectively eliminate imbalances arising from both shaft eccentricity and shaft misalignment. Furthermore, the introduction of ball mass asymmetry or balancing planes that are not equally spaced has little effect on the stable region provided that the machine is operating in the supercritical regime with typical values for the eccentricities[3]. Our investigations of the rotor run-up highlights the phenomena of non-synchronous motions near critical speeds and also the presence of increased vibrations as the balls desynchronize with each other in order to resynchronize with the rotor speed. To eliminate the non-synchronous motions completely one could envisage using a partitioned race [22] or a clamping mechanism where the balls would be fixed until supercritical rotation speeds are reached [1].

We plan to continue the present work by investigating the symmetry properties of the bifurcations which give rise to the balanced state. Normal form theory and the equivariant branching lemma can then be used to give explicit conditions for the stability of the bifurcating solutions. The global stability properties of the system will also be investigated through the use of Lyapunov functions and experimental work is currently in progress in order that we may verify our findings and determine the influence of other effects such as ball-race interactions and rotor flexibility.

6. ACKNOWLEDGEMENTS

The authors are grateful to Rolls-Royce plc. for financial support of this research.

APPENDIX

The system (4) can be non-dimensionalized by writing it with respect to the following non-dimensional state variables

$$\bar{X} = \frac{X}{R}, \quad \bar{Y} = \frac{Y}{R}, \quad \bar{\phi}_X = \phi_X, \quad \bar{\phi}_Y = \phi_Y, \quad \bar{\alpha}_i = \alpha_k \quad \text{for} \quad i = 1 \ldots 4,$$

and the dimensionless time

$$\bar{t} = \Omega_1 t.$$

We have taken the same race radii for all the balls $R_k = R$, and Ω_1 is the critical frequency associated with cylindrical whirl given by $\Omega_1 = \sqrt{k_{11}/M}$. In addition, we introduce the dimensionless parameters

$$\bar{\Omega} = \frac{\Omega}{\Omega_1}, \quad \bar{\epsilon} = \frac{\epsilon}{R}, \quad \bar{\chi} = \chi, \quad \bar{m} = \frac{m}{M}, \quad \bar{z} = \frac{z}{R}, \quad \bar{l} = \frac{l}{R},$$

$$\bar{J}_t = \frac{J_t}{MR^2}, \quad \bar{J}_p = \frac{J_p}{MR^2}, \quad \bar{k}_{11} = \frac{k_{11}}{M\Omega_1^2} \equiv 1, \quad \bar{k}_{12} = \frac{k_{12}}{M\Omega_1^2 R}, \quad \bar{k}_{22} = \frac{k_{22}}{M\Omega_1^2 R^2},$$

$$\bar{c}_{11} = \frac{c_{11}}{M\Omega_1}, \quad \bar{c}_{12} = \frac{c_{12}}{M\Omega_1 R}, \quad \bar{c}_{22} = \frac{c_{22}}{M\Omega_1 R^2}, \quad \bar{c}_b = \frac{c_b}{m\Omega_1 R^2}.$$

The result of the non-dimensionalisation procedure is the rescaling of the mass M, the race radius R and the stiffness k_{11} so that they each have value of one unit.

---

[3]For machine tools and aircraft gas turbines $\epsilon \simeq 1 \times 10^{-5}$

REFERENCES

[1] E. Thearle, A new type of dynamic-balancing machine, *Transactions of the ASME* **54**(APM-54-12) (1932) 131-141.

[2] J. Lee and W. K. Van Moorhem, Analytical and experimental analysis of a self compensating dynamic balancer in a rotating mechanism, *ASME Journal of Dynamic Systems, Measurement and Control* **118** (1996) 468-475.

[3] J. Chung and D. S. Ro, Dynamic analysis of an automatic dynamic balancer for rotating mechanisms, *Journal of Sound and Vibration* **228**(5) (1999) 1035-1056.

[4] M. Hedaya and R. Sharp, An analysis of a new type of automatic balancer, *Journal of Mechanical Engineering Science* **19**(5) (1977) 221-226.

[5] J. Chung and I. Jang, Dynamic response and stability analysis of an automatic ball balancer for a flexible rotor, *Journal of Sound and Vibration* **259**(1) (2003) 31-43.

[6] C.-P. Chao, Y.-D. Huang, and C.-K. Sung, Non-planar dynamic modeling for the optical disk drive spindles equipped with an automatic balancer, *Mechanism and Machine Theory* **38** (2003) 1289-1305.

[7] L. Sperling, B. Ryzhik, Ch. Linz and H. Duckstein, Simulation of two-plane automatic balancing of a rigid rotor, *Mathematics and Computers in Simulation* **58** (2002) 351-365.

[8] D. J. Rodrigues, A. R. Champneys, M. I. Friswell and R. E. Wilson, Automatic two-plane balancing for rigid rotors, *International Journal of Nonlinear Mechanics* (in press)

[9] J. Chung, Effect of gravity and angular velocity on an automatic ball balancer, *Proc. IMechE Part C: J. Mechanical Engineering Science* **219** (2005) 43-51.

[10] G. Genta, *Dynamics of Rotating Systems*. Springer, New York (2005).

[11] K. Green, A. R. Champneys, and N. J. Lieven, Bifurcation analysis of an automatic dynamic balancing mechanism for eccentric rotors, *Journal of Sound and Vibration* **291** (2006) 861-881.

[12] L. Sperling, F. Merten and H. Duckstein, Self-synchronization and automatic balancing in rotor dynamics, *International Journal of Rotating Machinery* **6**(4) (2000) 275-285.

[13] B. Ryzhik, L. Sperling, and H. Duckstein, Auto-balancing of anisotropically supported rigid rotors, *Technische Mechanik* **24** (2004) 37-50.

[14] K-O. Olsson, Limits for the use of auto-balancing, *International Journal of Rotating Machinery* **10**(3) (2004) 221-226.

[15] N. Van De Wouw, M. N. Van Den Heuvel, H. Nijmeijer and J. A. Van Rooij, Performance of an automatic ball balancer with dry friction, *International Journal of Bifurcation and Chaos* **20**(1) (2005) 65-85.

[16] E. Doedel, A. Champneys, T. Fairgrieve, Y. Kusnetsov, B. Sanstede, and X. Wang, AUTO97: Continuation and bifurcation software for ordinary differential equations, http://indy.cs.concordia.ca/auto/ (1997).

[17] I. Blekhman, *Vibrational Mechanics*. World Scientific (2000).

[18] L. Sperling, B. Ryzhik and H. Duckstein, Single-plane auto-balancing of rigid rotors, *Technische Mechanik* **24** (2004) 1-24.

[19] R. Horvath, G. T. Flowers and J. Fausz, Influence of nonidealities on the performance of a self-balancing rotor system, *Proc. of IMECE* (2005).

[20] B. Ryzhik, L. Sperling, and H. Duckstein, Non-synchronous motions near critical speeds in a single-plane auto-balancing device, *Technische Mechanik* **24** (2004) 25-36.

[21] D. J. Rodrigues, A. R. Champneys, M. I. Friswell and R. E. Wilson, Automatic balancing of a rigid rotor with misaligned shaft, *Applied Mechanics and Materials* **5-6** (2006) 231-236.

[22] K. Green, A. R. Champneys, M. I. Friswell and A. M. Muñoz, Investigation of a multi-ball, automatic dynamic balancing mechanism for eccentric rotors, *Phil. Trans. R. Soc. A* **366** (2008) 705-728.

Balancing of a north seeking device using partial, noisy and delayed measurements

I Bucher, A Elka
Dynamics Laboratory, Mechanical Engineering
Technion, Israel Institute of Technology, Israel

ABSTRACT

This paper describes a balancing procedure of a precise north-finding device seeking the earth axis of rotation. This information is obtained by combining the rotation of accelerometers and the natural rotation of earth to generate a measurable Coriolis acceleration. The basic working principle of this apparatus has been patented in the past, but it turns out that the useful signal level is extremely small and therefore even relatively small vibrations deteriorate the device's performance considerably. The main contribution to the unwanted vibration comes from synchronous mass unbalance, even when it is considered relatively small for normal applications. A new procedure that enabled us to produce an accurate balancing procedure in the presence of shaft asymmetry is proposed. The method is explained and demonstrated in a laboratory experiment.

INTRODUCTION

Consider a rotating disk on which an accelerometer is mounted as shown in Fig.1.

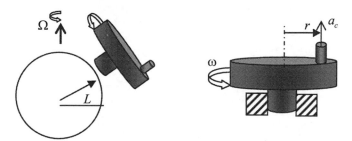

Figure 1: Left: north seeking device located at a certain latitude on the face of earth; Right: North seeking device showing the rotating disk and the sensor (in red)

It has been shown (e.g. [1]) that the combined effect of earth rotation at angular frequency of Ω [Rad/s] and the rotation of the sensor at a frequency ω, give rise to a useful signal due to the Coriolis effect, the sensor will thus measure:

$$a_c = -2\omega \cdot \Omega \cdot r \cdot \cos L \cdot \sin(\omega t + \phi_0) \frac{m}{s^2} \qquad (1)$$

Where ω - angular velocity of the disk, Ω - angular velocity of the Earth, L – latitude, r – sensor's distance from the disk center, ϕ_0 - unknown phase related to azimuth

Unfortunately, the acceleration, being measured by the sensor in Fig. 1, has additional components that can be attributed to mechanical and electrical imperfections. Some of these components appear at the same frequency as a_c and cannot be eliminated by filtering or averaging. One of the most disturbing forces appears due to the flexibility of the structure and it is excited by mass unbalance forces.

As often happens in reality, this device was not been designed to allow for an ordinary balancing procedure. The rotating part was made of machined surfaces that are not round and a key-phasor was difficult to produce since the rotating part is hidden.

The paper discusses the elastic response of the structure and its minimization by a suitable balancing means. In particular, a new influence coefficient based approach that is suitable for asymmetric rotors, had to be developed with unique signal processing, signal alignment and filtering.

The balancing procedure employed a proximity probe that measured the radial deviation of the rotating disc, as shown in Fig. 2. The Coriolis acceleration is measured in an orthogonal direction as illustrated below.

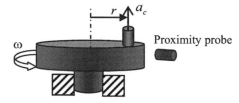

The disc is supported by a set of ball-bearings and it is driven by a DC motor. It is worth mentioning that the first critical speed appears at a speed which is considerably higher than the nominal speed of the device.

Figure 2: Sensor measuring the radial deviation and vibration

BALANCING A ROTATING MASS

Balancing is an important but routine task that is performed on many rotating machines to reduce the vibration.

Balancing is a well known procedure [2-6], but often it is assumed that the rotating part is perfectly axi-symmetric. Indeed, asymmetric rotors have been dealt with extensively in the literature [2-6]. Balancing, as was mentioned before, often

910

neglects low levels of rotor asymmetry without deteriorating the resulting balancing procedure too much. In some cases, the asymmetry level and the required balancing quality are beyond standard applications. In these cases a different procedure has to be employed as was explained in [2-3]. In these papers and especially in [2] it was stated that the ordinary influence coefficient method would not be suitable for asymmetric rotors. None of the cited works proposes a complete balancing procedure for damped asymmetric rotors nor have they shown an experimental application of such a system.

The importance of the present work is that it demonstrates by simple mathematical expressions why asymmetric rotors cannot be balanced by (say) a one (or multiple) run influence coefficient approach. It is clearly shown that several trial mass runs are required to compute the right balancing mass value.

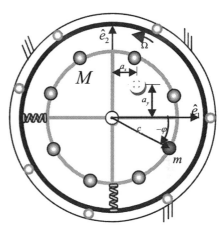

The present paper deals with an asymmetric rotating part that is illustrated in figure 2.

This device, as can be seen in Fig. 3 is not perfectly axis-symmetric and indeed when the natural frequencies for the modes in the x and y directions were measured, distinct differences were found. This is a clear indication for asymmetry.

Figure 3: Model of an unbalanced rotating mass.

PRECISE BALANCING OF AN ASYMMETRIC ROTOR

Consider a mass rotating at angular speed Ω. As shown in Fig. 2, the rotating part has a mass M whose centre of mass is not at the origin where the centre of rotation resides at

$$r_{cm} = a_x \hat{e}_1 + a_y \hat{e}_2 . \tag{2}$$

There is one trial mass m which is located at

$$r_m = c \cos \varphi \hat{e}_1 + c \sin \varphi \hat{e}_2 \tag{3}$$

The mass is mounted on a flexible shaft and bearings with stiffness k_x, k_y in the directions $\hat{e}_1, \hat{e}_2$ respectively. The squared natural frequencies of the rotating part are therefore:

$$\omega_x^2 = \frac{k_x}{M+m}, \quad \omega_y^2 = \frac{k_y}{M+m} \tag{4}$$

in the appropriate directions.

It is assumed that the system has some damping, d, in the stationary frame of reference, giving rise an asymmetric stiffness matrix. The none-dimensional damping ratio is computed according to [4,5,6-Eq.6.56]

$$\zeta = \frac{d}{\sqrt{2(M+m)(k_x+k_y)}} \tag{5}$$

The center of rotation is displaced by

$$r = x\hat{e}_1 + y\hat{e}_2 \tag{6}$$

The dynamics of the system in the presence of unbalance in body-fixed coordinates can be expressed as:

$$\begin{pmatrix} 1 & 0 \\ 0 & 1 \end{pmatrix}\begin{pmatrix} \ddot{x} \\ \ddot{y} \end{pmatrix} + \begin{pmatrix} 2\zeta\Omega & -2\Omega \\ 2\Omega & 2\zeta\Omega \end{pmatrix}\begin{pmatrix} \dot{x} \\ \dot{y} \end{pmatrix} + \begin{pmatrix} \omega_x^2-\Omega^2 & -2\zeta\Omega \\ 2\zeta\Omega & \omega_y^2-\Omega^2 \end{pmatrix}\begin{pmatrix} x \\ y \end{pmatrix} = \Omega^2\begin{pmatrix} \alpha_x \\ \alpha_y \end{pmatrix} \tag{7}$$

where

$$\alpha_x \triangleq \frac{Ma_x + mc\cos\varphi}{M+m} \quad ; \quad \alpha_y \triangleq \frac{Ma_y + mc\sin\varphi}{M+m} \tag{8}$$

The symbols have been defined in Fig. 2.

In steady state, the amplitude as measured in body-fixed coordinates, does not change in time, therefore the whirl orbit of the shaft centre of rotation can be computed from Eq.(7,8) by eliminating the time derivatives:

$$\begin{pmatrix} \omega_x^2-\Omega^2 & -2\zeta\Omega \\ 2\zeta\Omega & \omega_y^2-\Omega^2 \end{pmatrix}\begin{pmatrix} x \\ y \end{pmatrix} = \frac{\Omega^2}{M+m}\begin{pmatrix} Ma_x + mc\cos\varphi \\ Ma_y + mc\sin\varphi \end{pmatrix} \tag{9}$$

Assuming that $(m \ll M)$, Eq.(9) can be solved for the response amplitude that appears as static deflection in body-fixed coordinates:

$$\begin{pmatrix} x \\ y \end{pmatrix} = \begin{pmatrix} \beta_{11} & \beta_{12} \\ -\beta_{12} & \beta_{22} \end{pmatrix}\begin{pmatrix} a_x \\ a_y \end{pmatrix} + c\frac{m}{M}\begin{pmatrix} \beta_{11} & \beta_{12} \\ -\beta_{12} & \beta_{22} \end{pmatrix}\begin{pmatrix} \cos\varphi \\ \sin\varphi \end{pmatrix} \tag{10}$$

or

$$\begin{pmatrix} x \\ y \end{pmatrix} = \begin{pmatrix} x_0 \\ y_0 \end{pmatrix} + c\frac{m}{M}\begin{pmatrix} \beta_{11} & \beta_{12} \\ -\beta_{12} & \beta_{22} \end{pmatrix}\begin{pmatrix} \cos\varphi \\ \sin\varphi \end{pmatrix} \tag{11}$$

Where

$$\begin{pmatrix} \beta_{11} & \beta_{12} \\ -\beta_{12} & \beta_{22} \end{pmatrix} \triangleq \Omega^2\begin{pmatrix} \omega_x^2-\Omega^2 & -2\zeta\Omega \\ 2\zeta\Omega & \omega_y^2-\Omega^2 \end{pmatrix}^{-1} \tag{12}$$

It is clear from Eq.(11) that upon moving the trial mass along the circumference of the disk, i.e. changing φ between 0 and 360 degrees, Eq.(10) traces an ellipse in the x,y plane.

912

It is also clear that no less than 3 trial mass runs are necessary in order to obtain the influence coefficients in this case. The process of indentifying the necessary parameters and placing the balancing mass is explained below.

CARRYING OUT THE BALANCING PROCESS - THEORY

During the balancing process, the trial mass is placed at

$$r_i = c\left(\cos\varphi_i\,\hat{e}_1 + \sin\varphi_i\,\hat{e}_2\right) \tag{13}$$

Measuring the complex amplitude, U_i of the response at the operating speed, we have

$$x_i = \Re(U_i), \quad y_i = \Im(U_i) \tag{14}$$

Combining Eq.(11) and Eq.(14), a set of linear equations can be formed:

$$\begin{pmatrix} x_i \\ y_i \end{pmatrix} = \begin{bmatrix} 1 & 0 & \cos\varphi_i & \sin\varphi_i & 0 & 0 \\ 0 & 1 & 0 & 0 & \cos\varphi_i & \sin\varphi_i \end{bmatrix} \begin{pmatrix} x_0 \\ y_0 \\ \beta_{11} \\ \beta_{12} \\ \beta_{21} \\ \beta_{22} \end{pmatrix} \tag{15}$$

Each trial mass contributes 2 independent equations in the form of Eq.(15), thus the 5 parameters we seek would require at least 3 different trial mass locations so that a unique solution can be found. Indeed, combining several trial mass runs we can produce a least-squares solution of the right hand vector in Eq.(15).

Having estimated:

$$\begin{pmatrix} x_0 \\ y_0 \end{pmatrix}, \quad T \triangleq c\frac{m}{M}\begin{pmatrix} \beta_{11} & \beta_{12} \\ -\beta_{12} & \beta_{22} \end{pmatrix} \tag{16}$$

One can make use of Eq.(10) and Eq.(16) to compute:

$$\begin{pmatrix} s_x \\ s_y \end{pmatrix} = T^{-1}\begin{pmatrix} x_0 \\ y_0 \end{pmatrix} = \frac{M}{cm}\begin{pmatrix} \beta_{11} & \beta_{12} \\ -\beta_{12} & \beta_{22} \end{pmatrix}^{-1}\begin{pmatrix} \beta_{11} & \beta_{12} \\ -\beta_{12} & \beta_{22} \end{pmatrix}\begin{pmatrix} a_x \\ a_y \end{pmatrix} = \frac{M}{cm}\begin{pmatrix} a_x \\ a_y \end{pmatrix} \tag{17}$$

Since m is known, we have identified the unbalance and one should place a correction mass of

$$m\sqrt{s_x^2 + s_y^2} = \frac{M}{c}\sqrt{a_x^2 + a_y^2} \tag{18}$$

at

$$\varphi = \arctan\frac{s_y}{s_x} + 180^{\circ} \tag{19}$$

to eliminate the unbalance.

SIGNAL PROCESSING

The rotating disk being imperfect produces a signal due to the geometrical imperfection that is not related to the elastic response. This variation is measured al low speed and it has to be subtracted from the measured response at the speed of operation at which balancing takes place. An analogue optical device produced a pulse every turn, but this pulse is slightly distorted at higher speed and thus slight phase delay takes place.

Examining Fig.4 it is clear that sensor measuring the radial motion of the disk surface is influenced considerably by the imperfection of the structure. In particular it is clear that there is some time delay between the static and dynamic measurements.

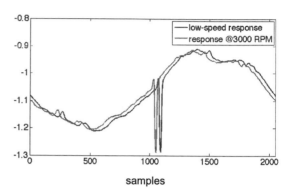

Figure 4: One cycle showing the 'static' response at low-speed and dynamic response at 3000 RPM.

In order to eliminate the static component from the measurement at 3000 RPM, these signals must be aligned. For this purpose a time shift parameter, τ is found by solving the optimization problem:

$$\tau = \arg\min_{\tau} \int_{0}^{T} \|y(t+\tau) - s(t)\| \, dt \qquad (20)$$

where $y(t+\tau), s(t)$ are the (time-shifted) dynamic and static response signals respectively. The difference between the displaced response and the static one is integrated over a complete cycle to produce the minimal time skew.

Indeed, this optimization produces aligned signals, as shown below:

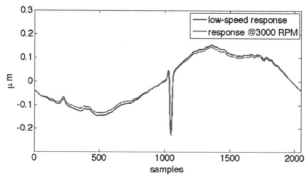

Figure 5: One cycle of the aligned 'Static' and dynamic responses

914

Once the static (measured at 100 RPM) response is subtracted from the measurement at 3000 RPM, the net effect of the unbalance force can be found.

It is clear from Fig. 6 that the alignment has produced a nearly pure sinusoidal response, as one should expect from the response to unbalance forces. The small misalignment, that is often overlooked, can result in considerable errors in the estimation of the vibrations and therefore the balancing procedure may not converge to produce sufficiently small residual vibrations.

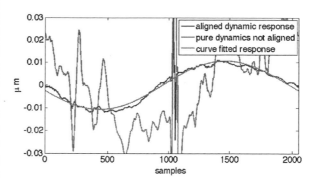

Figure 6: Aligned and misaligned response estimates and curve-fitted sine wave

Other unusual signal processing procedures (e.g. wavelet based filtering) and their dynamical interpretation will be described in the full paper alongside the complete balancing procedure.

EXPERIMENTAL VERIFICATION

The abovementioned procedure has been applied to the north seeking device that was described at the beginning. The device has 10 balancing holes spaced 36 degrees apart. The sensor depicted in fig. 2 was used to collect the data and the signal processing approach that was described above was used to clean the measurement. To the clean signal, a sinusoid of the form

$$p(t) = A \cos \Omega t + B \sin \Omega t \tag{21}$$

was curve fitted by a least squares procedure. Plotting the A vs. the B (same as x vis y) for all the trial masses, together with the curve fitted Eq.(11), we have

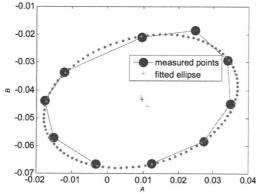

Figure 7: Measured response amplitudes (blue dots) and the fitted ellipse according to Eq.(11).

915

Having confirmed the model, the correction mass can be installed by using Eq.(16-19).

The parameters in Eq.(16) have been identified as:

$$\begin{pmatrix} x_0 \\ y_0 \end{pmatrix} = \begin{pmatrix} 0.0095 \\ -0.0439 \end{pmatrix}, \quad T = \begin{pmatrix} -0.0139 & 0.0237 \\ -0.0230 & -0.0070 \end{pmatrix}, \quad T^{-1} = \begin{pmatrix} -10.9244 & -36.9437 \\ 35.8668 & -21.6808 \end{pmatrix} \quad (22)$$

It is worth mentioning that the transformation that represents the influence coefficients has relatively large off-diagonal entries and therefore it is clear that a standard influence balancing procedure that does not take the asymmetry of the rotor in consideration would be far from optimal and may even diverge (see [2]).

Finally, using Eq.(18-19), having used a trial mass of 1.5 gram, it was found that a correction mass of 3.4 gram has to be placed at 220 degrees.

The measured vibrations before and after implementing the correction is depicted below.

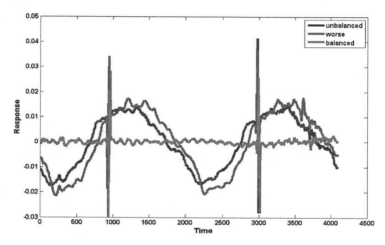

Figure 8: Measured vibrations before (blue green) and after (red) implementing the balancing correction mass

In order to examine the effectiveness of the balancing procedure, the driving motor's currents have been measured.

It can be clearly seen that once the balancing mass has been installed, a considerable synchronous current component has been removed and the system runs more smoothly.

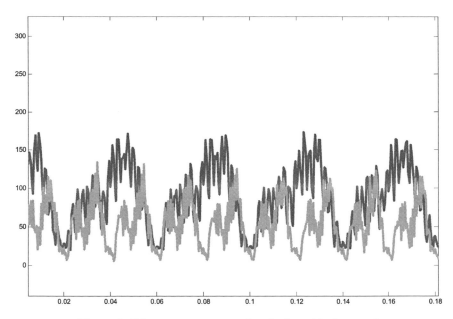

**Figure 9: Motor currents vs. time before (dark green)
and after (light green)balancing has been applied**

SUMMARY

The balancing of an asymmetric north finding device has been performed with satisfactory results with a new procedure. This method required several trial mass runs and as a bonus it provides accurate balancing results. It has been suggested that asymmetric rotors cannot be balanced by an ordinary procedure since the same trial mass can result in a different response amplitude change for different installation angles.

The method requires some signal processing and removal of static runout that has been proposed here with a special means to correct trigger time uncertainty.

ACKNOWLEDGEMENT

The authors would like to thank Azimut Technologies, the maker of the north-seeking device, for granting the authors the permission to publish these results.

REFERENCES

[1] Itzhack Y. Bar-Itzhack, Jacob Reiner and Michael Naroditsky, JOURNAL OF GUIDANCE, CONTROL, AND DYNAMICS Vol . 24, No. 2, March– April 2001

[2] Matsukura Y., Kiso M., Inoue T. and Tomisawa M., "On the balancing convergence of flexible rotors, with special reference to asymmetric rotors", *Journal of Sound and Vibration, Volume 63, Issue 3, 8 April 1979, Pages 419-428*

[3] Parkinson A. G. "The vibration and balancing of shafts rotating in asymmetric bearings", *Journal of Sound and Vibration, Volume 2, Issue 4, October 1965, Pages 477-501*

[4] Gasch R. Nordmann R. and Pfutzner H. "Rotordynamik", Springer , 2nd edition, 2006

[5] Genta G. *Dynamics of Rotating Systems*, 2005, Springer

[6] Lee C. W., *Vibration Analysis of Rotors*, Kluwer Academic Publishers, 1993.

Use of the co-variance matrix in rotor balancing

A I J Rix
Rolls-Royce plc., UK

S D Garvey, S Jiffri
Nottingham University, UK

ABSTRACT

Imbalance remains the most important source of forcing in virtually all rotating machines and efforts must be made on most rotors to correct the state of balance of the rotor after its initial manufacture. The distribution of imbalance on any given rotor must be measured and the information measured is necessarily incomplete whether the rotor is tested in a balancing machine or an in-situ balancing run. The complete state of imbalance of a rotor can be considered to be a vector sum of a known component (measurable) and an unknown (and unmeasurable) component. Standard practice is to assume that the unmeasurable component will contribute little to the vibration of the rotor and that it can be ignored. This paper suggests an alternative. Given a population of rotors and some knowledge of the manufacturing processes which are used to produce these, it is now possible, using modelling, to make some assessment of what the co-variance matrix will be for the distribution of imbalance vectors on that population. The co-variance matrix contains information about which patterns of imbalance are very likely and which are very unlikely to occur. It is possible to combine this information with the measured information from balancing tests to determine the most likely unknown component of the imbalance vector. This paper firstly explains how the above is achieved. Examples are presented which indicate that where such co-variance information exists, it is advantageous to use it.

1. INTRODUCTION

Manufacturing and build of rotors introduces a distribution of imbalance that is continuous along the rotor. The literature on rotor balancing is substantial (e.g. [1]) and reflects the importance of this operation. The most effective removal of these imbalances would be to apply correction masses either continuously or at every plane of significant mass such that the centre of gravity of every thin slice of the rotor normal to its axis of rotation lay on the axis. However, measurements from balancing tests can provide only limited information about imbalance. The limitations on obtaining information from such tests are due to measurement resolution and noise and the fact that in any one set of balancing tests, only a very small number of independent patterns of imbalance will cause measurable responses over a finite range of speeds.

When balancing tests are conducted with the rotor *in-situ* in the stator which will contain it for the remainder of its working life, the incompleteness of the information about the distribution of imbalance is not a significant issue since the user can be satisfied with reducing to near-zero, the excitation of each one of the independent patterns of response which can cause significant response. This can be done by applying imbalance corrections at a small number of planes (usually 2 or 3) and the appropriate corrections to add can be determined following age-old procedures such as that by Goodman [2]. Note that in the least-squares approach described in [2], the set of simultaneous equations produced during the balancing operation is over-determined – there are more equations than unknowns.

The most challenging problems in rotor balancing arise when balancing tests are conducted on the rotor in one dynamic environment (such as a balancing machine) and where the rotor will operate in a different dynamic environment when it is working. In this case, the patterns of imbalance distribution which contribute significant responses in the testing environment are different from those which contribute significant responses in the ultimate working environment. Garvey *et. al.* [3] outline model-based methods for the determination of which distributions of residual imbalance on a rotor will contribute significantly to *cost* and provide a methodology for the optimum correction of the state of imbalance of a rotor given knowledge of the expected distribution of operating condition parameters and knowledge of the relative importance of different vibration components. However, this *robust-balancing* approach relies on the determination of the original state of imbalance of a given rotor in sufficient detail in the first instance. Ehrich [4] describes a method for progressing towards this objective – achieving (it is claimed) the effects of high-speed balancing from low-speed tests. In fact, that method relies on some arbitrary intuitive judgements.

Based on the information from a balancing test alone (measurable data) the imbalance correction applied cannot take into account full information about the distribution of imbalance along the rotor and residual imbalances on the rotor will lead to response components when the rotor is fitted into a different stator. This may be one of the reasons that balancing is often described as being more of an art than a science.

The present paper proposes that additional information which is accessible to the manufacturer of rotors can be used to improve balancing outcomes. That additional information takes the form of a co-variance matrix. A co-variance matrix can be populated based on knowledge of the manufacture and build processes and the likely distributions of residual imbalance that they generate. This information is normally not measurable due to the tolerances involved, and the fact that the rotor has to be assembled in order for the assembly error influences on balance to be established. However this information may be arrived at theoretically.

Efforts to reduce this residual imbalance during the assembly process through processes such as selective orientation build are actually likely to make the co-variance matrix approach more effective because as the build process becomes more systematic, correlations between different components of imbalance are liable to increase. Selective alignment assembly processes in rotors are a good example of this; any assembly process, however careful, can never eliminate residual imbalance because of assembly and measurement tolerances which are many times higher than that of a balancing machine. However they are likely to cause underlying patterns to the imbalance distributions that can be exploited by the co-variance matrix approach which, once established, adds no additional cost to the process but a potentially significant benefit.

2.   WHAT IS A CO-VARIANCE MATRIX?

The co-variance matrix is the natural extension of variance, for two random variables. This is best illustrated with an example:

Consider a population of rotors with two discs assembled together with the imbalance at the two discs measured in the vertical plane only with their respective imbalances 'a' and 'b' as shown in figure 1:

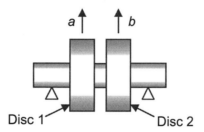

Figure 1: A rotor made up of 2 discs with imbalances a and b

For a single random variable, X, it is meaningful to talk about a population mean μ_x and a population variance σ_x^2, defined as follows:

$$\mu_x = \lim_{N \to \infty}\left(\frac{1}{N}\sum_{i=1}^{N} x_i\right) \qquad \sigma_x^2 = \lim_{N \to \infty}\left(\frac{1}{N}\sum_{i=1}^{N}(x_i - \mu_x)^2\right) \tag{1}$$

If the imbalances of the rotor are considered together, then there are two random variables ('A' and 'B') and then the quantity 'x' can be considered as a vector (**x**), therefore the population mean would also be a vector:

$$\mathbf{x} := \begin{bmatrix} a \\ b \end{bmatrix} \qquad \boldsymbol{\mu}_x := \begin{bmatrix} \mu_a \\ \mu_b \end{bmatrix} = \lim_{N \to \infty}\left(\frac{1}{N}\sum_{i=1}^{N}\begin{bmatrix} a_i \\ b_i \end{bmatrix}\right) \tag{2}$$

In this equation, the mathematical notation ":=" is used to indicate that a quantity on the left hand side of the equation is being defined. We can use these random vectors to produce the co-variance matrix, which we denote as **C**

$$C = \lim_{N \to \infty} \left(\frac{1}{N} \sum_{i=1}^{N} \left(\left(\begin{bmatrix} a_i \\ b_i \end{bmatrix} - \begin{bmatrix} \mu_a \\ \mu_b \end{bmatrix} \right) \times \left(\begin{bmatrix} a_i \\ b_i \end{bmatrix} - \begin{bmatrix} \mu_a \\ \mu_b \end{bmatrix} \right)^T \right) \right) \tag{3}$$

In the context of this paper the random vector, **x**, will represent a vector of discrete imbalances at different planes of the rotor and this vector will contain two entries for each plane. Then the covariance matrix can be expressed more generally as

$$C_x = \lim_{N \to \infty} \left(\frac{1}{N} \sum_{i=1}^{N} \left((\mathbf{x} - \mathbf{\mu}_x) \times (\mathbf{x} - \mathbf{\mu}_x)^T \right) \right) \tag{4}$$

In the context of balancing, the population mean, $\mathbf{\mu}_x$, is virtually always a vector of zeros.

3. WHAT DOES THE CO-VARIANCE MATRIX TELL YOU?

The leading diagonal of the co-variance matrix contains variances for individual random variables. In the present context, each random variable represents one component of imbalance at a given rotor plane. Each off-diagonal entry indicates the extent to which two random variables are correlated. Evidently (from (4)), the co-variance matrix is inherently symmetric. In the case of the simple example of the previous section, positive off-diagonal terms would occur if the imbalances arising on the two discs tended to have the same sign.

For any given vector of imbalances, **x**, the covariance matrix can be used to provide a dimensionless measure of *distance, h,* from the population mean, $\mathbf{\mu}_x$. The units of this measure are standard-deviations

$$h = (\mathbf{x} - \mathbf{\mu}_x)^T C_x^{-1} (\mathbf{x} - \mathbf{\mu}_x) \tag{5}$$

If the individual random variables within **x** follow a normal distribution, then it is found that these *distances* follow a "Chi-square" distribution. Then, for example, if **x** has 20 entries, the probabilities that h will be less than $\{2,5,10,20,50,100\}$ can be calculated to be (respectively)
$\{(1.1 \times 10^{-7}), (2.8 \times 10^{-4}), 0.032, 0.542, (1 - 2.2 \times 10^{-4}), (1 - 1.3 \times 10^{-12})\}$.

The eigenvalues of the co-variance matrix are themselves variances. These variances relate to random variables which are linear combinations of the original random variables. In the simple illustration of the previous section, it is clear that the complete imbalance information for the rotor can be expressed by giving the values of a and b. However, it could be expressed equally well by giving the values c and d where c might represent the average of $\{a,b\}$ and d might represent the difference between them. Formally, this relationship could be written:

$$\begin{bmatrix} c \\ d \end{bmatrix} = \begin{bmatrix} \frac{1}{2} & \frac{1}{2} \\ 1 & -1 \end{bmatrix} \begin{bmatrix} a \\ b \end{bmatrix} \quad , \quad \begin{bmatrix} a \\ b \end{bmatrix} = \begin{bmatrix} 1 & \frac{1}{2} \\ 1 & -\frac{1}{2} \end{bmatrix} \begin{bmatrix} c \\ d \end{bmatrix} \tag{6}$$

This concept extends easily to vectors of imbalance of arbitrary dimension. Thus, if $\mathbf{x}$ represents the state of imbalance of a rotor in terms of the components of imbalance in the x and y directions at each one of a number of rotor planes, then some other vector, $\mathbf{y}$, also represents the same state of imbalance provided that there is some invertible relationship between $\mathbf{x}$ and $\mathbf{y}$ such as

$$\mathbf{y} = \mathbf{U}^{-1}\mathbf{x} \quad , \quad \mathbf{x} = \mathbf{U}\mathbf{y} \tag{7}$$

A new co-variance matrix, $\mathbf{C}_y$, can be prepared from $\mathbf{C}_x$, by recognising that

$$\left(\mathbf{x} - \boldsymbol{\mu}_x\right)^T \mathbf{C}_x^{-1} \left(\mathbf{x} - \boldsymbol{\mu}_x\right) \equiv h \equiv \left(\mathbf{y} - \boldsymbol{\mu}_y\right)^T \mathbf{C}_y^{-1} \left(\mathbf{y} - \boldsymbol{\mu}_y\right) \tag{8}$$

In view of (7), we can conclude that

$$\mathbf{C}_y = \mathbf{U}^{-1}\mathbf{C}_x\mathbf{U}^{-T} \tag{9}$$

If the matrix, $\mathbf{U}$, is chosen such that its columns are eigenvectors of the matrix, $\mathbf{C}_x$, then with appropriate scaling, $\mathbf{U}$, is an orthogonal matrix and $\mathbf{C}_y$ is a diagonal matrix containing the eigenvalues of $\mathbf{C}_x$ along its diagonal.

The relevance of this material in the present context is now outlined. Given any two different vectors of rotor imbalance, $\mathbf{x}_1$ and $\mathbf{x}_2$, which would produce the same set of responses in a balance test, we can apply (5) to calculate the corresponding distances, h_1 and h_2, and the vector which returns the lower value of h is the more likely one to occur.

Illustration:
Assume that a large number of two-disc rotors from figure 1 have been produced and that the resulting co-variance matrix is:

$$\mathbf{C} = \begin{bmatrix} 0.0577 & 0.0768 \\ 0.0768 & 0.1024 \end{bmatrix} \tag{10}$$

If C is rewritten in terms of its eigenvalues and eigenvectors:

$$\mathbf{C} = \left(\begin{bmatrix} 0.6 \\ 0.8 \end{bmatrix} \times 0.16 \times \begin{bmatrix} 0.6 & 0.8 \end{bmatrix} \right) + \left(\begin{bmatrix} 0.8 \\ -0.6 \end{bmatrix} \times 0.0001 \times \begin{bmatrix} 0.8 & -0.6 \end{bmatrix} \right) \tag{11}$$

As the eigenvectors shown represent the imbalance at discs in the format [a, b], and the eigenvalues represent the variances of the corresponding independent random variables; it can be deduced that the magnitude of any imbalance in the direction [0.6, 0.8] is likely to be 40 times greater than the magnitude of any imbalance in the direction [0.8 -0.6] because:

$$40 = \sqrt{\frac{0.16}{0.0001}} \tag{12}$$

Now consider the contrived situation where we know the imbalance at disc 1 (a=24g.mm.) but not at disc 2 ('b' is unknown). The main value of the co-variance matrix is that some best estimate can be made for 'b'. Intuitively you might expect:

$$b \cong 24 \times \left(\frac{0.8}{0.6} \right) = 32 g.mm. \tag{13}$$

Formally, we get the correct answer by minimising J:

$$J := \begin{bmatrix} 24 & b \end{bmatrix} \mathbf{C}^{-1} \begin{bmatrix} 24 \\ b \end{bmatrix} \tag{14}$$

Omitting the formal mathematics (covered in section 7) the answer is $b = 31.944506$g.mm.

4.   POPULATING THE CO-VARIANCE MATRIX

Co-variance matrices can be generated in two ways, measurements or models. In general there is no possibility of trying to estimate $\mathbf{C}$ from measurements. If the measurement systems were good enough then $\mathbf{C}$ would not be required anyway.

There is often a good possibility of estimating $\mathbf{C}$ from modelling the manufacturing / assembly process. Simple examples of this would be modelling tilted and/or eccentric bores in shafts, or possible bends caused by the necessity to assemble the stages of a rotor in the horizontal plane.

5.   INFORMATION OBTAINED FROM BALANCING TESTS

When a rotor is spun during a balancing test, numerous different quantities can be measured – all of which are driven by the imbalance on the rotor. These quantities typically comprise bearing reaction forces and/or absolute rotor displacements but may also include displacements of the rotor relative to the stator, point velocities or accelerations on the stator and strains on the stator. Even if a large number of apparently-independent measurements are made, the fact is that the number of independent pieces of information is still small because, fundamentally, within the measuring resolution of the instruments, the signals are composed exclusively of contributions from a very small number of patterns of "significant residual imbalance". (The concept of "Modes of Significant Residual Unbalance" was presented in [3] and is discussed in depth there).

Thus, although a given rotor actually has balance errors at many locations along its length, the findings from any one set of balancing tests is unlikely to be able to distinguish between the actual distribution of imbalance and some notional set of (say) three discrete imbalances at prescribed planes. Fig. 2 illustrates.

The information obtained from a balancing test can always be summarised in the form:

$$\mathbf{A}\mathbf{x} = \mathbf{y} \tag{15}$$

Where:

A is a matrix that is known and describes the characteristics of the rotor.

x is the imbalance distribution along the rotor (as in figure 2a), with 2 real unknowns per plane for the 2 translational degrees of freedom.

y is the vector of measurements from the balancing ~~machine~~ test (as in figure 2(b)).

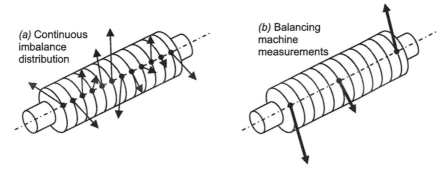

Figure 2: Rotor and imbalance vectors: balancing tests return only partial information

6. BALANCING USING MEASURABLE DATA ONLY – CURRENT PRACTICE

The least-squares methods attributed to Goodman [2] and used by many since then are addressing the problem of balancing in-situ where the measurements which can be made are a relatively direct indication of how well-balanced the rotor is. More specifically, there is no need in such cases to find information about resident components of imbalance which happen to contribute very little to the response measurands. Typically, that problem is <u>overdetermined</u> (because **x** need have only very few entries in it) and an optimal solution can be found as

$$\mathbf{x} = \left(\mathbf{A}^{\mathrm{T}}\mathbf{W}^{2}\mathbf{A}\right)^{-1}\left(\mathbf{A}^{\mathrm{T}}\mathbf{W}\right)\mathbf{y} \qquad (16)$$

where **W** is a weighting matrix – almost invariably diagonal and very often equal to the identity matrix. The problem of interest here is the opposite in one sense. The vector **x** has many entries in it (more than the number of independent measurements) and there is no unique solution for **x** from (15). We can, however find that solution having the minimum Euclidean norm as follows. Let

$$\mathbf{x} = \mathbf{A}^{\mathrm{T}}\mathbf{g} \qquad (17)$$

for some vector, **g** having the same dimensions as **y**. Substituting (17) into (15) provides for the determination of a unique **g** and thereafter, one possible solution to (15) can be determined from (17). This possible solution is denoted $\mathbf{x}_k$

$$\mathbf{x}_k = \mathbf{A}^{\mathrm{T}}\left(\mathbf{A}\mathbf{A}^{\mathrm{T}}\right)^{-1}\mathbf{y} \qquad (18)$$

925

We stress that this is not a unique solution for $\mathbf{x}$. Instead it simply parameterises a whole space of possible solutions $\mathbf{x}$ for equation (15) through

$$\mathbf{x} = \mathbf{x}_k + \mathbf{x}_u \tag{19}$$

where $\mathbf{x}_u$ is any vector satisfying

$$\mathbf{Ax}_u = 0 \tag{20}$$

The prevailing practice at present is simply to take $\mathbf{x}_u = \mathbf{0}$. Without knowledge of the co-variance matrix, this is the only reasonable assumption. However, as the following section shows, one can do rather better having knowledge of the co-variance matrix.

7. BALANCING USING THE MEASURABLE DATA AND THE CO-VARIANCE MATRIX – PROPOSED METHOD

The method proposed in this paper for using the co-variance matrix essentially provides a criterion for computing an appropriate vector $\mathbf{x}_u$. The method begins by developing an appropriate parameterisation for the space of all vectors $\mathbf{x}_u$ satisfying (20). Formally, this involves finding some matrix, $\mathbf{N}$, such that

$$\mathbf{AN} = 0 \quad , \quad \begin{bmatrix} \mathbf{A}^T & \mathbf{N} \end{bmatrix} \quad \text{full rank} \tag{21}$$

In MATLAB, this is achieved simply through $>> \mathrm{N} = \mathrm{null}(\mathrm{A})$. Then the space of all vectors $\mathbf{x}_u$ satisfying (20) is given by

$$\mathbf{x}_u = \mathbf{Nu} \quad , \tag{22}$$

where $\mathbf{u}$ is any arbitrary vector. One particular choice of the vector, $\mathbf{u}$, will result in minimising the distance h as described in (5). Combining (22), (19) and (5) results in

$$h := (\mathbf{x}_k + \mathbf{Nu})^T \mathbf{C}_x^{-1} (\mathbf{x}_k + \mathbf{Nu}) \tag{23}$$

Recall that by minimising h, here, we are discovering that vector $\mathbf{x}$ which is consistent with the measurements from the balancing test (equation (15)) and which is the least unlikely vector to satisfy that equation according to the information present in the co-variance matrix. The value of $\mathbf{u}$ which minimises h is found to be

$$\mathbf{u} = (\mathbf{N}^T \mathbf{C}_x^{-1} \mathbf{N})^{-1} (\mathbf{N}^T \mathbf{C}_x^{-1} \mathbf{x}_k) \tag{24}$$

It is beyond the scope of the present paper to prove that the most likely vector of imbalance which is consistent with the measurements should necessarily form the basis of the optimum imbalance correction. However, this can be shown [5]. It suffices here to conduct a simulated example to demonstrate that in most cases, the procedure provides an improvement

8.   SIMULATED EXAMPLE

This simulation used a simple rotor 1.2m in length comprising 12 uniform lengths of 0.1m each, illustrated in figure 3:

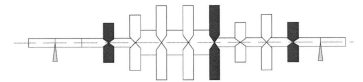

Figure 3: Example rotor

The shaft comprises two different shaft-diameters, 35mm and 100mm. The material properties are:

- Young's modulus=211E9 N/m^2, μ=0.3. ρ=7810 Kg/m^3.
- All discs are 0.04m thick, diameters 0.16m, 0.225m, 0.30m.
- Bearings indicted by triangles.
- Blue (dark) discs are balance locations
- The 3 balancing test response measurement locations are at the bearings and at disc 3 (counting from left).
- The first 3 natural frequencies of the rotor are at ≈20Hz, ≈74Hz and ≈192Hz

A co-variance matrix was populated by creating a large number of sample "imbalance distributions" constructed from combinations of sample distributions of random amplitudes that mainly followed the first four free-free modes plus some random noise. An example of each of these distributions (excluding the random one) is shown in figure 4 below:

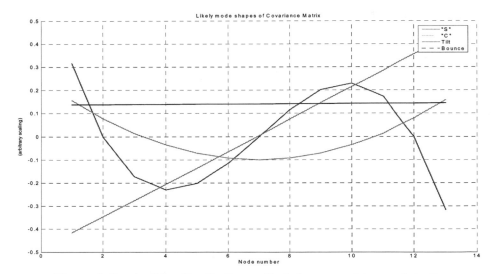

Figure 4: Some of the distributions of imbalance used to populate the co-variance matrix

The eigenvalues of the co-variance matrix indicate how many significant independent distributions of imbalance can exist in this population of rotors. These eigenvalues are graphed in figure 5 below:

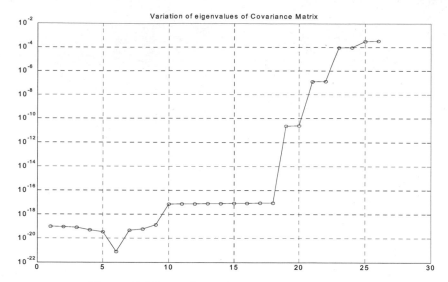

Figure 5: Eigenvalues of the co-variance matrix.

There are 26 eigenvalues because this model has 13 nodes and 2 degrees of freedom per node. If all eigenvalues were similar the co-variance matrix would not be useful; of particular importance is the large step change in value with only a few "most likely" eigenvalues that can be observed.

In order to simulate the process of balancing rotors both with and without the co-variance matrix,10,000 different imbalance distributions were generated and applied to the rotor in turn. Again these imbalance distributions were based on the first four free-free modes of the rotor and randomly combined together and with a further random distribution of much smaller amplitude.

The balance correction was calculated for each imbalance distribution using the methods described in section (6) (with x_u=0) and then again using the co-variance matrix method. The measure of success used to compare between the two methods was simply the sum of the squares of all rotor node translations.

For every one of the 10,000 simulated rotors the cost was calculated using the balance correction applied using different levels of knowledge of the imbalance distribution. These methods were:

1. Least-squares (current practice - as described in section 6)
2. Co-variance matrix method (proposed method)
3. Full balance distribution (unbalance known at every plane of the rotor, the best possible theoretical solution)

Institution of Mechanical Engineers
MECHATRONICS, INFORMATICS & CONTROL GROUP
TRIBOLOGY GROUP

AdvEnTech
group

Ninth International Conference on Vibrations in Rotating Machinery

Volume two

University of Exeter, UK

8-10 September 2008

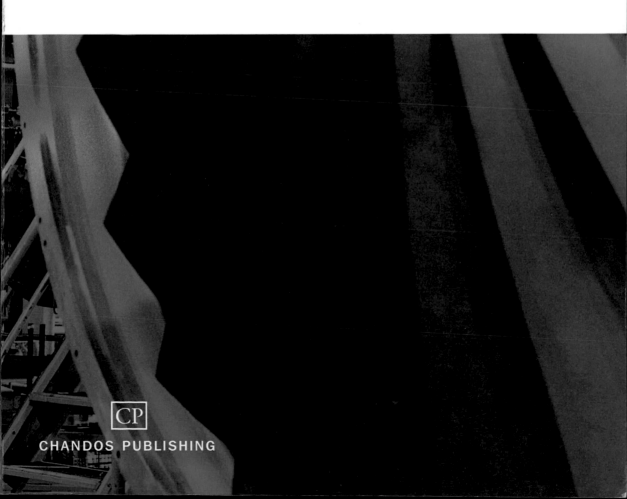

CHANDOS PUBLISHING

AUTHOR INDEX

Abele, E	465	Domes, B	15, 511
Amati, N	97, 761	Dufour, R	497
Anegawa, N	143	Dutt, J K	61
Arkkio, A	423, 449, 477		
		Ecker, H	371
Bachschmid, N	631	El Badaoui, M	573
Bash, T	85	Elka, A	909
Bonello, P	203, 671	El-Shafei, A	719
Brennan, M J	697	Ewins, D J	317
Bucher, I	909		
Burrows, C R	111	Feng, N	861
		Ferraris, G	497
Cade, I S	111	Fiori de Castro, H	359, 885
Callan, R	585	Friswell, M I	255, 645, 899
Carvalho Brito Jr., G	411	Fujiwara, H	143
Champneys, A R	899		
Chen, S L	585	Garvey, S D	255, 523, 919
Ciğeroğlu, E	185	Genta, G	33, 799
Clark, R E	305	Guendogdu, Y	171
Cole, M	215	Guillet, F	573
Combescure, D	289	Guo, H	709
Craig, M	585		
		Hahn, E	861
Das, A S	61	Hai, P M	203
De Lépine, X	97	Hameyer, K	437
Deckner, M	849	Hasch, B	75
Di Maio, D	317	He, Y Y	645
Dohnal, F	775	Helfert, M	49

Henke, M	171	Mace, B R	775
Holmes, R	697	Mahfoud, J	497
Holopainen, T P	423, 449	Maslen, E H	3
Hu, W	861	Matsushita, O	143
Huyanan, S	125	Memmott, E A	787
		Menq, C-H	185
Ibrahim, A	573	Morton, P G	383
Inman, D	85	Muñoz-Abella, B	657
Irretier, H	61		
Ivanenko, A	745	Nandi, A K	229, 709
		Neogy, S	229
Jiffri, S	919	Nicholas, J C	685
		Nordmann, R	49, 75, 465, 511
Kaletsch, C	511		
Kamesh, P	697	Okabe, A	143
Karpenko, E	761	Orbay, G	159
Karthikeyan, M	619	Orivuori, J	477
Kasarda, M	85	Özgüven, H N	159
Keogh, P S	111		
Kern, S	465	Paiva Okabe, E	331
Kill, N	267	Panin, S	745
Kirk, R G	85, 685	Panning, L	171
Knabel, J	603	Pennacchi, P	549, 631, 813
Koehler, R	511	Penny, J E T	255, 523, 645, 825
Kudo, T	143	Petermeier, B	733
		Peters, D	75, 511
Laiho, A	477	Petrov, E P	241
Lazarus, A	289	Powrie, H E G	585
Lee, C-W	277	Pugachev, A O	849
Lees, A W	255, 825		
Levecque, N	497	Quinn, D	85
Lindenborn, O	75		
Litvinov, D	489	Ramesh, K	787
Lucchesi Cavalca, K	331, 359, 885	Randall, R B	539, 573

Rémond, D 573

Ricci, R 549

Rix, A I J 919

Rodrigues, D J 899

Rongong, J A 745

Rubio, L 657

Sahinkaya, M N 111

Sawalhi, N 539

Sawicki, J T 645

Schiffler, A 465

Seo, Y-H 277

Seume, J 171

Siewert, C 171

Silvagni, M 761

Sims, N D 125, 305

Sinervo, A 477

Sinha, J K 565

Springer, H 371, 733

Staples, B 761

Stepanova, I 745

Stocki, R 603

Szolc, T 603

Talukdar, S 619

Tammi, K 477

Tanaka, M 871

Tanzi, E 631

Taura, H 871

Tauzowski, P 603

Tiwari, R 619

Tomlinson, G R 745

Tonoli, A 97, 761

van der Giet, M 437

Vania, A 347, 813

Violette, D 497

Wallaschek, J 171

Wang, L 585

Ward Franco de Camargo, L 885

Werner, U 395

Wilson, R E 899

Wood, R J K 585

Xie, L 49

Yusoff, A R 305

Zapoměl, J 837

Zenger, K 477

Zhang, N 861

Zhang, Q 709

This data showed that of the 10,000 simulated rotors, 88% were better balanced (produced a lower cost) with the co-variance matrix than by the pure least-squares method. In order to illustrate how much of an improvement the co-variance matrix made to each rotor, the ratio of the cost of the co-variance matrix balanced rotor to the "best possible solution" rotor was calculated in each case. For comparison, the ratio of the pure least-squares method to the best possible solution was also calculated for each rotor. The dataset produced was then analysed and the results are given in Table 1.

Table 1: Data analysis of ratios of costs / "best possible solution" cost.

	Min	Max	Mean	Std. Dev.
Least-squares Method	1.0	206.1	7.1	5.8
Co-variance Method	1.0	119.0	2.4	2.5

The minimum of 1.0 in both cases shows that occasionally both methods will, by chance, produce a result that is almost as good as the best possible. The maximum of the co-variance is significantly less than the least-squares, this shows that the "worst" rotors are always significantly improved by the use of the covariance method. The mean shows that the average result is significantly improved and the standard deviation shows that a significantly more consistent balance is achieved with the co-variance method.

Although it is clear that the co-variance method will have a positive effect on 88% of the rotors in this simulation, it is important to understand what it will do to the 12% of rotors that it had a negative effect upon. To assess this, a subset of the data used to produce Table 1 was selected which represented only the cases where the co-variance method proved to be less effective than least-squares. The data set analysis is shown in table 2.

Table 2: Data analysis of ratios of costs / "best possible solution" cost in the 12% of cases where the co-variance matrix method was not as good as the least squares.

	Min	Max	Mean	Std. Dev.
Least-squares Method	1.0	30.1	2.3	2.0
Co-variance Method	1.1	36.8	3.5	2.9

The maximums are relatively low compared to Table 1 which shows that it is only rotors that are near the "best possible solution" that can have their imbalance made worse by the covariance method, and the effect is minimal.

9.   CONCLUSIONS

1. This co-variance approach is applicable when the there are more independent influential planes of imbalance than can be measured on a balancing machine.

2. If a co-variance matrix can be produced that represents the likely arising imbalance distributions from the manufacturing / build process, it can be used to produce a "better" balanced rotor in most cases.

3. In a theoretical simulation the covariance matrix made a significant improvement in correcting 88% of the 10,000 imbalanced rotors simulated. The co-variance matrix had the greatest effect on the rotors that were the most out of balance. The 12% of cases where it was not as effective as the traditional methods were on rotors with relatively small imbalances, and the difference between the methods was small.

10.   REFERENCES

[1] W.C. FOILES, P.E. ALLAIRE, E.J. GUNTER, (1998) 'Review: Rotor balancing,' *Shock and Vibration*. Issue: Volume 5, Numbers 5-6. University of Virginia

[2] GOODMAN, T. P. (1964) 'A Least Squares Approach for Computing Balance Corrections,' *Journal of Engineering for Industry - Transactions of the ASME*, 86, pp. 273-279

[3] Garvey, S. D., Friswell, M. I., Williams, E. J., Lees, A. W., Care, I. D., "Robust Balancing for Rotating Machines", *Proc. Instn Mech. Engrs, Part C, Journal of Mechanical Engineering Science*, 2001, 216(C), pp. 1117-1130.

[4] Ehrich, F. F., "Pseudo-High Speed Balancing", *Transactions of the ASME, Journal of Engineering for Industry*, 1990, 112(4), pp. 418-426.

[5] Jiffri, S; Rix, A & Garvey, S.D. Enriching Readings from Rotor Balancing Tests using the Covariance Matrix. Manuscript submitted to IMechE, Part C, Journal of Engineering Science. March. 2008.